D. Readey

High-Temperature Ceramic-Matrix Composites II:

Manufacturing
and
Materials Development

Related titles published by the American Ceramic Society:

Design for Manufacturability of Ceramic Components (Ceramic Transactions Volume 50)
Edited by Asish Ghosh, Basavaraj Hiremath, and John Halloran
© 1995, ISBN 0-944904-88-2

Manufacture of Ceramic Components (Ceramic Transactions Volume 49)
Edited by Basavaraj Hiremath, Allan Bruce, and Asish Ghosh
© 1995, ISBN 0-944904-87-4

Functionally Gradient Materials (Ceramic Transactions Volume 34)
Edited by J. Birch Holt, Mitsue Koizumi, Toshio Hirai, and Zuhair A. Munir
© 1993, ISBN 0-944904-64-5

Structural Ceramics Joining II (Ceramic Transactions Volume 35)
Edited by Arthur J. Moorhead, Ronald E. Loehman, and Sylvia M. Johnson
© 1993, ISBN 0-944904-65-3

Low-Expansion Materials (Ceramic Transactions Volume 52)
Edited by David P. Stinton and Santosh Y. Limaye
© 1995, ISBN 0-944904-92-0

Advances in Ceramic-Matrix Composites (Ceramic Transactions Volume 38)
Edited by Narottam P. Bansal
©1993, ISBN 0-944904-69-6

Advances in Ceramic-Matrix Composites II (Ceramic Transactions Volume 46)
Edited by J.P. Singh and Narottam P. Bansal
©1994, ISBN 0-944904-80-7

For information on ordering titles published by the American Ceramic Society, or to request a publications catalog, please call 614-890-4700, or write to Customer Service Department, 735 Ceramic Place, Westerville, OH 43081.

High-Temperature Ceramic-Matrix Composites II:

Manufacturing
and
Materials Development

Edited by

A.G. Evans
Harvard University

R. Naslain
University of Bordeaux

Volume 58

Published by
The American Ceramic Society
735 Ceramic Place
Westerville, Ohio 43081

Proceedings of the Second International Conference on High-Temperature Ceramic-Matrix Composites, held August 21–24, 1995, in Santa Barbara, CA.

Copyright © 1995, The American Ceramic Society. All rights reserved.

No part of this book may be reproduced, stored in a retrieval system, or transmitted in any form or by any means, electronic, mechanical, photocopying, microfilming, recording, or otherwise, without written permission from the publisher.

Permission to photocopy for personal or internal use beyond the limits of Sections 107 and 108 of the U.S. Copyright Law is granted by the American Ceramic Society, provided that the base fee of US$5.00 per copy, plus US$.50 per page, is paid directly to the Copyright Clearance Center, 222 Rosewood Dr., Danvers, MA 01923, USA. The fee code for users of the Transactional Reporting Service for *Ceramic Transactions Volume 58* is 0-944904-99-8/95 $5.00+$.50. This consent does not extend to other kinds of copying, such as copying for general distribution, for advertising or promotional purposes, or for creating new collective works. Requests for special photocopying permission and reprint requests should be directed to the Director of Publications, The American Ceramic Society, 735 Ceramic Place, Westerville OH 43081, USA.

For information on ordering titles published by the American Ceramic Society, or to request a publications catalog, please call 614-890-4700.

Printed in the United States of America.

1 2 3 4–99 98 97 96 95

ISSN 1042-1122
ISBN 0-944904-99-8

Contents

Preface xi

CVI Processing of Ceramic-Matrix Composites 1
Theodore M. Besmann

Processing of CMCs from Novel Organometallic Precursors 13
G. Ziegler, J. Hapke, and J. Lücke

The Concept of Layered Interphases in SiC/SiC 23
R. Naslain

Development of Interfaces in Oxide and Silicate Matrix Composites 41
M.H. Lewis, M.G. Cain, P. Doleman, A.G. Razzell, and J. Gent

Interfacial Microstructure and Stability of BN-Coated Nicalon
Fiber/Glass-Ceramic Matrix Composites 53
John J. Brennan, Steven R. Nutt, and Ellen Y. Sun

Oxygen-Free Ceramic Fibers from Organosilicon Precursors and
E-Beam Curing 65
Hiroshi Ichikawa, Kiyohito Okamura, and Tadao Seguchi

Si-B-(N,C): A New Ceramic Material for High-Performance
Applications 75
H.-P. Baldus, G. Passing, D. Sporn, and A. Thierauf

Structural and Mechanical Characterization of Some Alumina- and
SiC-Based Fibers... 85
A.R. Bunsell, M.H. Berger, and N. Hochet

Design and Processing of All-Oxide Composites 95
Robert Lundberg and Lena Eckerbom

Dynamic Analysis of Pressure Infiltration Processes 105
Dhiman K. Biswas, Jorge E. Gatica, and Surendra N. Tewari

Fabrication of SiC-Matrix Composites Using a Liquid
Polycarbosilane as the Matrix Source 111
L.V. Interrante, C.W. Whitmarsh, and W. Sherwood

Development of a Two-Step, Forced Chemical Vapor Infiltration
Process 119
W.M. Matlin, D.P. Stinton, and T.M. Besmann

Carbon Fiber–Reinforced Silicon Nitride Composites by Slurry
Infiltration 125
   C. Grenet, L. Plunkett, J.B. Veyret, and E. Bullock

Reaction-Bonded Si_3N_4 Reinforced with Continuous SiC Fibers:
Processing and Interface Characteristics 131
   G.H. Wroblewska and G. Ziegler

Mechanism of Interface Formation in a Silicon Carbide Fiber–
Reinforced Magnesium Aluminosilicate 137
   Atul Kumar and Kevin M. Knowles

Filament Winding of SiC_{fiber}/Si_3N_4 Composites—Process and
Control of Fiber Distribution 143
   A. Kristoffersson, E. Laarz, R. Carlsson, and R. Lundberg

Interaction Between Capillary Flow and Macroscopic Silicon
Concentration in Liquid Siliconized Carbon/Carbon 149
   Frank H. Gern

Fiber-Reinforced Ceramic-Matrix Composite Fabrication by
Electrophoretic Infiltration 155
   S. Kooner, J.J. Campaniello, S. Pickering, and E. Bullock

Processing and Flexural Behavior of Alumina Multilayer Structures ... 161
   P. Letullier, K. Debray, E. Martin, J.M. Quenisset, and J.M. Heintz

Fiber Reinforcement of Reaction-Bonded Oxide Ceramics 167
   R. Janssen, J. Wendorff, and N. Claussen

Fabrication of Continuous-Fiber-Reinforced Ceramics with a
Nanosized Mullite Precursor 175
   O. Reese, B. Saruhan, B. Kanka, and H. Schneider

Processing and Microstructure of Continuous Fiber–Reinforced
$MoSi_2$-Based Composites 181
   A.R. Bhatti, A.J. Pritchard, and B. Mortimer

Development of Si-Ti-C-O Fiber-Reinforced SiC Composites by
Chemical Vapor Infiltration and Polymer Impregnation and
Pyrolysis 187
   S. Masaki, K. Moriya, T. Yamamura, M. Shibuya, and H. Ohnabe

Liquid Infiltration and Pyrolysis of SiC-Matrix Composite Materials... 193
   S. Casadio, A. Donato, C.A. Nannetti, A. Ortona, and M. Rescio

Interface and Mechanical Properties of Ceramic Fiber–Reinforced Silicon Nitride Composites Prepared by a Preceramic Polymer Impregnation Method 199
    K. Sato, H. Morozumi, A. Tezuka, O. Funayama, and T. Isoda

Advanced Refractory Ceramic-Matrix Composites from Polymer Precursor for Working Temperatures up to 1700°C 205
    R.A. Rabinovitch, S.L. Gershkohen, L.G. Polyak, N.M. Balagurova, V.I. Bezruchenko, N.A. Vygovskii, and E.A. Chernyshev

Fibers and Ceramic-Matrix Composites from Polymer Precursors: Methodology of Production and Properties 209
    R.A. Rabinovitch, N.M. Balagurova, S.L. Gershkohen, and E.A. Chernyshev

SiC- and Si_3N_4-Matrix Composites According to the Hot-Pressing Route 215
    K. Nakano, K. Sasaki, H. Saka, M. Fujikura, and H. Ichikawa

Rapid Densification of Carbon-Carbon Composites by Thermal-Gradient Chemical Vapor Infiltration 231
    I. Goleki, R.C. Morris, D. Narasimhan, and N. Clements

Characterizations of Carbon-Carbon Composites Elaborated by a Rapid Densification Process 237
    B. Narcy, F. Guillet, F. Ravel, and P. David

Optimization of the Fugitive Coating Thickness in Pressure-Infiltrated Mullite-Alumina Composites 243
    E.H. Moore, S. Shamasundar, and J.L. Kroupa

Effects of Carbon Coatings on SiC(Nicalon)-Pyrex Composites........ 249
    B. Mouhamath and N. Takeda

Effect of Polymer Addition on the Strength of Carbon Fiber–Reinforced Reaction-Bonded SiC-Matrix Composites 255
    Eiji Tani and Kazuhisa Shobu

Novel Organosilicon Precursors for Si-C Ceramics and Ceramic-Matrix Composites 261
    E.A. Chernyshev, S.A. Bashkirova, T.V. Tikchonovich, and V.V. Ivanov

Antioxidative Protective Coatings for Carbon Materials............... 267
    G.A. Kravetskii, V.I. Kostikov, A.V. Demin, and V.V. Rodionova

The Nanoscale Microstructure of 2-D C/C-SiC Ceramic Composites Processed via Silicon Capillary Impregnation...................... 275
    W. Braue, R. Pleger, and R. Weiss

Silicon Nitride Fiber Synthesis from Polycarbosilane Fiber by Radiation Curing and Pyrolysis Under Ammonia 281
    Seiji Kamimura, Kiyoshi Watanabe, Noboru Kasai, Tadao Seguchi, and Kiyohito Okamura

Effect of Rapid Heat Treatments on Electrical Properties of Polymer-Derived Ceramic Fibers... 287
    Masaki Narisawa, Koji Nakashiba, and Kiyohito Okamura

Reaction Mechanisms of SiC Fiber Synthesis from Radiation-Cured Polycarbosilane Fiber 293
    Masaki Sugimoto, Kiyohito Okamura, and Tadao Seguchi

Structure, Composition, and Mechanical Behavior at High Temperature of the Oxygen-Free Hi-Nicalon Fiber 299
    G. Chollon, R. Pailler, R. Naslain, and P. Olry

Structure and Thermal Evolution of SiC-Based Fibers with Low Oxygen Content 305
    G. Chollon, R. Bodet, R. Pailler, and X. Bourrat

SiC-Based Fibers with Low Free Carbon Content...................... 311
    A. Tazi Hémida, R. Pailler, R. Naslain, J.P. Pillot, M. Birot, and J. Dunoguès

Novel Synthesis and Characterization of Silicon Nitride Fibers........ 317
    Ulrich Vogt, Karl Berroth, and Georg Engeli

Ceramic Fibers for Functional Composite Materials 325
    T.M. Ulyanova and L.V. Titova

Environmental Effects on Creep and Stress Rupture Properties of Advanced SiC Fibers... 331
    H.M. Yun, J.C. Goldsby, and J.A. DiCarlo

SiC Fibers from Modified Polyorganosilanes: Production, Properties, and Application 337
    V.V. Ivanov, T.V. Tikchonovich, S.A. Bashkirova, L.G. Polyak, N.M. Balagurova, and E.A. Chernyshev

Models for the Thermostructural Properties of SiC Fibers 343
    J.A. DiCarlo, H.M. Yun, G.N. Morscher, and J.C. Goldsby

The Machining of CMC Materials Using YAG Laser 349
    I.P. Tuersley, A.P. Hoult, and I.R. Pashby

TEM and EDX Investigations of Experimental SiC_f-YMAS Composites 355
    F. Doreau, J. Vicens, and J.L. Chermant

Ceramic-Matrix Composites by a Preceramic Polymer Route 361
  Takeshi Isoda and Takemi Yamamura

Oxidation Kinetics of MgO-SiC Composites 371
  M.E.F. Carney and D.W. Readey

Effect of Interphase Carbon Thickness on Environmental Resistance
of Continuous Fiber–Reinforced Ceramic-Matrix Composites.......... 377
  James D. Cawley

Index 385

Preface

These proceedings are derived from presentations given at the Second International Conference on High-Temperature Ceramic Matrix Composites (HT-CMC-2). The first conference was held in Bordeaux, France, in 1993. Following the themes of that conference, the principal intent of these proceedings is to address essential engineering issues related to the large-scale implementation of CMCs.

CMCs have reached a critical stage in their development and application. The technological factors that govern progress include the development of design procedures and life prediction methodologies, as well as affordable manufacturing and process control strategies. This proceedings includes articles that address each of these issues.

Many of the basic mechanisms that control properties and performance are understood. These have been used to devise methods for calibrating the inelastic deformation that occurs in CMCs and thereby develop nonlinear constitutive laws capable of being implemented in finite element (FEM) codes. In consequence, design strategies are reaching maturity. Several articles give examples of nonlinear FEM calculations used for design.

Design criteria are based on a comparison of the stresses calculated by FEM with the ultimate tensile strength (UTS), measured at various orientations relative to the fiber direction. Fiber pull-out effects are also involved. In turn, the UTS and the pull-out lengths are predicated on the stochastics of fiber failure, modulated by frictional load transfer from the matrix. Several articles address the associated issues by introducing concepts of global and local load sharing. Of particular relevance to design are the statistical distributions of the UTS, including scale effects (if present).

The life expectancy of nonoxide CMCs is limited by stress oxidation, particularly upon cyclic loading. This effect is prevalent at intermediate temperatures and is similar to the "pest" effect found in other materials. The life limits are dictated by the chemistry of the constituents: the fibers, the matrix, and the fiber coatings. One of the major research themes on CMCs is concerned with the chemistry needed to suppress stress oxidation. A particular emphasis is on the development of fibers with stable high temperature chemistry having stable microstructures. These fibers are based primarily on SiC, Al_2O_3, and mullite. They are produced according to relatively low-cost manufacturing approaches, especially polymer precursor methods, because fiber costs are a predominant contributor to the overall manufacturing costs of CMC components. Several articles describe the status of fiber development.

The chemistry of the fiber coatings also has a crucial effect on the stress oxidation kinetics. These coatings must satisfy mechanical debonding and friction measurement, as well as have chemistries that resist rapid oxidation. For these reasons, C coatings have been replaced by BN, multilayer SiC/C, and oxide (*e.g.*, monazite) coatings. Articles that discuss these and other coating concepts are included in the proceedings.

Another approach to the stress oxidation problem is the development of oxide/oxide composites that are intrinsically resistant to embrittlement. These CMCs perform in accordance with different mechanisms than those that dictate the performance of nonoxide CMCs. These mechanisms and phenomena are described and the potential for these CMCs is discussed in several papers.

In addition to developing improved materials, a life prediction methodology is required. This development is facilitated by an analogy between stress oxidation and stress corrosion cracking. In accordance with this analogy, models that predict the crack growth rate as a function of the stress intensity and the oxidation kinetics are described. These models enable a lifing methodology to be defined and implemented.

Relatively high manufacturing costs are an impediment to large-scale commercial application of CMCs. The concern is being addressed through processing approaches compatible with near-net shape infiltration of fiber preforms, which also require minimal subsequent machining. Among the available approaches, those capable of high throughput by virtue of short cycle times and high-capacity furnaces are emphasized in the articles presented in these volumes. A substantial contribution to the overall CMC manufacturing cost derives from fiber manufacturing. This factor limits the fiber production approaches to those capable of yielding an affordable CMC, particularly the polymer precursor approaches noted above, hence the emphasis on these approaches in the articles within this proceedings

The proceedings is organized into two volumes that cover the above range of topics, from design to manufacturing. In both volumes, the invited articles are at the front and the contributed papers follow. The first volume includes articles that address design issues, CMC performance, and durability. The second volume contains articles concerned with manufacturing and processing as well as materials development issues, including new chemistries for fibers and interfaces.

A.G. Evans
R. Naslain

CVI PROCESSING OF CERAMIC MATRIX COMPOSITES

Theodore M. Besmann
Metals and Ceramics Division
Oak Ridge National Laboratory
Oak Ridge, TN 37831-6063
USA

ABSTRACT

Chemical vapor infiltration is a unique method for preparing continuous fiber ceramic composites that spares the strong but relatively fragile fibers from damaging thermal, mechanical, and chemical stress. The process is relatively complex however, requiring detailed phenomenological knowledge of the chemical kinetics and mass and heat transport to model. An overview is presented here of some of the current understanding of CVI and examples of efforts to optimize the processes.

INTRODUCTION

Chemical vapor infiltration (CVI) is simply chemical vapor deposition (CVD) on the internal surfaces of a porous preform and has been used to produce a variety of developmental materials. The greatest commercial interest is in continuous filament fibrous preforms, in which the high strength fibers can be aligned with the high stress directions. The preforms are infiltrated with carbon or ceramic, taking advantage of the relatively low stress CVD process, and resulting in carbon/carbon or ceramic matrix composites (CMCs).

Chemical vapor infiltration originated in efforts to densify porous graphite bodies by infiltration with carbon.[1] The technique has developed commercially such that half of the carbon/carbon composites currently produced are made by CVI (the remainder are fabricated by curing polymer impregnated fiber layups). The earliest report of CVI for ceramics was a 1964 patent for infiltrating fibrous alumina with chromium carbides.[2]

To the extent authorized under the laws of the United States of America, all copyright interests in this publication are the property of The American Ceramic Society. Any duplication, reproduction, or republication of this publication or any part thereof, without the express written consent of The American Ceramic Society or fee paid to the Copyright Clearance Center, is prohibited.

In CVI, reactants are introduced into a porous preform via either diffusion or forced convection, and the CVD precursors deposit the appropriate phase(s). As infiltration proceeds, the deposit on the internal surfaces becomes thicker. Thus after some length of time the growing surfaces meet, bonding the preform and filling much of the free volume with deposited matrix.

The major advantage CVI has over competing densification processes is the low thermal and mechanical stress to which the relatively sensitive fibers are subjected. CVD can occur at temperatures much more modest than the melting point of the deposited material, and therefore usually well below the sintering temperature. In addition, the process imparts little mechanical stress to the preform as compared to more traditional techniques such as hot-pressing.

There are a variety of techniques for infiltration classified as to whether they utilize diffusion or forced flow for transport of gaseous species, and whether thermal gradients are imposed.[3] Yet only two methods have been found to be practical. The most widely used commercial process is isothermal/isobaric CVI (ICVI),[4,5] which depends only on diffusion for species transport. These generally operate at reduced pressure (1-10 kPa) for deposition and transport rate control. Fixturing of the fibrous preforms is needed before initial densification to maintain the proper shape. Density gradients are minimized by a low reaction temperature, although in order to get economical densification rates, deposition is often sufficiently rapid to overcoat the outer surface before infiltration is complete. Interruption of the CVI process for periodic machining of parts is thus necessary for all but the thinnest parts, to open diffusion paths from the surface. Regardless, this diffusion-dependent process is still slow requiring several-week-long infiltration times. It is commercially attractive, however, because large numbers of parts of varying dimensions are easily accommodated in a single reactor.

The forced-flow/thermal gradient technique (FCVI) developed at Oak Ridge National Laboratory (ORNL) overcomes the problems of slow diffusion and restricted permeability, and has demonstrated a capability to produce thick-walled, simple-shaped components in times of the order of hours.[6] Its disadvantage lies in the extensive fixturing necessary to maintain both the thermal and pressure gradient, and as such has lagged ICVI in its application.

In a relatively recent development, CVI is being used to prepare low density materials for filtration applications. The 3M Company and DuPont Lanxide, Inc., are producing particulate filters for hot gas cleanup in advanced coal combustion systems to allow direct passage of combustion gases to a turbine.

ORNL with the Cummins Engine Company is developing filters for carbon particulate removal from diesel engine exhaust. The filters are prepared via ICVI within preforms with controlled porosity. In the case of the coal systems, these ~2 m long tubular filters have an infiltrated Nextel™ or Nicalon cloth structure which is coated with a slurry of fibers and then re-infiltrated to form the filtration surface.

Economic assessments of CVI processes have indicated that processing time (reactor cost) can be a major cost element, dominating others such as precursor cost.[7,8] Thus process optimization would most productively focus on reducing densification times while maintaining high and uniform density. This paper has therefore been devoted to a discussion of CVI with a focus on these issues.

Understanding of CVI has largely been the result of efforts to model the processes. Such models require representations of the physical system and chemical behavior. To accomplish this goal it is necessary to have accurate chemical kinetic relationships for deposition, appropriate mass transport expressions, thermal transport must also be considered for FCVI, and, finally, a means for describing pore distributions and how they will vary with densification.

CHEMICAL KINETICS

The CMC material systems of greatest interest utilize SiC as the matrix. SiC is typically deposited from methyltrichlorosilane (MTS), or similar chlorosilanes carried in hydrogen. While there has been a wealth of reports on the kinetics of SiC deposition from MTS, the results have been widely disparate.[9] Recent work has noted that SiC formation via the overall reaction is complex, and not a simple first order process, with the overall reaction written:[9-12]

$$CH_3Cl_3Si + H_2 \rightarrow SiC + 3HCl + H_2 \qquad (1)$$

Brennfleck et al[13] developed a first-order rate expression for SiC deposited from MTS in hydrogen that has been often used in CVI modeling. Such a simple approach has been shown to be incorrect, but still has resulted in predicted deposition rates not too distant from experiment. Figure 1 illustrates Brennfleck et al's[13] rate expression as compared to that of Besmann et al,[10] in which a reverse/etch rate relationship is used:

$$u = k_R C_R - k_P C_P \qquad (2)$$

where u is the deposition rate, k is the rate constant, C is the concentration, and the subscripts R and P signify reactant and product (HCl), respectively. As is seen in Fig. 1 and in experimental observations, the production of HCl has a strong effect on SiC deposition, particularly at lower temperatures. The effect is thus especially important for ICVI, which operates at ~1300 K, and for FCVI which utilizes temperatures between 1000 and 1500 K. Without accurately including the effect of HCl accumulation during infiltration the deposition rates will be markedly incorrect.

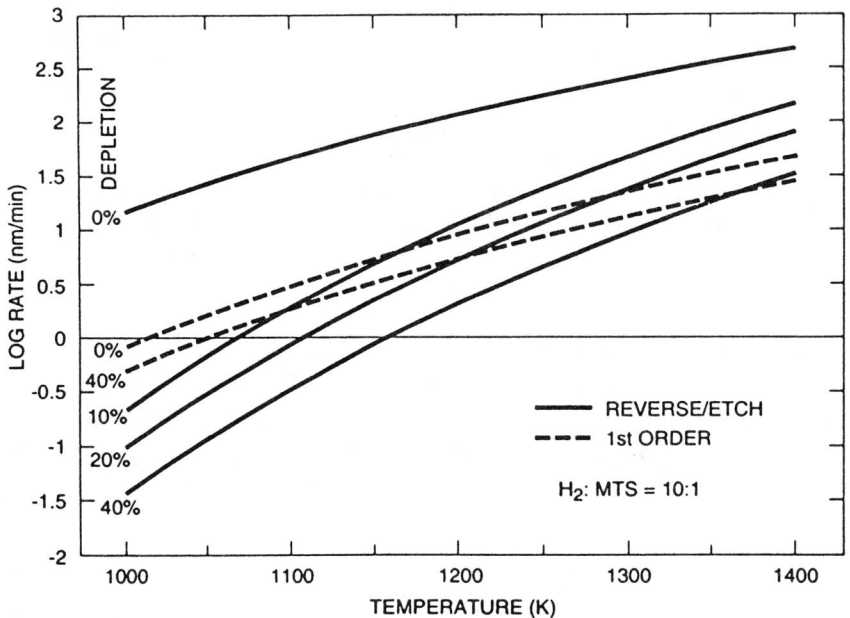

Fig. 1. HCl magnifies the effect of depletion on SiC deposition from MTS.

The effects of using the two relationships for predicting infiltration rates for ICVI was demonstrated by Sheldon and Chang.[14] Figure 2 is from their work and illustrates the substantial difference in optimal total pressure for densification of the coarser porosity in a preform, with the reverse/etch relationship resulting in over an order of magnitude lower pressure than does the first-order model. Experience confirms that lower pressures, approaching the 10 kPa range of Fig. 2, yield improved uniformity of densification.

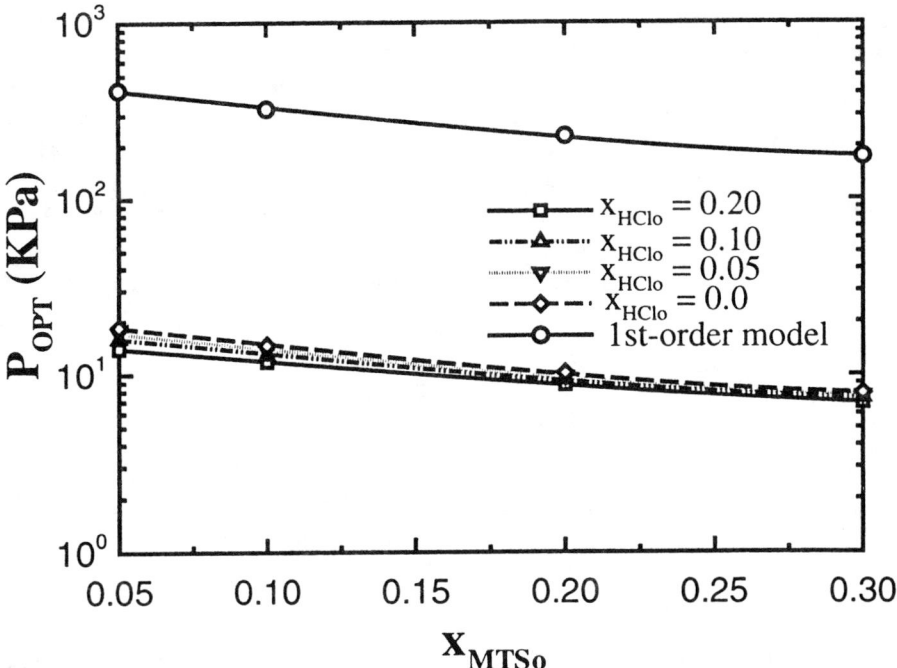

Fig. 2. Optimal pressure for ICVI infiltration of interbundle porosity based on the reverse/etch or first-order model at different initial HCl concentrations, X_{HClo}.[14]

MASS AND THERMAL TRANSPORT

Classical mass and thermal transport relationships serve for modeling these phenomena in CVI, complicated, however, by the fiber architecture. Most approaches use Darcy's Law for viscous flow, with Fickian and Knudsen diffusion as appropriate. The tortuosity, however, depends on the fiber architecture and may be anisotropic and permeability is best determined experimentally.[15] The Kozeny relationship has proven to be useful in this context.[16]

COMPETITIVE PHENOMENA

During CVI the primary objective is to maximize the rate of matrix deposition and minimize density gradients. Unfortunately, there is an inherent competition between the deposition reaction and the mass transport of the gaseous reactant and product species. Deposition reactions which are too rapid usually result in severe density gradients, where in ICVI there is essentially complete densification near the external surfaces and much lower densities in the interior regions. Alternatively, exceptionally slow deposition reactions require an uneconomically long time to densify a part. Relative rates of reaction and diffusion in ICVI, for example, can be expressed by a dimensionless parameter, the Thiele modulus, ϕ. The Thiele modulus is the ratio of the kinetic deposition rate to the diffusion rate:

$$\phi^2 = \frac{SuH^2}{VD_e} \qquad (3)$$

where S is the unit surface area, u is the first-order deposition rate, H is half the width of the preform (characteristic length), V is the molar volume of the matrix, and D_e is the effective diffusion coefficient. Although the deposition rate is not first-order, it can be treated that way at a fixed depletion, allowing use of the Thiele modulus in this form.

Chang et al[17] have derived a plot of final preform density as a function of the Thiele modulus for the first-order model of Brennfleck[13] for SiC deposition from MTS and the reverse/etch relationship of Besmann et al[10] for various HCl concentrations (Fig. 3). The HCl concentrations simulate depletion of the MTS precursor, although the sub-zero designation indicates the computations were performed assuming the initial porosity distribution before infiltration. It is apparent that first-order kinetics require larger Thiele modulus values for efficient densification, implying that faster diffusion or slower kinetics are needed. Assuming the reverse/etch model is correct, the system is more forgiving, with HCl buildup moderating the deposition rate. This effect is likely to account for the relative ease of infiltration using MTS.

MINIMIZATION OF INFILTRATION TIME

Non-random preform architectures result in a multi-modal pore size distribution that is not easy to quantify. This presents a particular challenge to densification and has resulted in recent approaches utilizing two steps, an initial stage in which the finer, intrabundle porosity is efficiently filled and a second stage which assumes the fine porosity is no longer available and seeks to rapidly fill the coarser, interbundle pores.

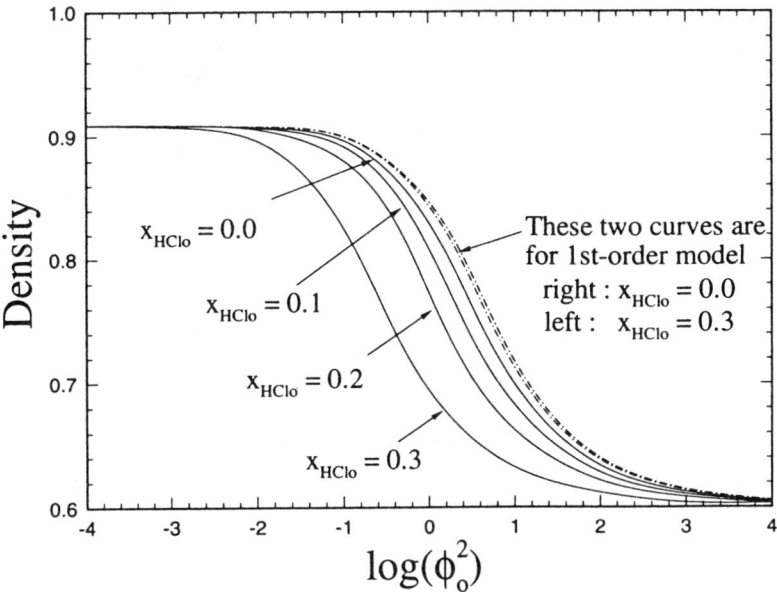

Fig. 3. Density vs. Thiele modulus for ICVI assuming a reverse/etch or first-order kinetic model.

Pore Size Distribution

The typical cloth layup for CVI generates a bimodal pore size distribution representing the intrabundle (within the fiber bundle or tow) and interbundle (between fiber bundles or tows in the cloth weave or between cloth layers) void space.[18-21] Starr[22] has estimated that in a 40 vol.% fiber, plain weave Nicalon™ (Nippon Carbon, Tokyo, Japan) cloth layup the fine and coarse porosity account for approximately equal fractions.

Minimizing ICVI Time

Chang et al[17,21] and Sheldon and Chang[14] have developed an approach to determining processing conditions which result in minimum infiltration times. The original work utilized the first-order kinetic model, however, they have recently found substantially better agreement with observations using the reverse/etch approach. Regardless of the kinetic relationships, however, two distinctly different sets of processing parameters are needed for the fine and coarse porosity. In ICVI the most efficient pore-filling protocol would infiltrate the fiber bundles uniformly, accepting some nonuniformity in the interbundle regions. After completion of bundle filling, then the conditions would be altered to most efficiently infiltrate the interbundle porosity. Their plot of the optimal pressures for each scale of porosity is shown in Fig. 4.

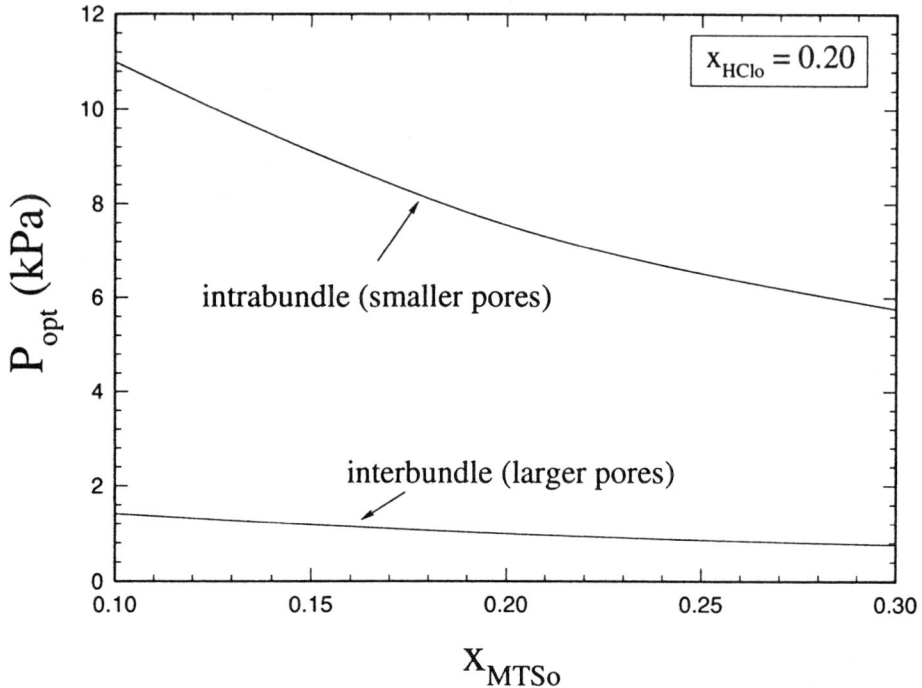

Fig. 4. Optimal pressures for ICVI within fiber bundles and in the interbundle volume.[23]

Minimizing FCVI Time

The computer model GTCVI[20] was utilized to develop FCVI conditions for filling the intra- and interbundle porosity in a Nicalon™ cloth layup. Infiltration of intrabundle porosity was assumed to be complete at 44 vol.% porosity for a preform with 40 vol.% fiber loading. Figure 5 shows the modeling results in which the thermal gradient was adjusted to allow uniform filling of the fine porosity within a 4.5 cm diameter disk 1.30 cm thick. This required increasing the cool-face temperature from what had been a standard 1025 K to 1225 K while maintaining the hot face at 1475 K. For the second stage, interbundle void filling, the cool-side temperature was reduced to 1025 K and the MTS concentration nearly doubled. The predicted densification profiles are shown in

Fig. 5 and agree with experiment to within 5%. The two-step process represents a 40% reduction in processing time.

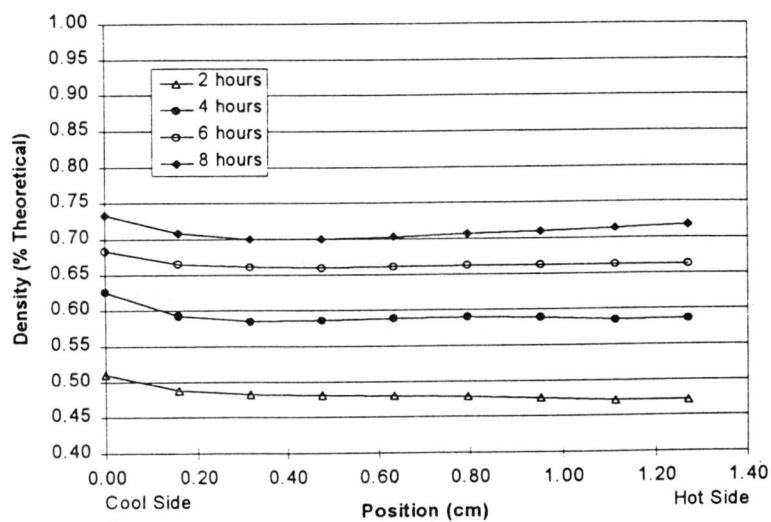

Fig. 5. Density vs. axial position for a 4.5-cm diameter disk during FCVI with an elevated cool-side temperature.

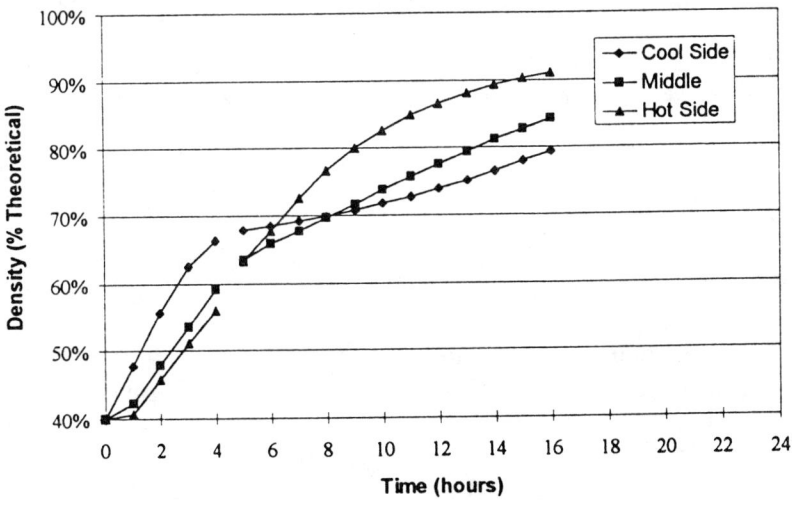

Fig. 6. Two-step FCVI process showing the change at 4 h to a lower cool-side temperature.

CONCLUSIONS

Our understanding of CVI processes has been substantially increased by the efforts to model the various systems. The complex, interactive phenomena in CVI are difficult to represent, however a combination of insightful modeling with key experimental efforts in delineating, for example, chemical kinetics has been very fruitful. There have been substantial improvement in processing times for FCVI, with potential for further advancement in this and other CVI techniques.

ACKNOWLEDGEMENTS

Research sponsored by the U. S. Department of Energy, Office of Fossil Energy, Advanced Research and Technology Development Materials Program, under contract DE-AC05-84OR21400 with Lockheed Martin Energy Systems, Inc.

REFERENCES

1. R. L. Bickerdike, A. R. G. Brown, G. Hughes, and H. Ranson. "The Deposition of Pyrolytic Carbon in the Pores of Bonded and Unbonded Carbon Powders"; pp. 575-583 in Proc. Fifth Conf. Carbon, Vol. I. Edited by S. Mrosowski, M. C. Studebaker, and P. L. Walker, Pergamon Press, New York, 1962.
2. W.C. Jenkin, "Method of Depositing Metals and Metallic Compounds Throughout the Pores of a Porous Body," U.S. Patent 3,160,517, Dec. 8, 1964,
3. W.J. Lackey and T.L. Starr, "Fabrication of Fiber-Reinforced Ceramic Composites by Chemical Vapor Infiltration: Processing, Structure and Properties"; pp. 397-450 in Fiber Reinforced Ceramic Composites. Edited by K.S. Mazdiyasni, Noyes Publications, Park Ridge, NJ, 1990.
4. E. Fitzer and R. Gadow, "Fiber-Reinforced Silicon Carbide," Am. Ceram. Soc. Bull. **65** [2], pp. 326-35 (1986).
5. R. Naslain and F. Langlais, "Fundamental and Practical Aspects of the Chemical Vapor Infiltration of Porous Substrates," High Temp. Sci. **27**, pp. 221-235 (1990)
6. T.M. Besmann, B.W. Sheldon, R.A. Lowden, and D.P. Stinton, "Vapor-Phase Fabrication and Properties of Continuous-Filament Ceramic Composites," Science **253**, pp. 1104-9 (1991).
7. T.M. Besmann, R.A. Lowden, D.P. Stinton, J.C. McLaughlin, B.W. Sheldon, T.L. Starr, and A.W. Smith, "Processing Science for Chemical Vapor Infiltration," Final Report, WL-TR-94-4044, U.S. Air Force Wright Laboratory, 1994.

8. Y.G. Roman and D.P. Stinton, "The Preparation and Economics of Silicon Carbide Matrix Composites by Chemical Vapor Infiltration," Materials Research Society proceedings, in press.
9. T.M. Besmann and M.L. Johnson, "Kinetics of the Low-Pressure Chemical Vapor Deposition of Silicon Carbide"; pp. 443-56 in Proceedings of the International Symposium on Ceramic Materials and Components for Engines (Las Vegas, NV, 1988). Edited by V.J. Tennery, American Ceramic Society, Westerville, OH, 1989.
10. T. M. Besmann, B.W. Sheldon, T.S. Moss III, and M.D. Kaster, "Depletion Effects of Silicon Carbide Deposition from Methyltrichlorosilane," J. Am Ceram. Soc. 75 [10], pp. 2899-903 (1992).
11. F. Loumagne, F. Langlais, and R. Naslain, "Kinetic laws of the chemical process in the CVD of SiC ceramics from CH_3SiCl_3-H_2 precursor," J. de Physique IV 3, pp. 527-533 (1993).
12. D. Lespiaux, F. Langlais, R. Naslain, S. Schamm, and J. Sevely, "Chlorine and Oxygen Inhibition Effects in the Deposition of SiC-based Ceramics from the Si-C-H-Cl System," J. Europ. Cer. Soc., pp. 81-88 (1995).
13. K. Brennfleck, E. Fitzer, G. Schoch, and M. Dietrich, "CVD of SiC-interlayers and their Interaction with Carbon Fibers and with Multilayered NbN-coatings"; pp. 649-55 in Proceedings of the Ninth International Conference on Chemical Vapor Deposition 1984 (Cincinnati, OH, 1984). Edited by McD. Robinson, C.H.J. van den Brekel, G.W. Cullen, J.M. Blocher, Jr., and P. Rai-Choudhury, Electrochemical Society, Pennington, NJ, 1984.
14. B.W. Sheldon and H.-C. Chang, "Minimizing Densification Times During the Final Stage of Isothermal Chemical Vapor Infiltration," Ceramic Trans. 42, pp. 81-93 (1994).
15. G.B. Freeman, T.L. Starr and T.C. Elston, "Transport Properties of CVI Preforms and Composites"; pp. 49-54 in Symposium Proceedings Vol. 168, Chemical Vapor Deposition of Refractory Metals and Ceramics (Boston, MA, 1989). Edited by T.M. Besmann and B.M. Gallois, Materials Research Society, Pittsburgh, PA, 1990.
16. T.L. Starr, "Model for Rapid CVI of Ceramic Composites"; pp.1147-55 in Proceedings of the Tenth International Conference on Chemical Vapor Deposition 1987 (Honolulu, HI, 1987). Edited by G.W. Cullen, Electrochemical Society, Pennington, NJ, 1987.
17. H.-C. Chang, T.F. Morse, and B.W. Sheldon, "Minimizing Infiltration Times During Isothermal Chemical Vapor Infiltration With Chlorosilanes," in preparation.

18. B.W. Sheldon and T.M. Besmann, "Reaction and Diffusion Kinetics During the Initial Stages of Isothermal CVI," J. Am. Ceram. Soc. **74** [12], pp. 3046-53 (1991).
19. J.K. Kinney, T.M. Breunig, T.L. Starr, D. Haupt, M.C. Nichols, S.R. Stock, M.D. Butts, and R. A. Saroyan, "X-ray Tomographic Study of Chemical Vapor Infiltration Processing of Ceramic Composites," Science **260**, pp. 789-92 (1993).
20. T. L. Starr, "Advances in Modeling of the Chemical Vapor Infiltration Process"; pp. 207-14 in Symposium Proceedings Vol. 250, Chemical Vapor Deposition of Refractory Metals and Ceramics II (Boston, MA, 1992). Edited by T.M. Besmann and B.M. Gallois, Materials Research Society, Pittsburgh, PA, 1992.
21. H.-C. Chang, T.F. Morse, and B.W. Sheldon, "Minimizing Infiltration Times during the Initial Stage of Isothermal Chemical Vapor Infiltration," J. Materials Proc. and Manuf. Sci. **2**, pp. 437-54 (1994).
22. T.L. Starr, "Gas Transport Model for Chemical Vapor Infiltration," J. Mat. Sci., in press.
23. B.W. Sheldon, personal communication.

PROCESSING OF CMCs FROM NOVEL ORGANOMETALLIC PRECURSORS

G. Ziegler, J. Hapke and J. Lücke
Institute of Materials Research, University of Bayreuth, D-95440 Bayreuth, Germany

ABSTRACT

Precursor synthesis in the system Si-C-N was improved regarding viscosity, wettability, long-term stability and ceramic yield. Novel organometallic precursors were developed in the system Si-Ti-C-N from mixtures of organometallic precursors for TiC/TiN and of various oligomeric precursors for Si-C-N (polycarbosilazanes). Thermal behavior of those polymer-derived ceramics was investigated in the temperature range up to 1400 °C. The composition of the resulting materials may be changed in a wide range (Si_3N_4-SiC-Ti(C,N)) via chemical modifications, maximum pyrolysis temperature and heating rate. The potential of those novel precursors for processing of fiber-reinforced composites and nano/micro composites as well as hybride composites is demonstrated.

INTRODUCTION

With the need and development of non-oxide based advanced ceramics for high temperature applications organometallic compounds have been investigated for their use as starting compounds for ceramic materials via pyrolysis [1]. Today, precursors exist for various ceramic matrices, e.g. carbides and nitrides of B, Al and Si [2, 3, 4]. Depending on their physical state (solid, liquid, gas) different processing techniques have been developed in order to use these organometallic compounds for fiber spinning, coatings, binders, high-purity powders as well as for infiltration of porous matrices (porous ceramics, fiber preforms).
Because of the high shrinkage from the organometallic (density ≈ 1 gcm^{-3}) to the inorganic state (density 2.4-3.4 gcm^{-3}) highly porous solids or powders are often obtained after pyrolysis. With the exception of vapour deposition (CVD / CVI) and fiber spinning [5, 6], only a few attempts have been successful in making dense and crack-free monolithic or composite materials [7, 8].
During the last years it was demonstrated that the pyrolysis of organometallic precursors, particularly based on Si-C-N, has a high potential for processing fiber-

To the extent authorized under the laws of the United States of America, all copyright interests in this publication are the property of The American Ceramic Society. Any duplication, reproduction, or republication of this publication or any part thereof, without the express written consent of The American Ceramic Society or fee paid to the Copyright Clearance Center, is prohibited.

reinforced composites. This procedure includes the liquid impregnation of suitable precursors with subsequent crosslinking to polymers and pyrolysis at temperatures higher than 700 °C. The liquid oligomer infiltration and pyrolysis method (**LOIP**) is a low temperature, pressureless process from which short processing time, low costs and the possibiltity of manufacturing complex-shaped parts are expected [9]. For this purpose **liquid** precursors are necessary which have to be optimized regarding viscosity, wettability and ceramic yield [10]. In order to control the resulting properties the constituent chemical elements should be variable over a wide range. At least appropriate amounts of the precursors have to be available.

Besides continuous fiber-reinforced ceramic composites advanced particulate-reinforced ceramic composites (micro- and/or nano-scaled) have a high potential [11]. In general, small grain sizes of the reinforcing component, down to the nanometer scale, and a homogeneous distribution of the elements are desired in order to improve strength, fracture toughness and high temperature characteristics. The pyrolysis of liquid precursors seems to be the most practicable way for the synthesis of hybride composites [12], which combine the toughening mechanisms of both, the fiber- and the particulate-reinforced composites. Pyrolysis of an appropriate precursor produces an amorphous residue from which at least two nanosized ceramic phases may crystallize which mutually retard their grain growth.

The aim of this work was to improve silazane-based precursors and to develop novel precursors by introducing further chemical elements, e.g titanium, into the molecules. The oligomers were modified regarding structure and chemical composition in order to combine low viscosity and high ceramic yield. Oxygen-free liquid precursors for amorphous Si-Ti-C-N materials with increased crystallization temperature are presented.

EXPERIMENTAL PROCEDURE

Polycarbosilazanes were synthesized at room temperature by ammonolysis of substituted chlorosilanes R_zSiCl_{4-z} (R = H, alkyl, $CH=CH_2$). The reactions

$$n\, R_zSiCl_{4-z} + NH_3 \text{ (excess)} \rightarrow [R_zSi(NH)_{(4-z)/2}]_n + n(4-z)\, NH_4Cl \qquad (1)$$

were carried out in dry toluene. After removing the ammonium chloride and solvent the liquid **polysilazanes** were obtained in 75 to 85 % yield (Tab. I, **PS**1-4). Titanium modified precursors were synthesized by reaction of low molecular weight silazanes with $Ti(NMe_2)_4$ (Me: CH_3) [12, 13] at temperatures of about 110-120 °C, resulting in **polytitanosilazanes** with various Si/Ti ratios (**PTS**1-2).

$$[RSiMe(NH)]_n + m\, Ti(NMe_2)_4 \rightarrow$$

$$[RSiMe\{NTi(NMe_2)_3\}_{m/n}(NH)_{1-m/n}]_n + m\, HNMe_2 \qquad (2)$$

Purity and structure of the precursors were determined spectroscopically and by chemical analysis. Precursor synthesis, characterization and processing were carried out in an inert gas atmosphere (Schlenk technique, glove box) to avoid oxygen impurities.
UD-reinforced ceramics were produced with carbon fibers (Torayca T800, Tenax HT12000) and silicon carbide fibers (Tyranno TYS1H08PX, Nicalon 607). Coatings of carbon and/or silicon carbide were applied by chemical vapour deposition. The fiber bundles were embedded into metal forms (10x8x4 mm, 40-60 vol%) and infiltrated under reduced pressure (5×10^{-2} mbar) with the **liquid** precursors. After curing at about 250 °C the samples were heated up to 1000 °C (1 K/min, flowing argon, tube furnace).
Crystallization as well as reaction behavior of the CMC components were investigated at temperatures up to 1400 °C (argon, 6 to 48 h). The samples were characterized by chemical analysis, thermoanalytical measurements, X-ray diffraction (powdered samples), density measurements and porosimetry as well as by SEM and bending tests.

RESULTS AND DISCUSSION

<u>Improvement of Precursor Synthesis in the System Si-C-N</u>

With respect to the infiltration process of fiber preforms, one essential result of the improved precursor synthesis in the system Si-C-N is that all precursors are liquid. The pure carbosilazanes are colorless to pale yellow with a viscosity value of 0.1 to 100 Pas depending on the starting silanes and the reaction temperature. The reactivity

Table I Structural units and ceramic yield ($\Delta m/m$) of silazane-based precursors
(* due to variation of flow and heating rate, Me: CH_3)

	-[R_1SiR_2-NR_3-]$_n$			
polysilazanes	R_1	R_2	R_3	$\Delta m/m$, %
PS1	H	Me	H	< 10
PS2	$CH=CH_2$	Me	H	< 10
PS3	H + $CH=CH_2$	Me	H	10-40*
PS4	H + $CH=CH_2$	$NHSiR_3$ + Me	H	70-80
polytitanosilazanes				
PTS1	H	Me	$Ti(NMe_2)_3$	67
PTS2	$CH=CH_2$	Me	$Ti(NMe_2)_3$	57

of the precursor (stability, curing temperature) is controlled by the type of functional groups (NH, SiH, $CH=CH_2$, see Tab. I) in the silazane chain. By varying the chemical substituents, the C/N ratio in the ceramic matrix can be changed over a wide range. With vinylic groups, carbon-rich residues are achieved whereas branched oligomers lead to a higher nitrogen content. Intra- or intermolecular mixtures of vinylic- and Si-H-substituents are responsible for high ceramic yields.

The precursors were analyzed by TGA. Only for the branched precursors the curing step (at about 250 °C, Fig. 1) is well separated from the ceramization step at about 550 °C (PS4, PTS1 and PTS2). If the precursor contains only rings and chains with no reactive groups at the silicon atom (PS1, PS2) the ceramic residue is poor. The possibility of crosslinking reactions (Si-H + $CH=CH_2$, PS3) and additionally branching the chains [-$Si(NH)_3$] (PS4) raises the yield up to 80% (see Fig. 1, Tab. I). PS3 exhibits a drastically higher mass loss in the TGA experiments than during pyrolysis in the tube furnace. This may be explained by higher heating rates and higher gas velocity in the thermobalance, indicating that hydrosilylation reactions could not take place.

Development of Polytitanosilazanes

The polytitanosilazanes **PTS** with R = H (PTS1) are obtained as solids, the polytitanosilazanes with R = -$CH=CH_2$ (PTS2) are highly viscous resins. In both cases the Si/Ti atomic ratio was approximately 3.5. The homogeneous distribution of Si and Ti in the precursor, which is necessary for the formation of homogeneous pyrolysis products and nanocomposites after crystallization, is caused by a statistical attachment of Ti along the polysilazane backbone. The polytitanosilazanes (PTS1 and PTS2) exhibit a much higher pyrolysis yield than the corresponding polysilazanes (PS1 and PS2), which were used as starting compounds (see Tab. I, Fig. 1). These linear and cyclic polysilazanes do not have the possibility of crosslinking reactions, therefore, they have more tendency to vaporize. In the case of PTS1 and PTS2, the pyrolysis process starts at about 300°C and is finished below 650 °C with relatively high pyrolysis yield of up to about 70 %. This means that the pyrolysis temperature can be reduced by approximately 50-100 °C in comparison to the branched and crosslinkable polysilazane PS4. The pyrolysis product of PTS2 contains a higher amount of excess carbon [12, 13]. The carbon is introduced by the vinyl group $CH=CH_2$ of the precursor. Therefore, further research was concentrated on PTS1.

With rising Si/Ti ratio the appearance of the products at room temperature changed from liquids (Si/Ti < 2.5) with increasing viscosity to solids (Si/Ti = 2.5-8) and back again to liquids (Si/Ti > 13) with decreasing viscosity (Fig. 2). The pyrolysis yield depends also strongly on the Si/Ti ratio in the precursor, with a maximum pyrolysis yield at Si/Ti ≈ 6-7 (see Fig. 2). The precursors with a Si/Ti ratio between 3.5 and 8 meet the requirements for the formation of hybride composites (combination of high pyrolysis (67-72%) yield and homogeneous distribution of Si and Ti). Since the mass loss of the precursors by evaporation is only approximately 10-20 % (see Fig. 1), most of the expensive Ti-compound is converted into the pyrolysis product.

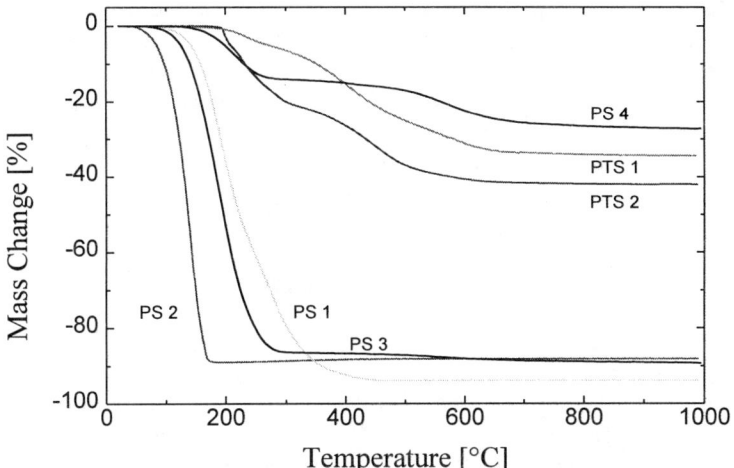

Figure 1 TGA of Several Si-C-N-Based Precursors (5 K/min, Flowing Argon)
PS1-PS4: Polysilazanes   PTS1-PTS2: Polytitanosilazanes

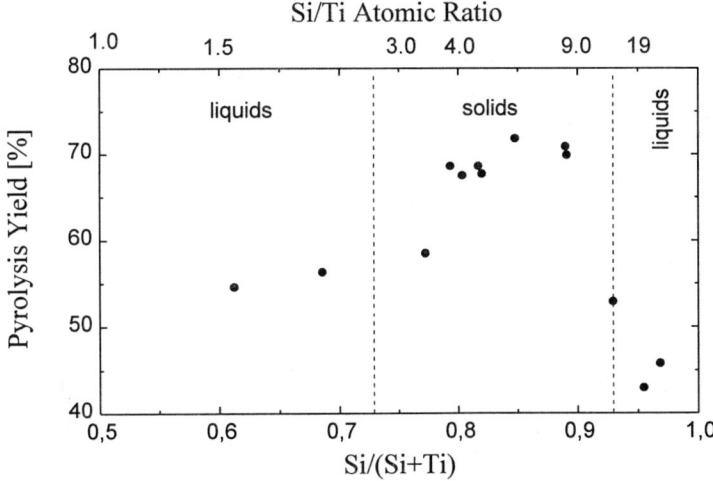

Figure 2 Pyrolysis Yield vs Si Content in the Polytitanosilazanes PTS1
(Tube Furnace, 1 K/min, 1000 °C, Flowing Argon)

Although the precursors are obtained as solids after cooling from the synthesis temperature to room temperature, they may be used for infiltration if the reaction is performed at lower temperatures (about 60 °C) and shorter reaction times. However, under these conditions the conversion is incomplete, and thus, the precursors contain unreacted species. Nevertheless, after infiltration of fiber preforms with liquid PTS1 at about 60 °C, the reaction can be completed by heating up to 110-120 °C. Further heat treament above this temperature converts the precursors into infusible solids or amorphous ceramics.

Processing of Continuous Fiber-Reinforced Si-C-N Ceramics

The polysilazanes, precursors for Si-C-N matrices, exhibit a good wettability for the fiber preforms as well as for the resulting porous ceramics. On repeating the infiltration and pyrolysis steps the calculated values of density and porosity are nearly achieved (Fig. 3a). With respect to the degree of filling, the content of solvent and the ceramic yield best results (that means a low number of repeating cycles) are achieved with **solvent-free** polysilazane PS4. No significant difference between the infiltration behavior of carbon or silicon carbide fibers, coated or uncoated, was observed (Fig. 3b). The open porosity is reduced from about 40-60 % in the as-received preforms to < 2 % after several infiltration and pyrolysis steps. The density reaches about 90 to 95 % of the theoretical

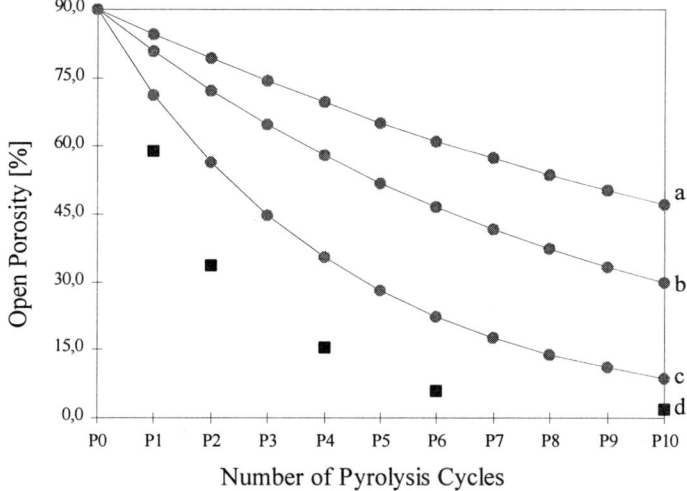

Figure 3a Calculated and Measured Porosity Values of a Model Sample
a) 50 % Solvent, 75 % Filling   b) 50 % Solvent, 100 % Filling
c) No Folvent, 100 % Filling   d) PS4 Infiltrated Short-Fiber Preform
Porosity = $(1-V_F) [1-\rho_P/\rho_M (1-\Delta m/m)]^n$

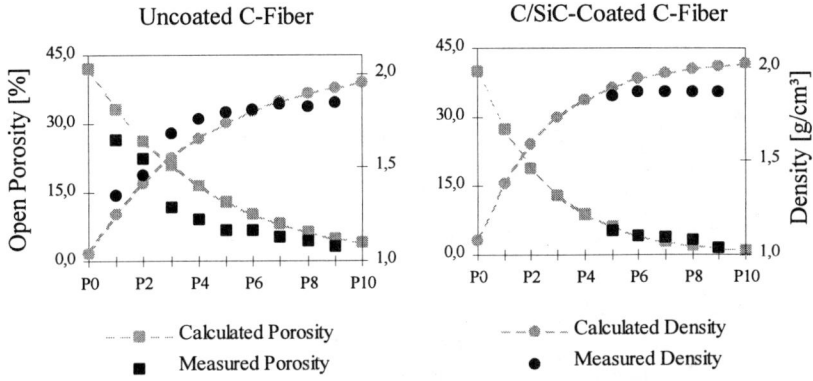

Figure 3b Densification of CMCs for Different Types of Fibers

Density: $\rho = \rho_F V_F + \rho_M (1-V_F)(1-[1-\rho_P/\rho_M (1-\Delta m/m)]^n)$

n Number of Infiltration/Pyrolysis Cycles, V_F Fiber Volume Content,
$\Delta m/m$ Pyrolysis Yield, ρ Density of Precursor (P), Matrix (M) or Fiber (F)

value indicating the existence of some porosity. The reason for this might be that the open pore structure reaches molecular dimensions before 'bottle-neck' pores inside the composite are completely filled.

During pyrolysis no fiber damage is observed. This is due to the low ceramization temperature (600 to 1100 °C) where no pressure is needed to achieve high ceramic yields. SEM investigations show that the ceramic matrix is built up homogeneously, and space filling between the fibers (down to 0.1 µm) is complete. Even voids between the fiber bundles (10 to 100 µm) are gradually filled. Nevertheless, in some cases larger voids (> 0.2 mm), originating from misalignment of the fiber bundles, may still exist.

In the bending tests strength values of up to 400 MPa, E-moduli of up to 200 GPa and strain values up to 2.5 % were obtained depending on the type of fibers and coating. The highest strength values of 400 MPa were achieved with C-coated SiC fibers. The best pseudoplastic behavior was observed with reinforcement with uncoated C fibers. It was experienced that small changes in precursor chemistry may change the stress-strain behavior dramatically.

Crystallization Behavior of Polysilazanes and Polytitanosilazanes

The crystallization behavior of the pyrolysis products was investigated by X-ray powder diffraction (Fig. 4 and 5). Additionally, precursor derived ceramic fibers used for preparation of the CMCs (Nicalon: Si-C-O and Tyranno: Si-C-Ti-O, 2 wt%-

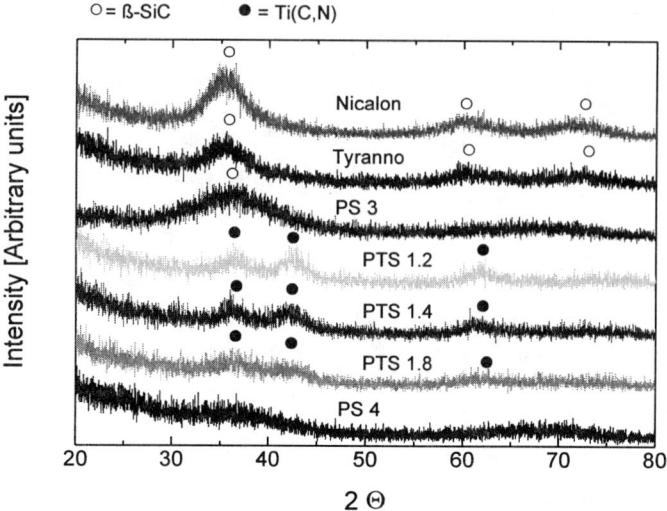

Figure 4 X-ray Powder Diffraction Pattern of Annealed Pyrolysis Products
(Graphite Furnace, 1200 °C, 48 h, 20 K/min, Flowing Argon)

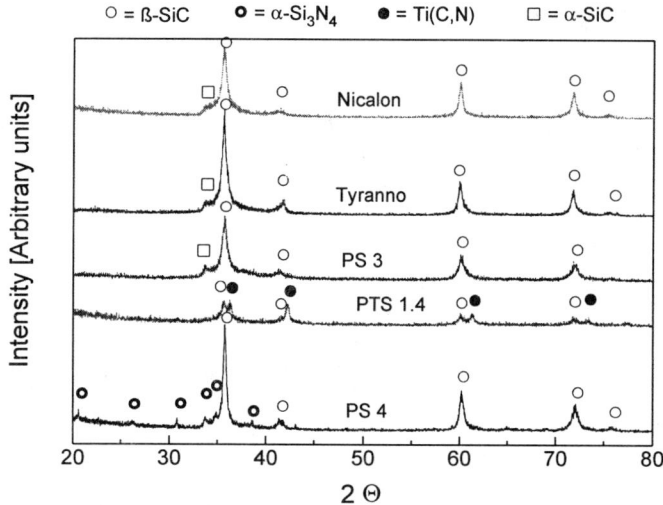

Figure 5 X-ray Powder Diffraction Pattern of Annealed Pyrolysis Products
(Graphite Furnace, 1400 °C, 48 h, 20 K/min, Flowing Argon)

Titan) were examined. After 48 h at 1200 °C the fibers already showed broad peaks of β-SiC interferences. Also the annealed product of the linear, but crosslinkable polysilazane (PS3) exhibits some formation of crystalline β-SiC, however, less than the Nicalon and Tyranno fibers. In opposite to these materials, the pyrolysis product of the branched and crosslinkable polysilazane PS4 is mainly amorphous. The annealed pyrolysis products of polytitanosilazanes PTS1.2/1.4/1.8 (Si/Ti = 2,4,8) consist of nanostructured Ti(C,N) embedded in an amorphous Si-C-N-matrix. Reflections of SiC or Si_3N_4 were not detected.

After 48 h at 1400 °C the formation of β-SiC with small amounts of α-SiC was detectable in the polysilazane-based materials as well as in the fibers (Fig. 5). Additionally, it should be mentioned, that after annealing at 1400 °C the pyrolysis product of precursor PS4 is more crystalline than that of precursor PS3, which is contrary to the annealing results at 1200 °C. This indicates that small differences in the structure of the precursors are responsible for a different pyrolysis behavior regarding temperature of decomposition as well as maximum rate of formation and growth of nuclei. An important result is that in the examined polytitanosilazane-based materials the amount of crystalline β-SiC phase is drastically reduced in comparison to pure silazane-based materials. Besides crystalline silicon carbide, only small amounts of silicon nitride are observed in annealed PS4 indicating that decomposition reactions of the amorphous ceramic predominate.

CONCLUSIONS

In the case of Si-C-N precursors the C/N ratio in the ceramic matrix can be changed over a wide range by varying the chemical substituents. With vinylic groups carbon-rich residues are achieved, whereas branched oligomers lead to a higher nitrogen content. Intra- or intermolecular mixtures of vinylic- and Si-H-substituents are responsible for high ceramic yields.

Polytitanosilazanes with varying Si/Ti ratio were developed. Viscosity and pyrolysis yield can be controlled by the Si/Ti ratio, nature and molecular weight of the polysilazane as well as by the degree of conversion. By incorporating titanium in the Si-C-N precursor the pyrolysis temperature may be reduced by about 50-100 °C in comparison to the branched and crosslinkable polysilazanes. The precursors with a Si/Ti ratio between 3.5 and 8 meet the requirements for the formation of fiber-, nano- and hybride-composites (high pyrolysis yield, homogeneous distribution of Si and Ti). These precursors exhibit a high pyrolysis yield between 67 and 72 %.

The precursors developed are well suited for infiltration of fiber preforms. During pyrolysis no fiber damage was observed. By repeating infiltration/pyrolysis of fiber preforms high density values have been achieved (porosity < 2 %). Moreover, controlled porosity values can be adjusted.

The annealed Si-Ti-C-N materials consist of nano-structured Ti(C,N) embedded in the amorphous Si-C-N matrix (1200 °C, 48 h, Ar). At higher temperatures (> 1400 °C) the matrix starts crystallizing to β-SiC. In the polytitano-based materials with

Si/Ti=4 the crystallinity of β-SiC is drastically reduced compared to pure silazane-based materials as well as to Nicalon and Tyranno fibers.

ACKNOWLEDGEMENTS

The authors thank the Deutsche Forschungsgemeinschaft and the Bayerisches Wirtschaftsministerium for financial support.

REFERENCES

[1] D. Seyferth, G.H. Wiseman, High-Yield Synthesis of Si_3N_4/SiC Ceramic Materials by Pyrolysis of a Novel Polyorganosilazane, J. Am. Ceram. Soc. **67**, C-132, 1984
[2] Y.D. Blum, K.B. Schwartz, R.M. Laine, Preceramic Polymer Pyrolysis, J. Mat. Sci. **24**, 1707, 1989
[3] D. Mocaer, R. Pailler, R. Naslain, Si-C-N Ceramics with a High Microstructural Stability Elaborated from the Pyrolysis of New Polycarbosilazane Precursors, J. Mat. Sci. **28**, 2615, 1993
[4] D. Bahloul, M. Pereira, P. Goursat, N.S.C.K. Yive, R.J.P. Corriu, Preparation of Silicon Carbonitrides from an Organosilicon Polymer I, J. Am. Ceram. Soc. **76**, 1156, 1993
[5] S. Yajima, Y. Hasegawa, J. Hayashi, M. Timura, Synthesis of Continuous Silicon Carbide Fiber with High Tensile Strength and High Young's Modulus, J. Mat. Sci. **13**, 2569, 1978
[6] T. Yamamura, Development of High Tensile Strength Si-Ti-C Fiber Using an Organometallic Polymer Precursor, Polym. Prep. **25**, 8, 1984
[7] R. Riedel, G. Passing, H. Schönfelder, R.J. Brook, Synthesis of Dense Silicon-Based Ceramics at Low Temperatures, Nature **355**, 714, 1992
[8] M. Seibold, P. Greil, Thermodynamics and Microstructural Development of Ceramic Composite Formation by Active Filler-Controlled Pyrolysis (AFCOP), J. Eur. Ceram. Soc. **11**, 105, 1993
[9] M. Keuthen, P. Gerstel, B. Martin, G. Ziegler, Carbon Fiber Reinforced Si_3N_4/SiC Ceramics by Polymer Pyrolysis, Proc. 8th Intern. Conf. on Modern Materials & Technologies (CIMTEC), Florence, 1994, in press
[10] J. Lücke, M. Keuthen, G. Ziegler, Development of New Silazanes for Infiltration/Pyrolysis Processing of Composites, Proc. Intern. Conf. Ceram. Proc. Sci. Technol., Friedrichshafen, 1994, in press
[11] K. Niihara, New Design Concepts of Structutal Ceramics - Ceramic Nanocomposites, J. Ceram. Soc. Jap. **99**, 974, 1991
[12] J. Hapke, P. Gerstel, M. Keuthen, G. Ziegler, Formation of Nonoxide-Based Composites from Liquid Precursors, eds: P. Duran, J.F. Fernadez, Proc. Third Euro-Ceramics, Madrid, 149, 1993
[13] J. Hapke, G. Ziegler, Synthesis and Pyrolysis of Liquid Organometallic Precursors for Advanced Si-Ti-C-N Composites, Adv. Mater. **7**, 1995, in press

THE CONCEPT OF LAYERED INTERPHASES IN SiC/SiC

R. Naslain
Laboratoire des Composites Thermostructuraux
UMR 47 CNRS-SEP-UB1, Domaine Universitaire
33600-Pessac, France

Abstract

SiC/SiC composites exhibit a non-brittle behavior when an interphase is used to control the fiber-matrix bonding. It is proposed here that a material with a layered crystal structure or a layered microstructure might be the best way to accomplish the interphase functions. This generalized form of the layered interphase concept is illustrated with examples of model and real composites comprising : (i) interphases of anisotropic pyrocarbon, turbostratic BN, Ti_3SiC_2 and phyllosiloxides (a new class of all-oxide layered materials deriving from micas) for the interphases with layered crystal structures and (ii) multilayered $(PyC-SiC)_n$ interphases for those with a layered microstructure. Correlations between the nature of the interphase, the mechanical behavior and the effect of the environment are discussed.

1- Introduction

The mechanical behavior of SiC/SiC composites is closely related to the fiber-matrix (FM) bonding. It is brittle when that bonding is too strong and tough when it has been weakened enough during processing, e.g. by introducing a thin interphase between the two constituents. In the SiC/SiC composites prepared fifteen years ago, the fibers, e.g. the Nicalon Si-C-O fibers[*] , were directly embedded in the SiC-matrix by chemical vapor infiltration (CVI) and the materials were mainly brittle. Fortunately,

---

[*] from Nippon Carbon, Japan

these fibers were no longer stable beyond about 1100°C, their surface usually exhibiting an excess of carbon and oxygen (presumably as a mixture of free carbon and silica) introduced in the fibers during processing. Although this surface layer was very thin (typically 10 nm) it weakened the FM-bonding sufficiently to yield some non-linearity in the stress-strain curve and to suggest that a deliberate addition of more carbon in the FM interfacial zone might enhance the phenomenon. The research further devoted to SiC/C/SiC composites has indeed clearly established that a thin layer of **pyrocarbon** (PyC) deposited on the fiber surface prior to the infiltration of the SiC-matrix was an efficient and elegant way of tailoring the FM-bonding [1-4].

Up to now, SiC (Nicalon)/PyC/SiC (CVI) composites exhibited two weak points for applications at high temperatures in oxidizing atmospheres. First, the Nicalon fibers are too limited in thermal stability. This weak point can now be considered as overcome with the development of near-stoichiometric oxygen-free SiC fibers (e.g. Hi-Nicalon[*]) [5]. The second weak point, i.e. the sensitivity of the PyC-interphase to oxygen, is much more serious taking into account the residual porosity and the microcracking of the matrix. Recent progress has been made in this field including : (i) the use of gas-tight seal-coatings consisting usually of glass-former materials or/and (ii) the reduction of the PyC-interphase thickness [6-7]. However, the poor oxidation resistance of the PyC-interphase, particularly at 500-700°C remains the key issue in the development of SiC/PyC/SiC composites.

Attempts have been made to extend the concept of layered interphases to materials exhibiting an oxidation resistance bettter than that of PyC. Hex-BN has been a first logical choice. It has almost the same crystal structure and its oxidation starts at a higher temperature ($\approx$ 900°C vs $\approx$ 500°C) with formation of a condensed oxide, B_2O_3 [8, 9]. Very recently, the concept has been extended to multilayered interphases such as the $(PyC-SiC)_n$ interphases in which PyC is partly replaced by SiC [10]. Finally, layered oxides although they have not yet been used in SiC/SiC composites also have some interesting potential [11, 12].
The aim of the present contribution is to depict, on the basis of the data which have been accumulated on SiC/SiC composites, a generalized form of the layered interphase concept in which the stratification of the

---

[*] an experimental fiber from Nippon Carbon, Japan

interphase is either at the sub-nm scale (as in pyrocarbon or hex-BN) or at the sub-µm scale (as in (PyC-SiC)$_n$.).

2- Layered Interphase Concept

In SiC/X/SiC composites, the interphase X has to simultaneously accomplish different functions including : (i) FM load transfer, (ii) deflection of the matrix microcracks (the so-called fuse function) parallel to the fiber axis, (iii) absorption of part of the thermal residual stresses (TRS) resulting from differential thermal expansion during processing, (iv) a function of diffusion barrier and finally (v) the interphase material has to be chemically compatible at high temperatures with both the fiber and the matrix, as well as with the atmosphere. Some of these functions are contradictory. The load transfer function supposes a strong enough FM-bonding whereas the fuse function requires a rather weak FM-bonding. In a similar manner, the chemical compatibility with the SiC-based fibers and the SiC-matrix suggests the use of a non-oxide X-material whereas that with oxidizing atmospheres favors the choice of oxide X-materials. Finally, the absorption of the radial TRS requires a compliant material.

We think, on the basis of our experimental data, that the use of a material X exhibiting either a **layered crystal structure** or (and) a **layered microstructure**, in which the layers are weakly bonded to one another might be the best way to accomplish the fuse function, assuming that the layers are in a first approximation oriented parallel to the fiber surface (fig. 1). Under such conditions, a matrix microcrack propagating in mode I will be easily deflected parallel to the fiber axis (with partial debonding of the interface(s) between the layers). Such a situation can be imagined at different scales. First, it can occur at the nm-scale, the layers being either single atomic planes (as in PyC or hex-BN) (fig. 1b, 2b) or sheets of more complex entities (as in micas). Unfortunately, the number of materials exhibiting a layered crystal structure with weak interlayer bonding and which are simultaneously refractory, resistant to oxidation and chemically compatible with SiC and Nicalon-type fibers at high temperatures, is extremely limited. However, if the concept is extended to the µm-scale, i.e. to a material with a **layered microstructure**, the choice becomes much larger. Under such an assumption, the interphase consists of a stack of thin films of different nature oriented parallel to the fiber surface (fig. 2d-f). In such multilayered interphases, referred to as (X-Y)$_n$ where X-Y is the sequence which is repeated, easy crack propagation paths parallel to the

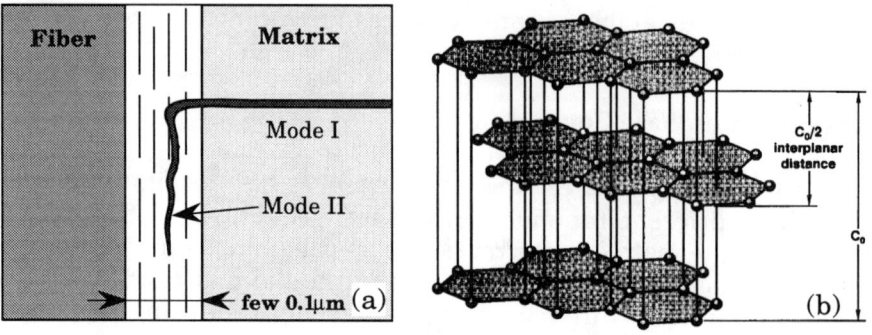

Fig. 1 : Deflection of a matrix microcrack within a layered structure interphase (a), crystal structure of graphite (b).

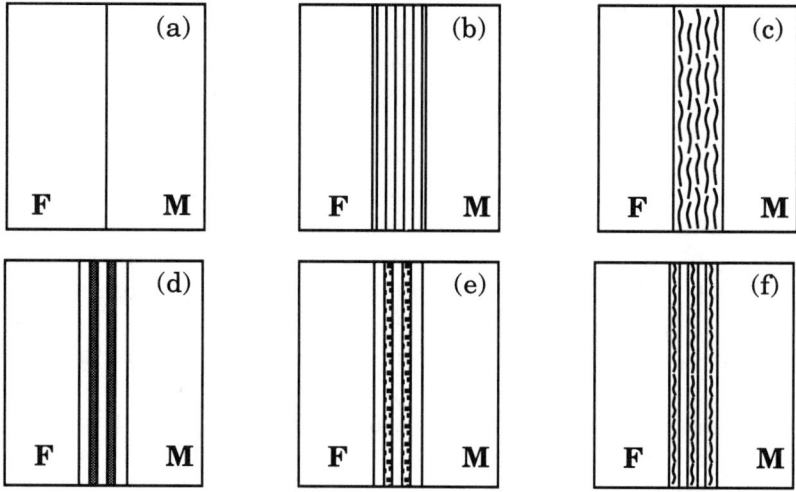

Fig. 2 : Weak fiber-matrix interface (a) and various forms of layered interphases : ideal (b) and defective (c) layered crystal structure interphase, multilayered (X-Y)n interphase with weakly bonded X, Y phases (d), with a phase exhibiting a porous microstructure (e) or a defective microtexture (f) (schematic).

fiber surface can be introduced in a variety of manners including : (i) weak interfaces between X and Y requiring that X and Y are only weakly bonded (as it occurs between glassy silica and anisotropic carbon), (ii) combining stiff/strong material X (e.g. dense SiC) with a compliant/weak material Y (e.g. porous SiC or porous refractory oxide such as ZrO_2) or (iii) intercalating thin films of a material X with a layered crystal structure (e.g. PyC) between thicker films of a stiff/strong glass-former material Y (such as SiC). In this latter case, even if X is sensitive to oxidation, it can well be protected by the glassy oxidation products of Y, if the X-layers are thin enough as established recently by computer simulation for model 1D-SiC/PyC/SiC composites [7]. Here, various processing parameters can be used to tailor the properties of the interphase including the nature and the number of the X-Y sequences as well as the thicknesses of X and Y.

All multilayered interphase materials do not necessarily fulfil **simultaneously** the fuse and the load transfer functions. As an example, a thin film of monocrystalline graphite (or mica) with the atomic planes parallel to the fiber surface, would probably behave as a poor interphase (fig. 2b). The first matrix microcrack deflected in such an ideal multilayered medium would debond the atomic layers over a long distance (owing to the weak FM-bonding) lowering the load transfer capability of the interphase. Thus, we think that **defects should be present** in order that a matrix microcrack does not undergo a single mode I/mode II deflection but a cascade of deflections and branching at a sub-µm scale within the bulk of the interphase (fig. 2c). Reducing the size of the carbon planes and folding somewhat these planes (e.g. introducing some defects), make crack propagation more difficult. Finally, introducing some roughness at the nm-scale in thin films stacked in the interphases (X-Y)$_n$ with a layered microstructure might play a similar role. Under such conditions, layered interphases containing structural/microstructural defects are expected to exhibit simultaneously good fuse and load transfer capabilities.

In composites fabricated with non-oxide fibers, it is highly preferable that the fibers do not see the atmosphere (which is often oxidizing) when the composites undergo microcracking at high temperatures. To fulfil this requirement : (i) the bonding between the fiber and the interphase should be stronger than the shear strength of the interphase and (ii) crack deflection should preferentially occur "far" from the fiber surface. Such a requirement can be achieved by tailoring the interphase.

Tailored interphases can be fabricated from liquid (e.g. sols or organometallics) or gaseous precursors, according to methods commonly used in thin film technology. Chemical vapor deposition (CVD) or chemical vapor infiltration (CVI) are particularly appropriate inasmuch as the composition of a gas flow can be easily changed to yield deposits with different compositions or/and microstructures. Examples will now be given of attempts made to apply the generalized layered interphase concept to SiC/SiC composites during the last few years at LCTS.

3- SiC/PyC/SiC Composites

In SiC/PyC/SiC composites, the interphase is a sub-μm film of turbostratic pyrocarbon deposited onto the fiber from a hydrocarbon. Three parameters can be adjusted during processing, namely, the microtexture of the interphase, its thickness and the strength of the fiber-interphase bonding, in order to control the balance between the fuse and load transfer functions or/and the oxidation resistance.

Pyrocarbon can be deposited in a **variety of microtextures** ranging from isotropic to rough laminar which is strongly anisotropic, depending on the nature of the precusor and the deposition conditions. The anisotropy of a pyrocarbon is commonly characterized by the so-called extinction angle, Ae, whose value ranges from Ae < 4° for isotropic carbon to Ae >18° for rough laminar. The use of hydrocarbons such as propane (C_3H_8) or propylene (C_3H_6) favors the deposition of strongly anisotropic pyrocarbons [4, 9, 13, 14] whereas that of methane is more suitable for CVI. Dupel et al. have shown that anisotropic pyrocarbon (rough laminar) could be deposited in model pores by pulse CVI from C_3H_6 or C_3H_8 [13]. The extinction angle (and thus the anisotropy) remains constant all along the pore (total length : 20 mm ; cross-section : 2 x 0.06 mm^2), when 950 < T < 1000°C for P = 3 kPa and a residence time t_R = 10 s (fig. 3a). Furthermore, the lattice fringe TEM-image of the deposit (fig. 3b) corresponds quite well to the defective layered crystal structure interphase depicted in fig. 2c. The carbon planes are of relatively large size (5-10 nm) and roughly parallel to the substrate surface and to one another but they also contain defects, i.e. they are undulated. 1D-SiC (Nicalon)/PyC/SiC minicomposites were prepared from single Nicalon fiber tows with such anisotropic interphase by pulse chemical vapor infiltration (P-CVI) and tensile tested at ambient [14]. The results suggest two conclusions (fig. 4 and table I). First, for a given interphase

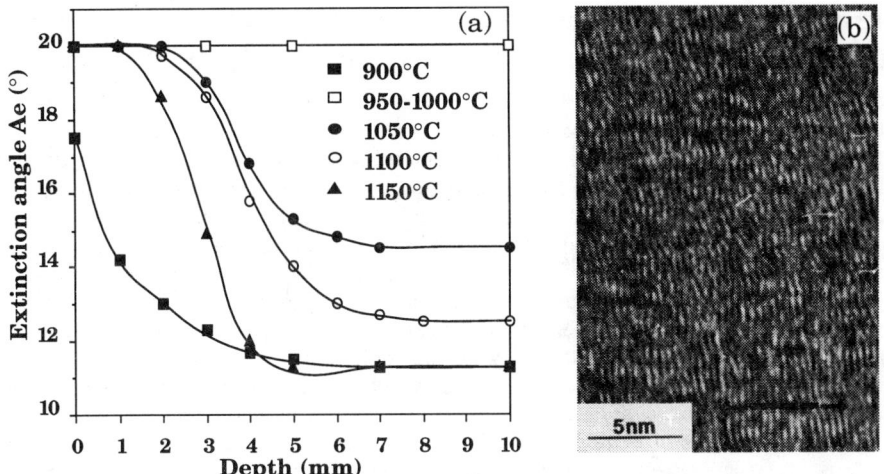

Fig. 3 : Pyrocarbon deposited in a 60 μm model pore from C3H8 by pulse-CVI (P = 3 kPa ; tR = 10 s) : (a) variations of Ae vs depth within half pore length, (b)TEM carbon 002 lattice fringe image, according to [13].

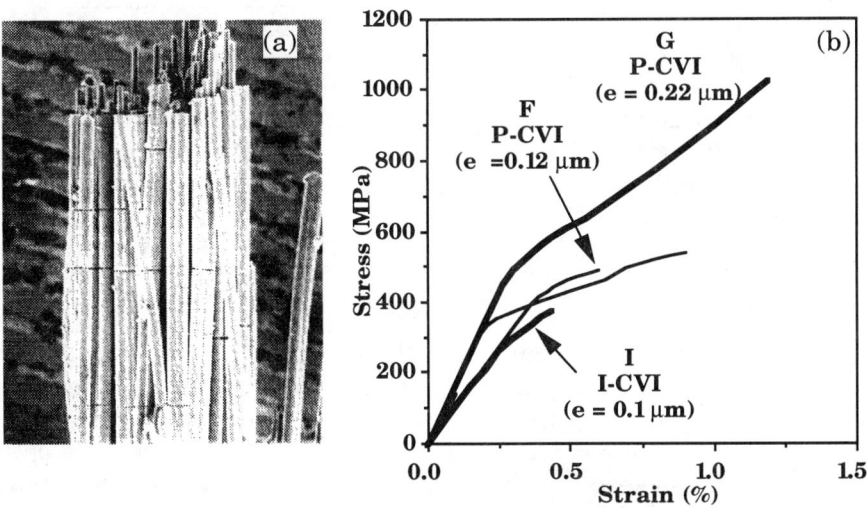

Fig. 4 : Model 1D-Nicalon/PyC/SiC minicomposites : (a) failure surface and matrix microcracking , (b) tensile curves at ambient, according to [14].

microtexture (e.g. that obtained by P-CVI), the microcrack interspacing, d_S, and the interfacial strength τ_c (calculated from d_S), on the one hand, the extension of the non-linear stress-strain domain as well as the failure strength, on the other hand, strongly depend on the interphase thickness e. The best results are achieved for e ≈ 200 nm. A similar conclusion has been reported by Lowden et al. for 2D-Nicalon/PyC/SiC composites prepared by forced-CVI (F-CVI) with an ex-propylene PyC-interphase [4, 9]. The optimum interphase thickness might be related to the TRS-absorption function of the interphase which requires enough compliant material to be effective. Second, for a given value of e, i.e. ≈ 100 nm, the minicomposite H prepared by P-CVI exhibits a better mechanical behavior than that prepared by conventional I-CVI (sample E), as expected from its anisotropic interphase.

Another important parameter is the strength of the bonding between the fiber and the PyC-interphase with respect to the intrinsic shear strength of the PyC layered interphase [1]. When the former is low with respect to the latter, the fiber/PyC interface is the weakest link and both matrix microcrack deflection and debonding occur at that interface (material of "type A" in [1]). In other words, the layered interphase does not play its role of fuse since there is a weaker fuse elsewhere. Such a situation, observed in composites fabricated with as-received Nicalon fibers, yields stress-strain tensile curves exhibiting a plateau-like feature (sample I in fig. 5a) [10]. It might be related to an additional glassy silica/anisotropic carbon dual layer formed during processing at the fiber surface, with a low bonding between silica and carbon, as mentioned in section 2. In such "type A" composites, there is a limited number of matrix microcracks and debonding occurs over very long distances (fig. 5 inset (a)). As a result, the load transfer capability is rapidly lowered explaining thus the plateau-like feature of the σ-ε curve. Conversely, the situation becomes very different when the fiber is strongly bonded to the PyC-interphase, as achieved by a proper pre- treatment of the Nicalon fibers. Under such conditions, the layered crystal structure interphase can now fully play its role of fuse since it is the "weakest" link in the interfacial zone ("type B" composite in ref. [1]). Since the PyC-interphase deposited by CVI contains numerous defects (fig. 2c), each matrix microcrack follows a tortuous path within the interphase with numerous branching and formation of many nm-size cracks (fig. 5, inset (b)). As load is increased more matrix microcracks are formed, saturation being achieved very near failure. The corresponding stress-strain curve (sample J in fig. 5) no longer exhibits a plateau-like

Table I : Characteristics of 1D-Nicalon/PyC/SiC composites[14].

Sample reference	Interface thickness (nm)	Failure[3] Stress (MPa)	Failure Strain (%)	d_s	τ_c[4] (μm) (MPa)
D [1]	50	300	0.26	200	-
E [1]	90	470	0.65	180	18
F [1]	120	549	0.64	70	43
G [1]	220	939	1.30	30	91
H [1]	1000	470	0.97	260	10
I [2]	100	408	0.46	70	37

[1]prepared by P-CVI ; [2]prepared by I-CVI ; [3]load at failure divided by overall cross-section area ; [4]calculated calculated from ds

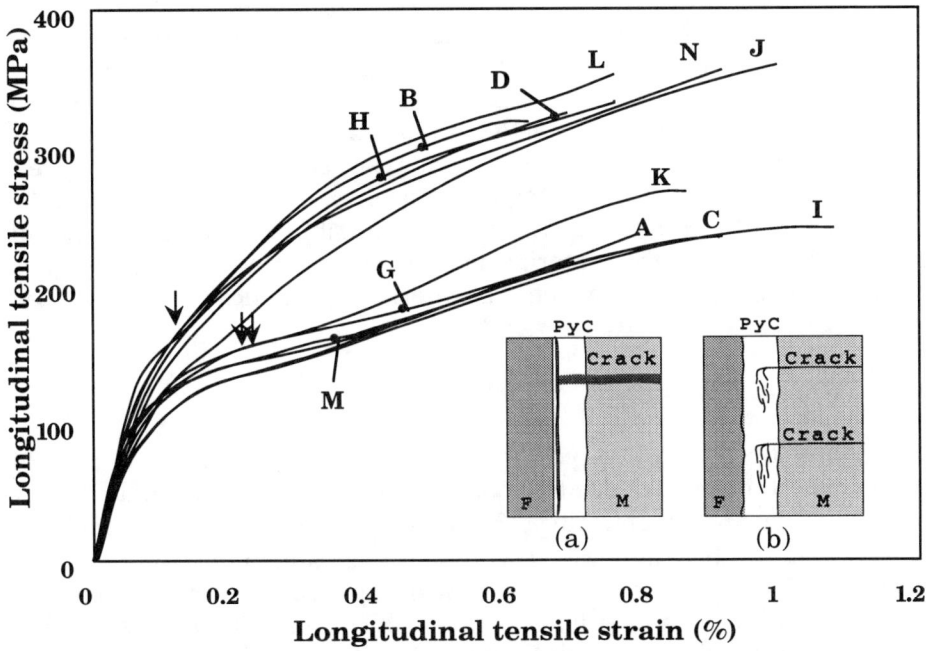

Fig. 5 : Tensile stress-strain curves at ambient of 2D-Nicalon/PyC/SiC composites (samples I and J) and 2D-Nicalon/(PyC-SiC)n/SiC composites. The insets show the crack deflection modes : (a) weak fiber/PyC bonding, (b) strong fiber/PyC bonding, according to [10].

feature, the interphase keeping a good load transfer capability although it becomes more and more damaged. In other words, its load transfer capability remains high [1, 10].

The oxidation of 1D-Nicalon/PyC/SiC composites has been studied experimentally and modelled [7, 15]. The basic phenomena which are involved are schematically depicted in fig. 6a. As oxygen reacts with the PyC-interphase, an annular pore is formed around each fiber. Thus, oxygen has to diffuse along that pore to react with PyC. During this diffusion oxygen also reacts with the SiC pore walls to form silica scales which can eventually plug the pore. When temperature is low and the PyC-interphase is thick, the formation of silica is very slow. Oxidation is thus an in-depth phenomenon and the load transfer capability of the interfacial zone is destroyed. Conversely, when temperature is high and the PyC-interphase is thin (typically ≈ 0.1 µm), the rate of formation of silica is fast, the pore is rapidly plugged by silica and the oxidation is limited to near the composite surface (fig. 6b). In other words, the composite exhibits a self-healing behavior. There is thus an advantage in reducing the thickness of the PyC-interphase, particularly when the composite is exposed to an oxidizing atmosphere at high temperatures. Additionally, a coating consisting of a glass-former material (such as SiC) deposited on the composite surface also contributes to limit the flux of oxygen diffusing in the material. Finally, the protective effect of silica might become insufficient when the material is periodically microcracked (load cycling).

4- SiC/BN/SiC Composites

Hexagonal BN has a layered crystal structure very similar to that of graphite and it can be deposited in a turbostratic form from e.g. BF_3-NH_3 or BCl_3-NH_3. It is chemically compatible with SiC in the temperature range of interest here and thus it is another interesting interphase material for SiC/SiC composites. Hex-BN has two important advantages. First its oxidation starts at a higher temperature ($\approx 850°C$ vs $450°C$) and second, it yields an oxide, B_2O_3, which remains liquid over a broad temperature range (i.e. from $\approx 500°C$ to $\approx 1100°C$) and is known for its healing properties (fig. 7a) [1, 8, 19].

The feasibility of the infiltration of hex-BN within a porous fiber preform has been already demonstrated for both BF_3-NH_3 [16-18] and BCl_3-NH_3 [19-22]. Furthermore, 2D-Nicalon/BN/SiC composites exhibit the non-

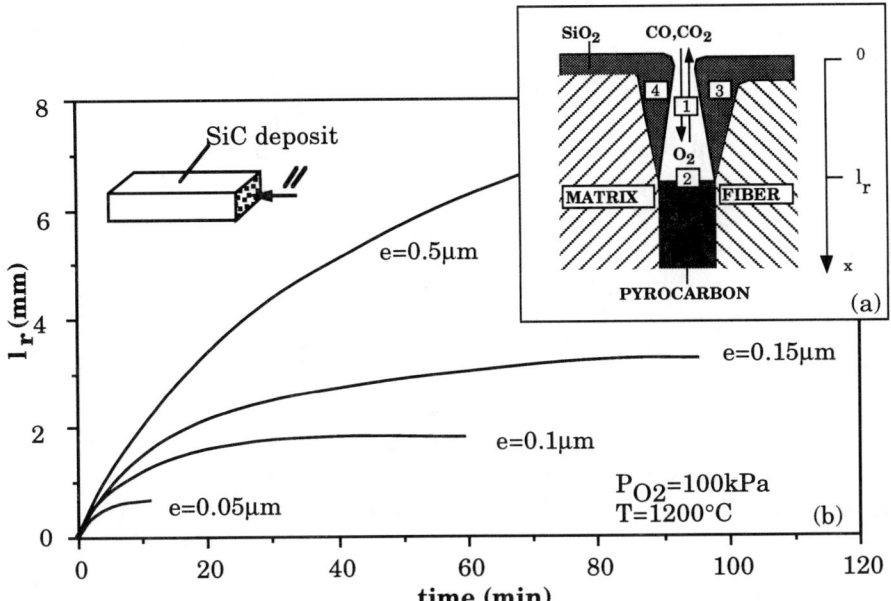

Fig. 6 : Oxidation of 1D-Nicalon/PyC/SiC model composites : (a) basic phenomena, (b) simulated curves of gasification of the PyC interphase, according to [7] and [15].

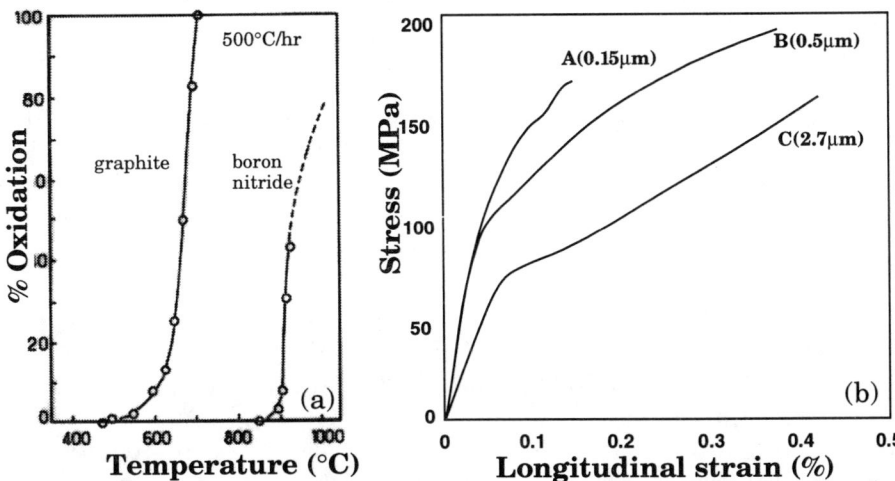

Fig. 7 : Nicalon/BN/SiC composites : (a) comparative oxidation kinetics of carbon and boron nitride [21], (b) tensile stress-strain curves at ambient of 2D-Nicalon/BN/SiC composites with various interphase thicknesses [18].

linear stress-strain tensile behavior expected for materials with a layered crystal structure interphase, as well as a clear effect of the interphase thickness (fig. 7b). Conversely, composites with a fully optimized mechanical behavior (i.e. similar to that reported in fig. 5 sample J for 2D-Nicalon/PyC/SiC composites) are not yet obtained in a routine manner (as an example, in fig. 7b, sample A undergoes an early failure), owing to processing difficulties. First, the BX_3-NH_3 precursors (with X = F, Cl) are more corrosive than hydrocarbons for Nicalon fibers. Second, it is more difficult to achieve a strong bonding between the fiber and hex-BN owing to the occurrence of additional carbon/silica interphases at the fiber surface. Third, it is also more difficult to properly control the microtexture of the layered hex-BN interphase.

Oxidation tests on model 1D-Nicalon/BN/SiC microcomposites performed under tensile static loading (at a stress level high enough to cause matrix microcracking), have clearly shown the potential of these materials in the intermediate temperature range (i.e. 500 - 800°C) where their 1D-Nicalon/PyC/SiC counterparts are particularly vulnerable. Under a stress level of 600 MPa, the lifetime of the microcomposites is dramatically improved when the PyC interphase is replaced by a hex-BN interphase presumably due to the healing of the SiC-microcracks by a boron oxide-based glass [20].

5- SiC/(PyC-SiC)$_n$/SiC Composites

2D-Nicalon/(PyC-SiC)$_n$/SiC (CVI) composites with n ranging from 1 to 4, have been prepared from Nicalon fabrics, in which the overall interphase thickness was maintained constant and equal to 0.5 µm (table II and fig. 8). Their tensile mechanical behavior at ambient was compared to that of the composites (samples I and J) fabricated with a homogeneous 0.5 µm PyC-interphase [10].
Although the layered microstructure of the (PyC-SiC)n interphase (corresponding to type f in fig. 2) was not fully optimized (i.e. the SiC sublayers exhibit a coarse grain and a non-uniform thickness ; fig. 8), the stress-strain curves (fig. 5) suggest the following remarks. First, all the composites prepared from a given fiber type, i.e. as-received Nicalon fibers or pre-treated Nicalon fibers, behave the same way whatever the interphase microstructure. When the composites have been prepared from as-received Nicalon fibers, the fiber-interphase bonding is weak (see inset (a) in fig. 5), as already mentioned for the composite (sample I) with a homogeneous PyC interphase, and the tensile curves exhibit the classical

plateau-like feature. Conversely, when the composites have been fabricated from pre-treated fibers, the fiber-interphase bonding is strong (inset (b) in fig. 5), as for the 2D-Nicalon/PyC/SiC sample J, and the curves show a regular convex curvature almost up to failure with intensive matrix microcracking in 0° tows and no plateau-like feature. Second, replacing part of the compliant PyC phase by the stiff SiC phase (in sample L, the overall thickness of PyC is only 0.2 µm whereas it is 0.5 µm in sample J) does not degrade the mechanical behavior (the toughness of the material is even improved).

It is anticipated from the discussion presented in section 3 (fig. 6) that the oxidation resistance of the composites with the layered $(PyC-SiC)_n$ interphase would be higher than that of the composites with the homogeneous PyC interphase of same overall thickness. As a matter of fact, in the former the overall amount of carbon in the interphase is lower and the thickness of each PyC sublayer is reduced by one order of magnitude (0.05 µm in sample L vs 0.5 µm in sample J). Thus the protective effect of silica resulting from the oxidation of the SiC sublayers might be more effective at least at high enough temperatures.

6- Other Layered Interphases

Ti_3SiC_2 has been reported to exhibit a layered crystal structure with sheets of Ti_6C units alternating with planes of silicon atoms, to crystallize as hexagonal platelets and to show some anisotropy in its mechanical properties. A preliminary study has been done to assess the potential of this material as an interphase [22, 23]. First, Ti_3SiC_2 is stable up to about 1300°C and compatible with SiC. It can be deposited from $TiCl_4$-$SiCl_4$-CH_4-H_2 at ≈ 1100°C but under very narrow deposition conditions and never as a pure phase. It grows as hexagonal platelets which are unfortunately oriented perpendicular to the substrate (and not parallel, as expected). Finally, the oxidation of Ti_3SiC_2 starts at temperatures as low as 400°C with kinetics only slightly lower than that for the oxidation of TiC, a result showing that the silicon content is too low to give enough silica for providing a protective effect. Thus Ti_3SiC_2 despite its "layered" crystal structure might not be an appropriate interphase material.

The phyllosilicates, e.g. micas, are well known layered crystal structure materials in which the elementary layers (consisting of condensed tetrahedral (Td) and octahedral (O_d) sheets) are only weakly bonded to

Table II : Nature of the fibers and interphases in 2D-Nicalon/SiC composites [10].

Samples	Nature of fabrics	Nature of the interphase and thicknesses of sub-layers (in µm)
A	NT[1]	F[3] /PyC /SiC /PyC M[4]
B	T[2]	0.1 0.3 0.1
C	NT	F /PyC /SiC /PyC/SiC/PyC /M
D	T	0.1 0.1 0.1 0.1 0.1
G	NT	F /PyC/SiC/PyC/SiC/PyC/SiC/PyC/ M
H	T	0.05 0.1 0.05 0.1 0.05 0.1 0.05
I	NT	F /PyC /M
J	T	0.5
K	NT	F /PyC/SiC/PyC/SiC/PyC/SiC/PyC/M
L	T	0.05 0.05 0.05 0.1 0.05 0.15 0.05
M	NT	F /PyC /M
N	T	0.1

[1] as-received   [2] pretreated   [3] fiber   [4] SiC-matrix

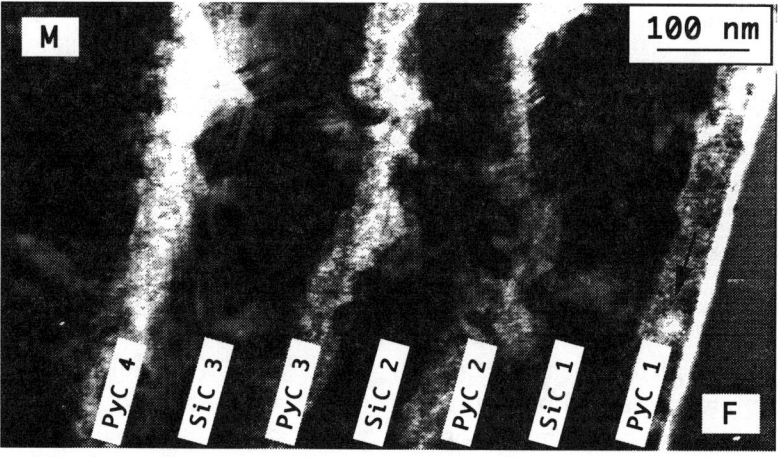

Fig. 8 : TEM-images of a $(PyC-SiC)_n$ multilayered interphase (sample G in fig.5), according to [10].

one another (fig. 9). As a result, they exhibit a low shear strength and a turbostratic disorder as in pyrocarbon or turbostratic BN. All these features suggest that they could be ideal interphase materials, totally insensitive to oxidation. Unfortunately, most phyllosilicates contain OH$^-$ ions responsible for their decomposition at low temperatures (typically 500-700°C) with destruction of the layered crystal structure. In some micas, such as fluorophlogopite, KMg$_3$(SiAl)O$_{10}$F$_2$, all the OH$^-$ ions are replaced by F$^-$ ions with a significant improvement in the thermal stability. Such a material has been already used as interphase in glass-ceramic matrix composites [11, 24]. However, its decomposition at high temperatures yields an evolution of fluorine bearing corrosive species.

A new family of all-oxide (i.e. containing no OH$^-$ or F$^-$ ions) materials isostructural with the micas and called phyllosiloxides, has been recently prepared according to a combined sol-gel/non-aqueous solvothermal treatment approach [12]. As an example, the phyllosiloxide corresponding to phlogopite has the formula KMg$_2$Al(Si$_4$)O$_{12}$ (fig. 9c). It is stable up to ≈ 950°C at atmospheric pressure and beyond 1350°C under a pressure of 2 GPa (these limits being compatible with CVI and hot-pressing). Some preliminary experiments performed on model materials have suggested that such a phyllosiloxide can act as an interphase material but numerous processing aspects still have to be solved.

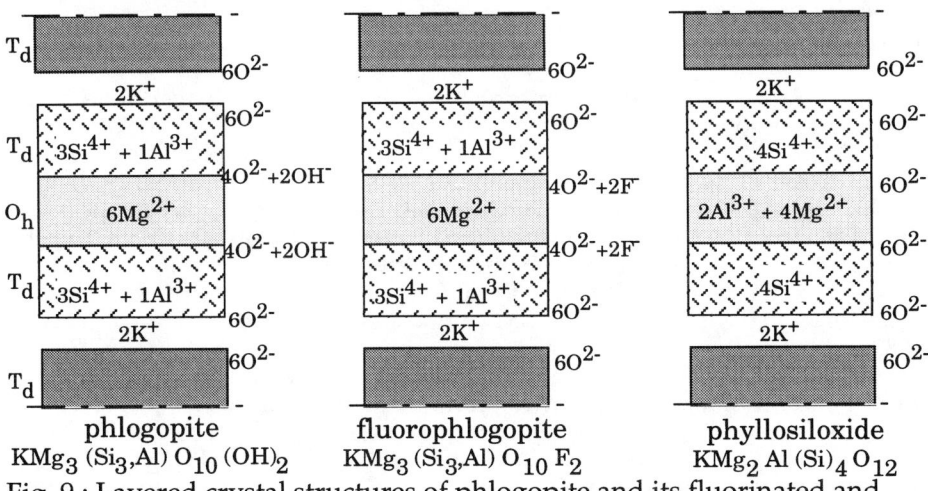

Fig. 9 : Layered crystal structures of phlogopite and its fluorinated and phyllosiloxide derivatives.

7- Conclusions

From the data and the discussion presented in sections 3-6, it appears that materials with a layered crystal structure or/and a layered microstructure act as efficient interphases in SiC/SiC composites, when the layers have been oriented parallel to the fiber surface. Materials with a layered crystal structure (e.g. pyrocarbon, turbostratic BN or micas) provide the simplest interphases but are very limited in number. Among them pyrocarbon constitutes the most effective interphase but its use in oxidizing atmospheres can be questioned. Turbostratic BN has a better oxidation resistance but its infiltration is not yet fully optimized. The feasibility of micas (or derivatives) as interphases in SiC/SiC composites is still far from being established. Finally, the extension of the layered interphase concept to materials exhibiting a layered microstructure may provide the best way for tailoring the interphase in SiC/SiC composites.

Acknowledgements

This synthesis has been written on the basis of research conducted at LCTS, SEP and LCS (for the phyllosiloxides) during the last few years. The author acknowledges the contribution among others of C. Droillard, X. Bourrat, J. Lamon, S. Prouhet, A. Guette, G. Camus, C. Racault, F. Langlais, all from LCTS, P. Reig and G. Demazeau from LCS, as well as S. Goujard and J.M. Jouin from SEP. He is also indebted to F. Lamouroux and J. Forget for the preparation of the manuscript.

References

1- R. Naslain, "Fiber-matrix interphases and interfaces in ceramic-matrix composites processed by CVI". Composite Interfaces, 1 [3] 253-286 (1993).
2- D. Cojean, "Composites SiC/C/SiC : relation entre les propriétés mécaniques et la microstructure des interphases", PhD Thesis, Univ. Pau, Sept. 12, 1991.
3- J.M. Jouin, J. Cotteret and F. Christin, "SiC-SiC interphase : case history", in Proc. Designing Ceramic Interfaces II (S.D. Peteves, ed.), pp. 191-203, Office for Official Publications of European Communities, Luxembourg, 1993.
4- R.A. Lowden and D.P. Stinton, "The influence of the fiber-matrix bond on the mechanical behavior of Nicalon/SiC composites", ORNL-TM 10667, December 1987.
5- M. Takeda, J. Sakamoto, A. Saeki, Y. Imai and H. Ichikawa, "High performance silicon carbide fiber Hi-Nicalon for ceramic-matrix composites", 1995 ACerS Cocoa Beach Conf., Ceram. Eng. Sci. Proc. (in press).
6- F. Lamouroux, G. Camus, R. Naslain and J.J. Thébault, "Kinetics and mechanisms of oxidation of 2D-woven C/SiC composites. I- Experimental approach", J. Amer. Ceram. Soc., 77 [8] 2049-2057 (1994).

7- L. Filipuzzi and R. Naslain, "Oxidation mechanisms and kinetics of 1D-SiC/C/SiC composite materials, 2- Modelling", J. Amer. Ceram. Soc., 77 [2] 467-480 (1994).

8- R. Naslain, O. Dugne, A. Guette, J. Sèvely, C. Robin-Brosse, J.P. Rocher and J. Cotteret, "Boron nitride interphase in ceramic matrix composites", J. Amer. Ceram. Soc., 74 (1991) 2482-2488.

9- R.A. Lowden, "Characterizaton and control of the fiber-matrix interface in ceramic matrix composites", ORNL-TM-11039, March 1989. Available NTIS, US Dept. Commerce, Springfield, Virginia.

10- C. Droillard, "Elaboration et caractérisation de composites à interphases séquencées PyC/SiC", PhD Thesis, n° 913, Univ. Bordeaux, June 19, 1993.

11- G. Beall, S. Daves, K. Chyung, K. Gadkaree and S. Hoda, "Fiber reinforced composites comprising mica matrix for interlayer", Europ. Patent, EP 366234, August 22, 1989.

12- P. Reig, G. Demazeau and R. Naslain, "Phyllosilicates de synthèse et procédé pour leur préparation", Europ. Patent, 93403098.2-, December 20, 1993.

13- P. Dupel, X. Bourrat and R. Pailler, "Anisotropic pyrocarbon obtained at low temperatures by pulse-CVD/CVI : structural characterization", Carbon, in press.

14- P. Dupel, "CVD/CVI pulsée du pyrocarbone. Application aux matériaux composites thermostructuraux", PhD Thesis, n° 927, Univ. Bordeaux, May 24, 1993.

15- L. Filipuzzi, G. Camus, R. Naslain and J.Thébault, "Oxidation mechanisms and kinetics of 1D-SiC/C/SiC composite materials : 1- an experimental approach", J. Amer. Ceram. Soc., 77 [2] 459-466 (1994).

16- H. Hannache, J.M. Quenisset, R. Naslain and L. Héraud, "Composite materials made from a porous 2D-carbon-carbon preform densified with boron nitride by chemical vapour infiltration", J. Mater. Sci., 19 (1984) 202-212.

17- H.O. Pierson, J. Composite Mater., 9 (1975) 228-240.

18- S. Prouhet, G. Camus, C. Labrugère, A. Guette and E. Martin, "Mechanical characterization of Si-C(O) fiber/SiC (CVI) matrix composites with a BN-interphase", J. Amer. Ceram. Soc., 77 [3] 649-656 (1994).

19- V. Cholet and L. Vandenbulcke, "Chemical vapor infiltration of boron nitride interphase in ceramic fiber preforms : discussion of some aspects of the fundamentals of the isothermal vapor infiltration process", J. Amer. Ceram. Soc., 76 (1993) 2846-2858.

20- S. Jacques, A. Guette, F. Langlais, R. Naslain and S. Goujard, " High temperature lifetime in air of SiC/C (B)/SiC microcomposites prepared by LPCVD", this volume.

21- J. Economy and R. Lin, "Boron nitride fibers", in "Boron and Refractory Borides (V.I. Matkovich, ed.), pp. 552-564, Springer, Berlin, 1977.

22- C. Racault, F. Langlais and R. Naslain, "Solid state synthesis and characterization of the ternary phase Ti_3SiC_2", J. Mater. Sci., 29 (1994) 3384-3392.

23- C. Racault, F. Langlais, R. Naslain and Y. Khin, "On the CVD of Ti_3SiC_2 from $TiCl_4SiCl_4CH_4H_2$ gas mixtures, 2- An experimental approach", J. Mater. Sci., 29 (1994) 3941-3948.

24- T.T. King and R.F. Cooper, "Ambient temperature mechanical response of alumina-fluoromica laminates", J. Amer. Ceram. Soc., 77 [7] 1699-1705 (1994).

DEVELOPMENT OF INTERFACES IN OXIDE AND SILICATE MATRIX COMPOSITES

M. H. Lewis and M. G. Cain, University of Warwick, Coventry, UK.
P. Doleman, A. G. Razzell and J. Gent, Rolls-Royce plc., Derby, U.K.

ABSTRACT

Silicate and oxide matrix CMCs are being developed for application in advanced gas turbines. High-performance Silicate/Nicalon CMCs have been characterised mainly as materials for interface, process and mechanical modelling due to their limited thermal and oxidative stability. Saphikon (Al_2O_3) monofilaments have been used in the development of interphase chemistry and processing via vapour and liquid-precursor methods. Prototype Al_2O_3-matrix CMCs have been fabricated and exploration of alternative fibre/interphase chemistries conducted via reactivity studies up to 1600°C.

INTRODUCTION

Long fibre CMCs are now accepted as being an essential part of a design strategy for critical engineering components within energy-conversion systems. One of the most demanding applications is that of the gas turbine in view of the high component temperatures which are required to demonstrate a significant gain in efficiency and performance. The design of CMCs is being considered for combustors, turbine aerofoils, shroud rings, exhaust and reheat components.[1] Since the 1940s there have been dramatic increases in thrust/weight ratio and turbine entry temperature via improvements in metallic alloys such that directionally solidified or single crystal 'superalloy' turbine blades are operating at material temperatures of 1000°C. The introduction of air cooling, in the 1960s, enabled metallic blade operation at gas temperatures up to 1600°C. To achieve the 21st century targets of a 15:1 thrust/wt.ratio and entry temperature of 1800°C, it is clear that over 800°C of cooling is necessary (Fig.1). Much greater efficiency would result from the elimination of the compressed air, which does not participate in the combustion process; an estimated 16% efficiency gain would occur for uncooled components. Further benefits of reduced air cooling are in the combustor area, with the reduction in NO_x emission which is a current problem in civil aero and stationary gas turbines.[1]

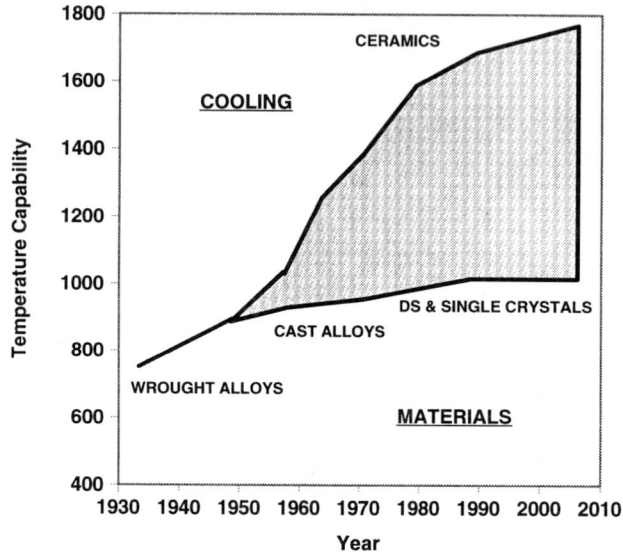

Fig.1. The evolution of turbine materials and their temperature capability.

The high thermal and mechanical stability required in many turbine components, frequently in oxidising environment, means that currently-available CMCs do not match the required property targets. A general philosophy within a broad research programme has been initially to use existing fibres as a basis for development of model interfaces, matrix constitution, fabrication and CMC properties over a range of temperature. This paper describes the development of high performance silicate matrix CMCs, which have limited thermal stability, and then focuses on the development of interphases in oxide/oxide CMCs which have potential for operation at temperatures above 1200°C.

HIGH PERFORMANCE SILICATE-MATRIX COMPOSITES

Processing, Microstructure and Mechanical Properties

The earlier projects have resulted in high strength model CMCs with Si-C-O fibres (e.g. - Nicalon, Nippon Carbon, Japan) in glass (borosilicate based) or glass-ceramic (MAS) matrices. Typical stress-strain data (Fig.2 a,b) exhibits high ultimate bend strength in the range 1-1.5 GPa for U.D. fibre volumes of ~50%, illustrating the retention of pristine fibre strength. Critical features of the fabrication cycle necessary to avoid fibre surface mechanical damage during hot pressing depend on maximising viscous-flow matrix densification, delaying pressure application until above the glass softening temperature and using a pressing isotherm above that for rapid crystal nucleation.[2,3] To facilitate this, in the MAS glass ceramic, an off-stoichiometric composition is used near to a eutectic between the primary crystallising phase (cordierite - $2MgO.2Al_2O_3.5Al_2O_3$) and enstatite $MgSiO_3$). For the borosilicate a modified 'Pyrex' composition has been developed to suppress surface nucleation of SiO_2 (cristobalite) during the heating cycle. Both matrices have low thermal expansions; the borosilicate as a stable glass and the MAS as a tailored phase mixture which crystallises slowly at the pressing isotherm or on post-pressing heat treatment.

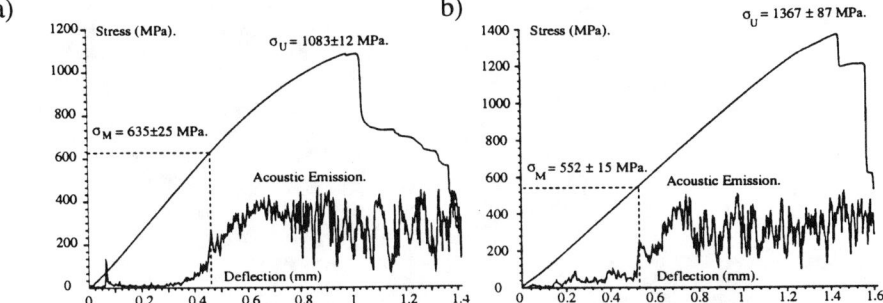

Fig.2. Bend stress-strain curves and associated acoustic emission traces for: (a) borosilicate glass and (b) MAS glass-ceramic matrix/Nicalon 607 CMCs.

An important additional pre-requisite for high strength is the use of interfaces formed by precoating, the most convenient form being the pyrocarbon-coated Nicalon 607, which inhibits thermochemical surface degradation. These carbon-rich interfaces (Fig.3) provide the required ratio of ultimate stress/matrix microcracking stress which dictates the level of notch-sensitivity, i.e. the ability to relax stress concentrations without fibre failure. The values of microcracking stress are most effectively controlled by the interfacial shear stress τ after debond and τ varies with precoating thickness and thermal cycle during fabrication. Typical values for pyrocarbon-coated Nicalon in borosilicate glass and MAS matrices are listed in Table I. These have been measured from 'push-down' tests using an instrumented, SEM-based, indentor system[4,5] and the values of debond energy G_i result from mode II (shear) initiation and hence are above the estimated model limits for debond

($G_i/G_{fibre} \sim 0.25$-0.5, depending on fibre/matrix modulus mismatch) based on mode I opening.

Table I - Interface micromechanical parameters for pyrocarbon coated Nicalon in glass and glass ceramic matrices, together with matrix microcracking stress (σ_m) and ultimate stress (σ_u)

	Processing Temperature and hold time	$G_i(Jm^{-2})$	τ(MPa)	σ_m(MPa)	σ_u(MPa)
NL607/ borosilicate glass	950°C/20 min 1100°C/20 min	20±10 20±7	158±74 51±9	>1000 ~ 600	1290±30 1380±50
NL607/MAS	1200°C/2hrs	12±5	48±15	665±78	1168±41

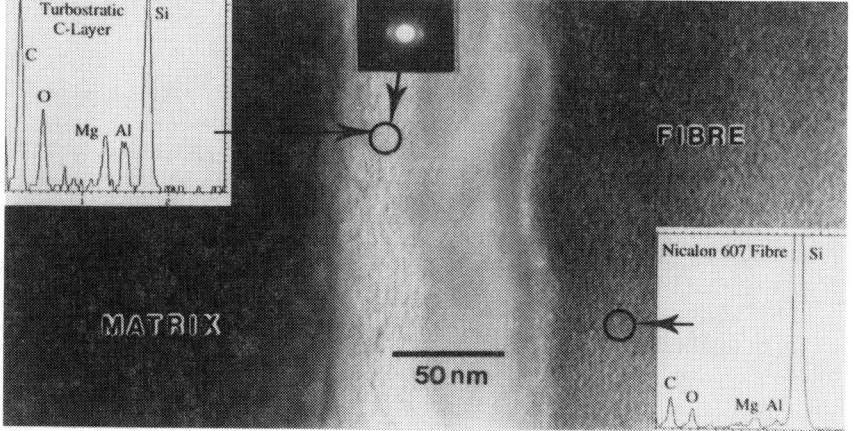

Fig.3. Carbon-rich interface microstructure in a precoated Nicalon fibre/MAS CMC (Courtesy R. L. Cain, Univ. of Warwick)

It is clear that, for the small precoating thicknesses (10-20nm), there is a processing time and temperature dependent change in interface structure and properties, consistent with the observed interdiffusion of matrix elements and some thickening of the carbon rich layer produced by SiC oxidation. An example, for the matrices studies here, is imaged in Fig.3.

High Temperature Thermal and Mechanical Stability.

The well-established generic problem for carbon-rich interfaces in oxidising conditions above ~500°C results from selective interface oxidation either by channelled reaction from exposed fibre ends or via matrix cracking or porosity.[6] This is reflected in an intermediate temperature dip in strength and corresponding rise in τ, with a reversion to brittle failure consistent with a progressive bridging of the interface with a strongly-bonded SiO_2 oxidation film. For glass ceramic matrices the channelled reaction is suppressed above ~1000°C by passive oxidation at fibre ends, preventing oxygen ingress, such that interface oxidation is limited by very slow diffusion of O_2 or reaction products through the silicate matrix conferring adequate protection for long exposure (< 100 hours at 1200°C). An important observation is the retention of composite (and hence fibre) strength in this protected state, indicating that fibre surface oxidation is critical for strength-loss.[6]

This passivating oxidation is useful as a pre-treatment in studies in high temperature creep deformation at stresses below that for matrix microcracking. The borosilicate glass matrix UD composite is a model material in which fibres undergo elastic strain during load transfer from a visco-elastic creeping matrix in the temperature range below its softening point (400-550°C) and may be simulated by a spring/dashpot mechanical element.[7] The glass-ceramic composites undergo the

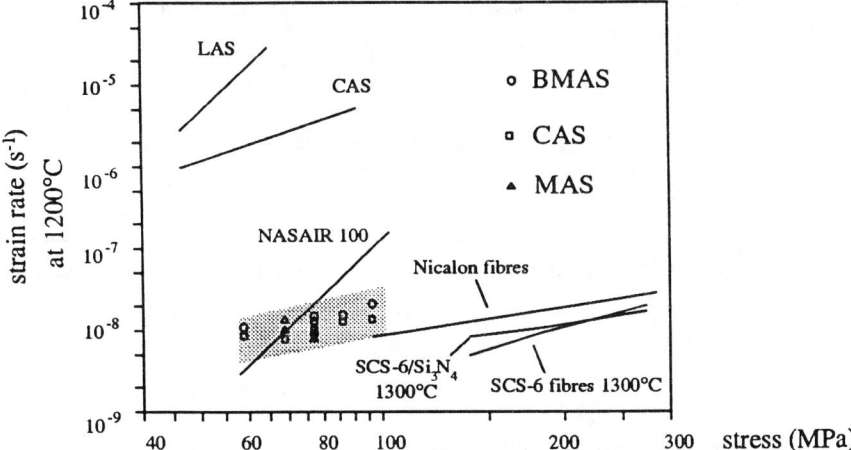

Fig.4. Comparison of creep rates for glass-ceramic matrix composites (GCMCs - shaded data points) with various fibres and matrices.

same transient creep response up to a near steady state strain rate dominated by fibre creep.[7] Fig. 4 demonstrates the matrix-insensitivity of creep rate for a range of matrix chemistry and the similarity to isolated fibre creep under normalised stress conditions. Also shown for comparison is the related behaviour for SiC monofilament reinforced Si_3N_4, typical glass-ceramic creep rates and a single crystal superalloy. The latter comparison emphasises the surprisingly good matrix-independent creep properties for this class of CMC up to ~1200°C. Above this temperature fibre instability and creep rate presents an application limit. In addition, stress-rupture will be dominated by fibre oxidation problems at stress levels above the microcracking thresholds.

OXIDE CMCs

Interphase Design Strategies.

The preliminary interface study has been conducted on Al_2O_3 fibres, mainly using single crystal sapphire (Saphikon Inc., USA), c̲ axis, monofilaments which have the required thermal stability for interphase synthesis and CMC fabrication. Interphases have been formed by precoating using vapour or liquid precursors (LP) followed by heat-treatment to develop preferred structures. Interphase chemistry has been selected on the basis of thermodynamic prediction of reactivity with fibre and matrix and the probability of matrix microcrack-induced debonding at either matrix/interphase or fibre/interphase. The following strategies have been explored:-

(i) Simple oxides (e.g. ZrO_2) with microporosity[8] within a columnar PVD layer or containing thermally-induced microcracks as an aid to debond since the Al_2O_3/ZrO_2 interfaces does not appear to have an intrinsic energy that satisfies the predicted G_i/G_f ratio.

(ii) Complex oxides which either have low intrinsic G_i values with Al_2O_3 or have a layered crystallography with weak layer-plane bonding. The former is exemplified by the rare earth phosphates[9] ($LaPO_4$ has good high temperature stability) and the latter by the β-aluminas (e.g. $ReAl_{11}O_{18}$ form stable couples with α-Al_2O_3).[10]

Interphase Synthesis and Microstructure.

The main effort has been directed at liquid or sol-precursor methods for interphase synthesis in view of their relative simplicity and cost and the potential for rapid continuous coating. Initial studies have utilised dip coating of short lengths of monofilament and selected processes have been scaled up to generate continuously coated filaments for CMC fabrication using a system shown schematically in Fig.5.

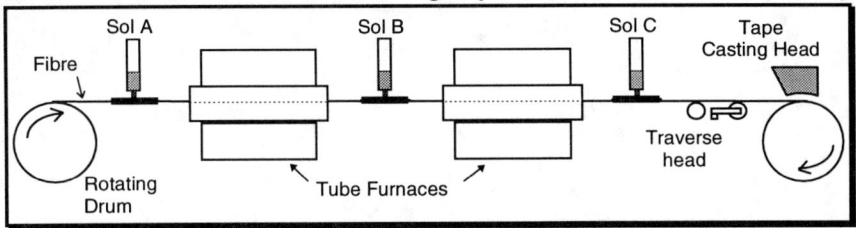

Fig.5. Experimental system for continuous coating and preforming of monofilament arrays.

The system can be adapted for varying thicknesses of single coating, for bi-layer chemistries which may include an outer layer of matrix composition to facilitate densification, or to form an interphase debond layer by in-situ reaction. Examples of LP coatings are:-

(i) ZrO_2 or Al_2O_3 - based compositions from <u>commercial sols</u>; the best pyrolysed coatings were obtained using a flame hydrolysed sol with the addition of a non-dispersed fine powder to minimise cracking during the drying/calcining stage.

(ii) <u>Solution precursors</u> (such as the acetates with appropriate dopant nitrates etc) as a basis for ZrO_2, Al_2O_3 or phosphates formed by simple polymerisation/condensation reactions during pyrolysis. These reactions require relatively high temperatures (up to ~1400°C) and multiple dipping/pyrolysis is necessary for thick coatings (>1μm).

Vapour-deposited coatings have been produced by R.F. magnetron sputtering or laser ablation from small area targets, to effect stoichiometric transfer of the required oxide, assisted by a variable atmosphere of low-pressure gas. Rotating filament holders enabled uniform coating thickness in the PVD process and differing microstructures (scale of columnar crystal structure, porosity, degree of crystallinity) were produced with variable operating condition (power level, substrate bias etc).

Coating microstructures, after inclusion within Al_2O_3 matrices initially using hot-pressed tape-cast matrix lamellae, are exemplified in Fig.6. ZrO_2-based interphases may be produced with variable porosity and columnar grain structure (Fig.6a). In Fig.6b the βAl_2O_3 plate-shaped crystals which constitute the interphase exhibit stability in αAl_2O_3 at the sintering temperature (1550°C) and evidence for interlayer fracture from surface-induced cracking. Fig.6c shows a βAl_2O_3 phase formed by in-situ reaction via out-diffusion of the β-stabilising cations from a doped

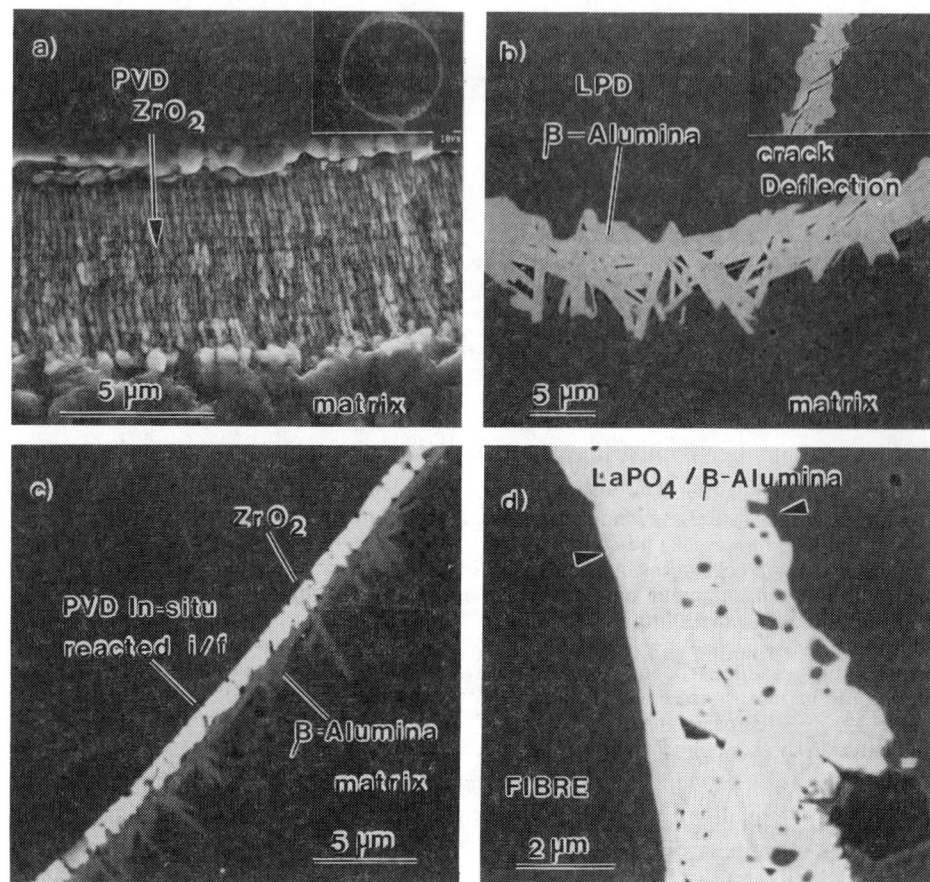

Fig.6. Interphase microstructures in Saphikon/Al_2O_3 CMCs; (a) PVD ZrO_2 with inset showing uniform coating coverage, (b) βAl_2O_3 via either powder Boehmite/La nitrate slurry, Al-acetate/La nitrate solution or Al sec.-butoxide/La nitrate solution precursors with inset indent initiated crack deflection within interphase (c) bilayer with reaction-synthesised βAl_2O_3, (d) $LaPO_4$

prelayer which forms a filament protective reaction/abrasion barrier. The phosphate interfaces are not always stable with respect to Al_2O_3 and in figure 6d has partially reacted to form La-stabilised β-alumina during sintering at 1550°C. An intrinsically low-energy interface between $LaPO_4$ and Saphikon may result from a discontinuity in the oxygen framework and a predominantly low density O-La-O bonding similar to that in the layered β aluminas.

Model CMC Fabrication and Mechanical Behaviour

Matrix-indent-initiated cracking may be used as a qualitative indicator of interface debonding, as illustrated above. Quantitative measurement of G_i and τ has been attempted using indentation tests on fibre ends in U.D. sections. An instrumented SEM based microindent system has been used in 'push-through' tests on transverse sections which are necessary because of the very high indent loads required for 'push-down' tests on the large diameter monofilaments. The shear stress-displacement traces, exemplified in Fig.7, exhibit sharp debonding events at various levels above that for subsequent frictional sliding, usually much higher than for the pyrocarbon interfaces previously studied[11] in Textron SiC/Si_3N_4 CMCs and reproduced for comparison in Fig.7. Debonding of some oxide interfaces is not possible within the indentor load capacity (~20N) and is sensitive to porosity in ZrO_2 and crystal morphology in β alumina; it is possible to produce a preferred orientation of layer plane in the latter interphases, parallel to the filament surface, by sintering/reaction heat treatment in the isolated fibre state.

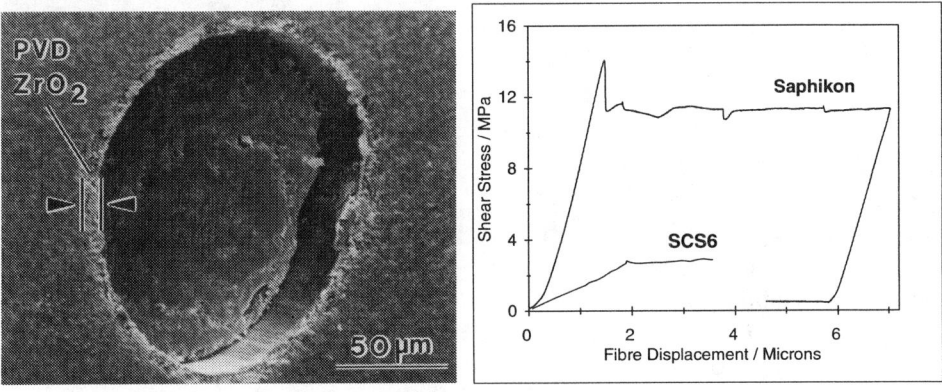

Fig.7. Monofilament 'push-through' indent traces for Saphikon and SCS6(SiC) with ZrO_2 and C interfaces, respectively (SEM image illustrates debond)

Prototype U.D. composites have been fabricated by hot-pressing parallel arrays of Saphikon between tape cast sheets of alumina. The best densities and filament distributions have been obtained by direct tape casting onto the wound filaments as a post-coating process illustrated in Fig.5. The single layer preforms are stacked and the tape-casting binder removed during the temperature rise to hot-pressing (typically 20 MPa at 1500°C). Some interlayer porosity remains and at higher pressures a plasticity of the Saphikon occurs in the soft lateral direction (normal to the c axis). The hot-pressed tiles (100 x 20 x 4 mm) may be machined for bend or tensile test specimens; typical bend stress-strain and fracture behaviour is shown in

Fig.8. CMC-like multiple stage stress-strain response is achieved with limited 'pull-out' observable on fracture surfaces (Fig.8b) for a ZrO_2-based interphase material. A comparative study is in progress for the alternative interphase chemistries and a modelling of mechanical behaviour based on interface micromechanical data.

ALTERNATIVE FIBRE/INTERPHASE/MATRIX COMBINATIONS

Saphikon is a good model filament for CMC process development and modelling of high temperature deformation. It has obvious limitations of high plastic anisotropy, high cost and large, non weavable, diameter. Alternative melt-grown filament chemistries are under development in UK programmes[1] together

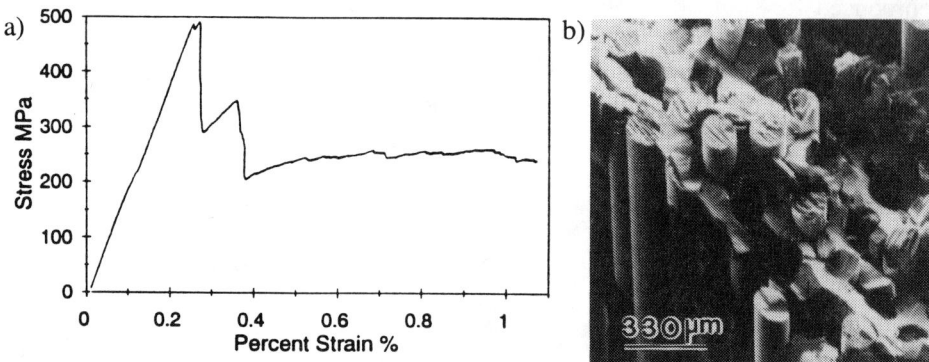

Fig.8. Saphikon/ZrO_2/Al_2O_3 CMC, bend stress/strain curve and fracture surface after 50hrs at 1400°C.

with polycrystalline and multiphase or eutectic fibres from sol-based materials. The objectives are to produce smaller fibres, at acceptable cost, with stability and good creep resistance above 1200°C. Initial research has examined phase compatibility of complex oxides, which may be components of novel fibres, and high temperature reactivity with potential interfaces. Model reaction couples have been fabricated either using mixed particle aggregates or by liquid-precursor deposition of oxides onto single and polycrystalline substrates. Preliminary reactivity data, obtained from X-ray diffraction and analytical electron microscopy, is exemplified in Fig.9.

There is encouraging stability for spinel and garnet phases with potential interphases. The more complex oxides have a greater plastic isotropy than αAl_2O_3 with lower mean creep rates dictated by dislocation glide/climb in the monofilament state and, with control of grain size, may have acceptable diffusional creep rates in the polycrytsalline fibre state. Stable diphasic fibres, such as garnet-alumina with a major garnet phase in off-eutectic composition is a possible combination with a phosphate or β alumina interphase and αAl_2O_3 matrix to form a CMC with superior thermal/mechanical stability for critical component application at acceptable cost.

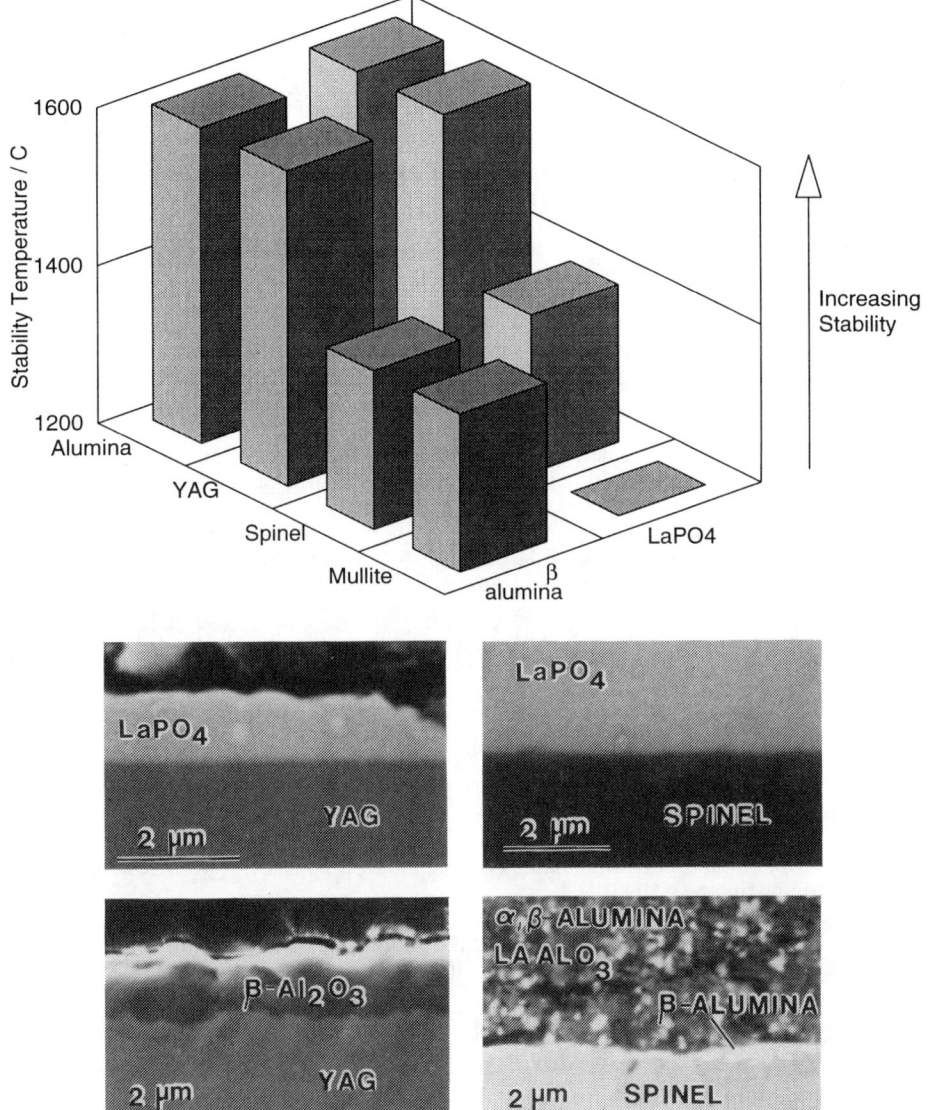

Fig.9. Measured temperature levels for phase compatibility with examples of interface microstructures.

ACKNOWLEDGEMENTS

Financial support for the broad programme on CMCs has been gratefully received within the EPSRC/DTI 'LINK' initiative and from Rolls Royce plc and the EEC (BRITE programme BREU 4610).

REFERENCES

1. M.H. Lewis, E.G. Butler and P.M. Marquis, 'Ceramic Materials for Gas Turbine Application' in Ceramic Materials and Components for Engines, Proc. 5th Int. Symposium, Shanghai (ed. D.S. Tan) in press 1995.

2. A. Chamberlain, M.W. Pharaoh and M.H. Lewis, 'Novel Silicate Matrices for Fibre-Reinforced Ceramics', Ceram.Eng.Sci.Proc. 14, 939-946 (1993).

3. A. Chamberlain, A.M. Daniel, M.W. Pharaoh and M.H. Lewis, 'Fracture Mechanical Behaviour and Interface Properties in Glass and Glass-Ceramic Matrix Composites', Proc. HT-CMC 1 - Bordeaux, ed. R. Naslain, J. Lamon and D. Doumeingts, 321-328 (Woodhead Publ. 1994).

4. M.H. Lewis, A.M. Daniel, A. Chamberlain, M.W. Pharaoh and M.G. Cain, 'Microstructure - Property Relationships in Silicate-Matrix Composites', J. Microscopy 169, 109-118 (1993).

5. A.M. Daniel, S.T. Smith and M.H. Lewis, 'A Scanning Electron Microscopy-Based Microindentation System', Rev.Sci.Instrum. 65, 632-640 (1994).

6. M.W. Pharaoh, A.M. Daniel and M.H. Lewis, 'Stability of Interfaces in CAS-Matrix/Nicalon SiC Fibre Composites', J.Mat.Sci.Lett.12, 108-113 (1993).

7. S. Sutherland, K.P. Plucknett and M.H. Lewis, 'High Temperature Mechanical and Thermal Stability of Silicate Matrix Composites', Comp.Eng. (in press).

8. A. G. Evans, F. W. Zok and J. Davis, "The Role of interfaces in fiber-reinforced brittle matrix composites" Composite Science and Technology, 42, 3-24 (1991).

9. P.E.D. Morgan and D.B. Marshall, 'Oxide/Oxide Composites', 365 MRS Symposium Proceedings (in press 1995).

10. M.H. Lewis, A.M. Daniel and M.G. Cain, 'Interface Characterisation Using an SEM-Based Microindentor' 365 MRS Symposium Proceedings (in press 1995).

11. A.G. Razzell and M.H. Lewis, 'Silicon Carbide/SRBSN Composites', J. Microscopy 169, 215-223 (1993).

INTERFACIAL MICROSTRUCTURE AND STABILITY OF BN COATED NICALON FIBER/GLASS-CERAMIC MATRIX COMPOSITES

John J. Brennan, United Technologies Research Center, East Hartford, CT 06108
Steven R. Nutt, Dept. of Materials Science, Univ. of Southern Calif., Los Angeles, CA 90089
Ellen Y. Sun, Oak Ridge National Lab, Oak Ridge, TN 37831

ABSTRACT

Glass-ceramic matrix composites with dual-layer CVD coated SiC/BN interfaces applied to Nicalon SiC fibers have been developed. These composites have been shown to possess high strength and good fracture toughness to temperatures of 1200°C. Tensile creep, tensile fatigue, and long-time tensile stress-rupture tests were performed, with the results correlated to the stability of the fiber/matrix interfacial microstructure and chemistry.

INTRODUCTION

Fiber-reinforced glass-ceramic matrix composites are prospective materials for high-temperature, lightweight, structural applications.[1] In recent years, research in this type of ceramic matrix composite has concentrated on the fiber/matrix interface and the relationship of the interfacial microstructure, chemistry, and bonding to the resultant composite mechanical properties and interface stability.[2-4] It has been found from these studies that polymer derived SiC type fibers such as Nicalon (Nippon Carbon Co.), that contain excess carbon and oxygen over stoichiometric SiC, form a thin (20-50nm) carbon rich fiber/matrix interfacial layer when incorporated into glass-ceramic matrices at elevated temperatures. The formation of this weak interfacial carbon layer is responsible for the high toughness and strength observed in these composites, but is also responsible for composite embrittlement and concurrent strength and toughness degradation when either stressed at elevated temperatures in an oxidizing environment or thermally aged in an unstressed condition in oxidizing environments. This embrittlement and strength degradation is a result of oxidation of the carbon layer and its replacement by a glassy oxide layer that is bonded strongly to both the fiber and matrix, thus inhibiting matrix crack deflection at the fiber/matrix interface.

It is thus imperative that the fiber/matrix interface in these types of ceramic matrix composites be controlled, or "engineered", so that relatively weak interfacial bonding exists for matrix crack

To the extent authorized under the laws of the United States of America, all copyright interests in this publication are the property of The American Ceramic Society. Any duplication, reproduction, or republication of this publication or any part thereof, without the express written consent of The American Ceramic Society or fee paid to the Copyright Clearance Center, is prohibited.

deflection while maintaining oxidative stability. An approach to accomplish this is to utilize coatings on the fiber surfaces that are applied before composite processing.[5-11] Not only must these interfacial coatings be weak and oxidatively stable, they must also be resistant to matrix and/or fiber interdiffusion so that interfacial reactions do not occur during composite processing or during subsequent thermal exposure. The utilization of dual layer SiC over BN fiber coatings to accomplish this is the subject of this paper.

EXPERIMENTAL PROCEDURE

The glass-ceramic matrix utilized for the present study was barium-magnesium aluminosilicate (BMAS), formulated to yield the barium osumilite phase ($BaMg_2Al_3(Si9Al_3O_{30})$) on crystallization. The reinforcement was a polycarbosilane derived Si-C-O fiber (Nicalon NLM 202) with a dual layer SiC over BN coating. The coatings were deposited continuously on fiber tows by atmospheric chemical vapor deposition (CVD) at a temperature of ~1000°C, with BN and SiC coating thicknesses of ~300 and 200nm, respectively. The BN was deposited using a proprietary precursor chosen to yield a composition of ~42at% B, 42% N, and 15% C. From scanning Auger analysis, the oxygen content of both the BN and SiC layers was less than 2%. The measured tensile strength of the coated Nicalon fibers of 2.22±5.6 GPa indicated essentially no loss in fiber strength during coating.

Composites (10cm x 10cm) were fabricated by hot-pressing a layup of 0/90° oriented unidirectional fiber plus matrix powder plies at a maximum temperature of ~1450°C for 5 minutes under 6.9 MPa pressure, yielding an essentially fully dense matrix with a fiber volume of ~45%. After hot-pressing, the composite panels were machined into dogbone tensile samples (2.54cm gage length, 0.5cm gage width) and then heat-treated ("ceramed") in argon at 1200°C for 24 hrs to crystallize the BMAS matrix to the barium osumilite phase.

Although a variety of mechanical property measurements were done, including flexural strength and flexural creep vs temperature, the composites to be discussed in this paper were subjected to uniaxial tensile testing at temperatures of 20°, 1100°, and 1200°C, tensile creep testing at 1100°C, and tensile fatigue and stress-rupture testing at 1100° and 1200°C. All testing was conducted in air. After composite testing, the fracture surfaces were examined using scanning electron microscopy (SEM), and the composite microstructures characterized by transmission electron microscopy (TEM) of polished composite cross-section replicas and ion beam thinned foil sections.

RESULTS AND DISCUSSION

As-Fabricated Composite Microstructure

A typical microstructure of a 0/90° ply layup BMAS matrix/SiC over BN coated Nicalon fiber composite is shown in Figure 1. The light micrographs show the overall composite microstructure, while the TEM replica shows the details of the SiC/BN interfacial coatings. The

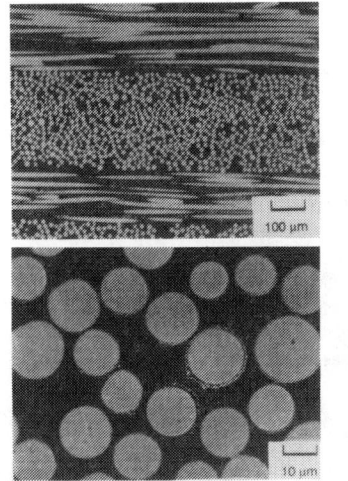

Fig. 1 - Microstructure of 0/90° BMAS Matrix/SiC/BN Coated Nicalon Fiber Composite

particular composite shown in Fig. 1 is in the as-pressed condition, and from the TEM micrograph it can be seen that the BMAS matrix is primarily glassy, with just a few crystals of mullite present that tend to precipitate out on cooling from the fabrication temperature. Figure 2 shows the interfacial microstructure of a composite after "ceraming", or matrix crystallization, and a high resolution TEM image of the BN/Nicalon fiber interfacial region. From Fig. 2a, it can be seen that a small amount of residual matrix glassy phase exists after ceraming, particularly at the interface between the SiC coating and the matrix. During ceraming, the matrix grains of barium osumilite grow along preferred crystallographic directions in lathlike configurations, often spanning interfiber distances. In composites where the SiC/BN layers remained tightly adhered to the Nicalon fibers after composite fabrication, no reactions or interlayers were observed at either the SiC/BN interface or the BN/Nicalon interface (Fig. 2b). This image shows a total absence of

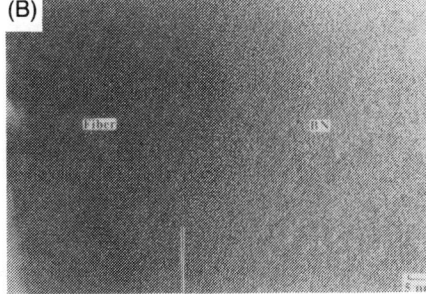

Fig. 2 - BMAS Matrix/SiC/BN/Nicalon Fiber Interfacial Microstructure (a) and HRTEM Image (b) of Nicalon Fiber/BN Interface.

the dual sublayers of carbon-rich and silica-rich material between the BN coating and Nicalon fiber that has been observed to form in CVI SiC matrix composites with BN coated Nicalon fibers.[8,9]

In BMAS matrix composites where the SiC layer becomes debonded from the BN layer during composite fabrication, however, changes do occur within the BN and at the BN/Nicalon fiber interface. As shown in Fig. 3, if the SiC layer cracks or debonds and allows the BMAS matrix to

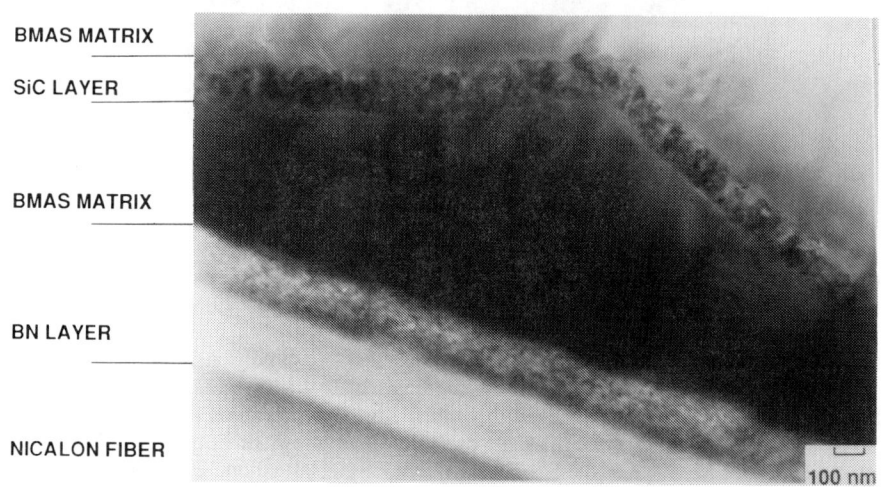

Fig. 3 - SiC Layer Debonding in BMAS Matrix/SiC/BN Coated Nicalon Fiber Composite

contact the BN layer during composite fabrication, coarsening of the turbostratic BN layer occurs. From EDS and scanning Auger analyses, a significant amount of matrix element (Si, Al, Mg, Ba, O) diffusion was found in the coarsened BN. For a BN layer thickness of ~400nm, as shown in Fig. 3, the BN is coarsened through only about one-half of its thickness, and there is no evidence of any interaction between the BN and the Nicalon fiber. However, if the BN layer is substantially thinner, the total BN layer thickness (~200nm) is coarsened (crystallized), with subsequent carbon plus silica sublayers formed at the BN/Nicalon interface, as seen in Fig. 4. These layers are formed by the carbon condensed oxidation of the Nicalon fibers, as initially described by Cooper and Chyung[3]:

$$SiC_{(s)} + O_{2(g)} \rightarrow SiO_{2(s)} + C_{(s)} \qquad (1)$$

This reaction is enhanced by the diffusion of oxygen from the BMAS matrix through the thin BN layer during composite fabrication. The thicker BN layer in Fig. 3 prevented oxygen diffusion from reaching the fiber surface during composite fabrication, thus limiting the carbon condensed oxidation reaction of the Nicalon fibers. From mechanical property measurements, the crystallization of the BN was found to result in lower strength and toughness composites, due to the crystallized BN being a less effective crack deflecting interfacial layer than turbostratic BN.

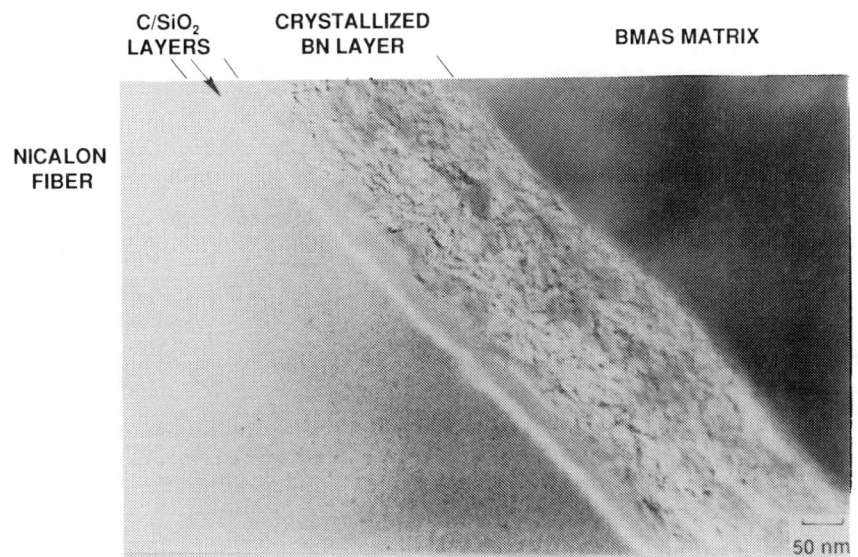

Fig. 4 - Crystallized BN Layer with C/SiO$_2$ Sublayer Formation at Nicalon Fiber Surface

Heat-Treated Composite Microstructure

A BMAS matrix composite with SiC/BN coated Nicalon fibers was subjected to thermal aging in air at 1200°C for 500 hrs. From TEM examination of the composite after thermal aging, no significant changes were found in the Nicalon fibers, the SiC or BN coatings, or the BMAS matrix, compared to the as-ceramed composite. However, throughout much of the composite a subtle dual C/SiO$_2$ nano-scale layer formation was found at the BN/Nicalon fiber interface, as shown in Fig. 5. The SiC coating appears to be well-bonded to the BN layer, and no BN coarsening is apparent, implying that matrix elements did not penetrate the SiC overlayer, and that the formation of the dual sublayers did not occur by oxygen diffusion from the matrix through the BN layer. Since it was found that the dual sublayers in the heat-treated composite tended to be more pronounced near the composite surface, the oxygen source for this reaction is thought to be the aerobic environment surrounding the composite.

Tensile Test Results

The BMAS matrix composites (0/90°) were subjected to uniaxial tensile testing at 20°, 1100°, and 1200°C. Figure 6 shows typical stress-strain curves for these tests, along with the fracture surface of a composite tested at 1100°C. In all cases, the composites exhibited elastic deformation upon initial loading, manifested as the linear part of the curves, from which the elastic modulus values were determined. The stress at which the curve deviated from linearity marked the proportional limit (PL) stress, which is normally attributed to the initiation of microcracking in the matrix. As shown in Fig. 6, and Table I, the ultimate tensile strength and strain-to-failure of the composites increased with temperature, while the elastic modulus decreased.

The proportional limit stress also decreased slightly with temperature.

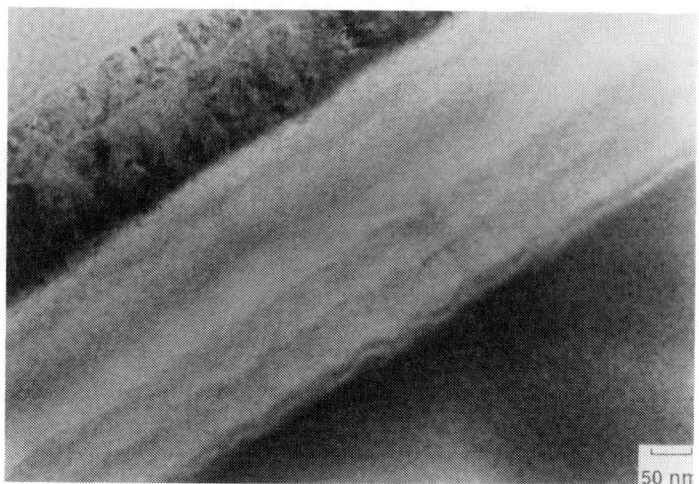

Fig. 5 - Interfacial Microstructure of Thermally Aged (1200°C, 500 hrs) Composite

Table I - Tensile Properties for BMAS Matrix/SiC/BN Coated Nicalon Fiber Composites (0/90°)

Temperature	UTS (MPa)	PL (MPa)	E (GPa)	ε_f - %
20°C	237	95	132	0.63
1100°C	272	88	73	0.70
1200°C	304	70	59	0.88

The fracture surfaces of the composites were quite fibrous in nature, regardless of the test temperature, with debonding occurring primarily within the BN layer, as can be seen in Fig. 6b. At 1200°C, the fracture surface appeared somewhat glassy and bubbly, indicating that oxidation of the SiC fibers and the SiC/BN coatings occurred after composite failure. From TEM examination of the composite microstructures after high-temperature tensile testing, no distinct changes were observed at the fiber/coating(s)/matrix interfaces, or in the Nicalon fibers or BMAS matrix.

Tensile Creep Testing

Tensile creep testing was performed on the BMAS matrix composites at 1100°C in air, at stress levels of 103 and 138 MPa. The stress level of 103 MPa was above the proportional limit stress of 88 MPa measured from 1100°C tensile testing. The 103 MPa tensile creep strain vs time curve is shown in Fig. 7, along with a photograph of the as-tested sample. An ~50 hr transient creep mode (the data gap was caused by an electronic malfunction) was followed by a very low steady state creep regime that lasted until the test was terminated without failure at 266 hrs. The measured creep rate was 8.32×10^{-4} %hr^{-1}, which corresponds to a rate of 2.3×10^{-9} s^{-1}. The creep behavior of the sample crept at 138 MPa was almost identical.

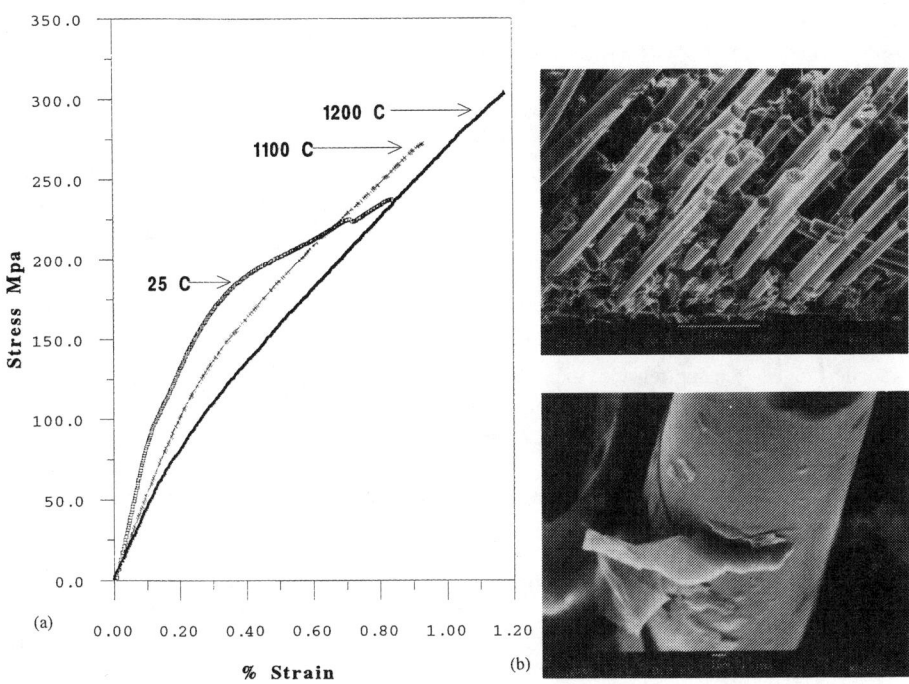

Fig. 6 - Composite Tensile Stress-Strain Behavior and Fracture Surfaces

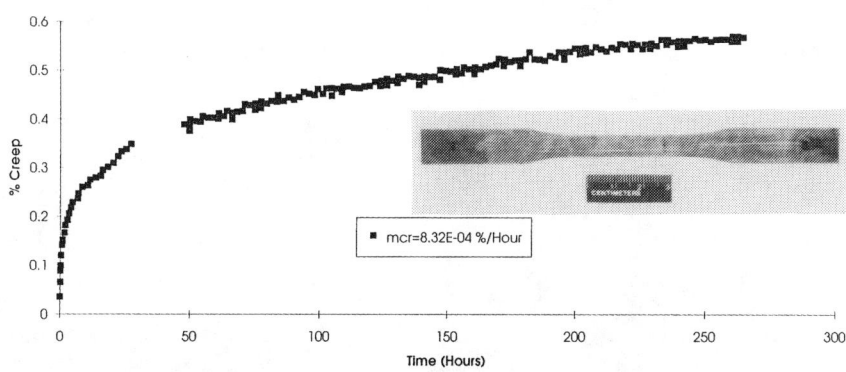

Fig. 7 - Composite Tensile Creep Response at 1100°C, 103 MPa.

Microstructural TEM analysis of the composite crept at 1100°C, 103 MPa, for 266 hrs in air revealed that no significant changes occurred in fiber/matrix interfacial regions in either the 0° or 90° plies, although the beginnings of a very subtle nano-scale C/SiO_2 dual layer formation at the BN/Nicalon fiber interface was noted. This dual layer is similar to, but much less distinct than, the C/SiO_2 dual layer found for the 1200°C, 500 hr, air aged composite, as was shown in Fig. 5. Even though this composite was stressed above the 1100°C proportional limit stress of 88 MPa, no matrix microcracking or fiber/matrix debonding could be seen. One change that was noted, however, was that evidence of spinodal decomposition of the BMAS matrix was seen that consisted of regions of crystalline tubules of hexacelsian (BAS), surrounded by a siliceous glass. It appeared these regions initiated within the intergranular glassy phase that was present in the as-ceramed composites. The fact that the spinodal decomposition of the BMAS matrix was not seen in the 1200°C, 500 hr aged sample, but was in the 1100°C, 266 hr crept sample, indicates that it may have been influenced by the application of stress on the matrix.

Tensile Fatigue Testing

Tensile fatigue (tension-tension) experiments were conducted on BMAS matrix composites at 20°C, 1100°C, and 1200°C, under a maximum stress of 103 MPA (minimum 10 MPa), and at 1100°C under a maximum stress of 138 MPa (minimum 14 MPa). The frequency was between 2-3 Hz, which resulted in a 10^5 cycle runout time of between 9 and 13 hrs. All of the samples fatigued at 103 MPa survived 10^5 cycles without failure. The one sample fatigued at 138 MPa failed after ~13,000 cycles.

Microstructural TEM analysis of the fatigued samples revealed several interesting features. In the samples fatigued at a maximum stress of 103 MPa, cracks in the matrix occurred in both the 0° and 90° plies. The cracks invariably were deflected and/or arrested at the fiber/matrix interface, usually within the BN interfacial layer. This is shown in Fig. 8 for the sample fatigued at 20°C for a 0° ply region. At elevated temperature, a similar crack pattern was seen, except that in the 90° plies, an oxidation product formed in the interfacial cracks, as shown in Fig. 9 for the sample fatigued at 1100°C. While the interfacial region in Fig. 9a was unaffected by the fatigue testing, that shown in Fig. 9b has a crack that is partially filled with a mottled appearing oxidation product. Interestingly, the BN layer on the matrix side of the crack in Fig. 9b is not oxidized,

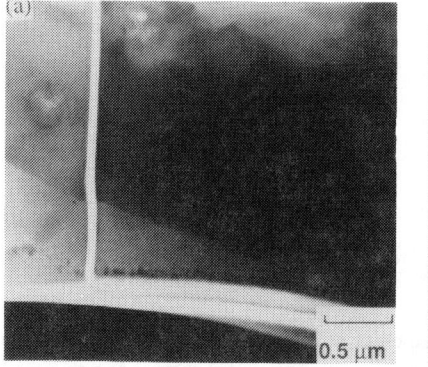 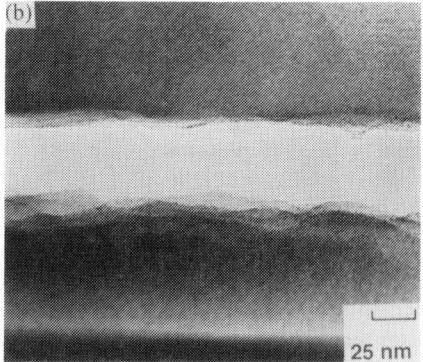

Fig. 8 - Crack Deflection in 0° Plies After RT Tensile Fatigue Testing at (a) BMAS Matrix/Coated Nicalon Fiber Interface and (b) Within BN Interfacial Layer.

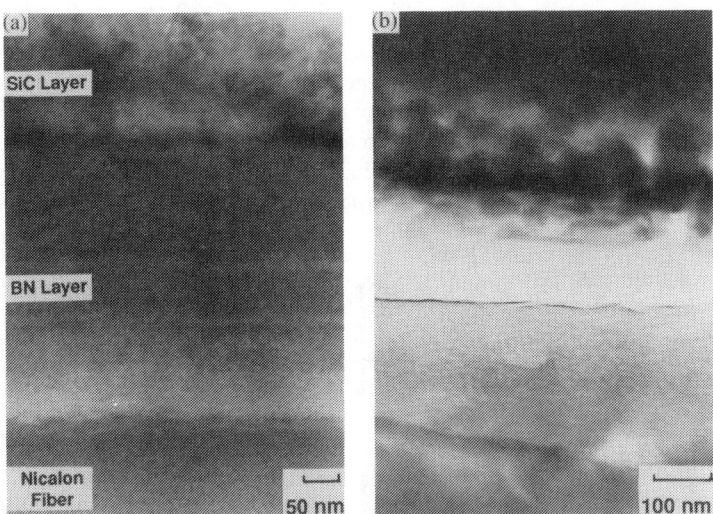

Fig. 9 - Composite Interfacial Microstructure (90° Plies) After 1100°C, 103 MPa Tensile Fatigue

with compositional analysis by EDS and PEELS revealing that the oxidation product contained only silicon and oxygen (no boron or carbon). It is apparent that different mechanisms are occurring in the microstructural evolution of the BMAS matrix composites depending on whether the stress state at elevated temperatures is monotonic or cyclic. Under monotonic loading (creep), the matrix cracks that may have formed did not propagate through the composite cross-section due to crack closure near the composite surface from oxidation reaction products. Under cyclic loading, however, crack-wake contact prevented accumulation of oxidation product, thus precluding the possibility of crack-healing by reaction products near the composite surface. Thus, oxidation of the Nicalon fibers occurred throughout the composite, although predominantly in the 90° plies. At higher fatigue stress levels (138 MPa), crack formation was more prevalent and progressed into the 0° plies, leading to composite fracture prior to the 10^5 cycle runout.

Tensile Stress-Rupture Testing

BMAS matrix composites with SiC over BN coated Nicalon fibers were subjected to long-time tensile stress-rupture testing at 1100° and 1200°C in air. The composite tested at 1200°C consisted of a 0/±45/90° ply layup configuration, and was loaded to a stress level of 69 MPa, which, for this ply layup configuration, was higher than the 1200°C fast fracture proportional limit stress of 50 MPa. The composite remained at a stress of 69 MPa at 1200°C in air for 11,725 hrs, which is ~1.35 yrs, when it finally fractured. The fracture surface, which is shown in Fig. 10, was quite fibrous except for a region around the periphery of the 0.2" x 0.1" (0.51 x 0.25 cm) composite cross-section in the gauge. Figure 11 shows a cross-section of an edge region of this composite with reacted fibers and matrix porosity quite evident within 150-200μm of the composite surface. It appeared that the oxidation of the Nicalon fibers occurred more rapidly in the 90° plies than in the 0° plies, in that the SiC/BN coatings on the fibers in the 0° ply next to the composite surface were intact within ~150μm of the composite surface, while those in the 90°

plies were visibly reacted to a depth of at least 250µm. Thus, "pipeline" diffusion that initiates at cut ends of the fibers exposed at the composite surface, with oxygen transport occurring along the fiber/coating(s)/matrix interface, is evidently faster than "transverse" diffusion of oxygen through the matrix. TEM replica analysis of the composite (0° ply) near the surface and away from the surface reaffirmed that, while the fibers and matrix near the surface were severely reacted, the microstructure of the bulk of the composite, including the SiC/BN fiber coatings, was unaffected by this severe environmental and mechanical exposure.

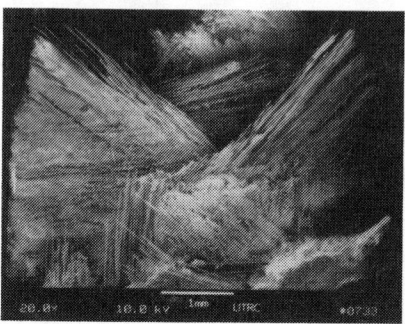 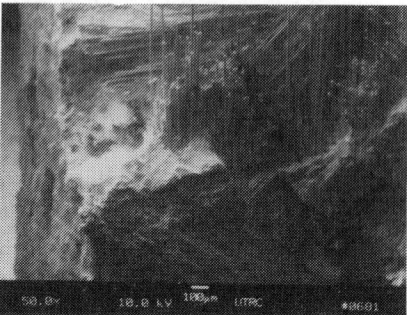

Fig. 10 - Fracture Surface of 0/±45/90° Composite After 1200°C, 69 MPa, 11,725 hr Tensile Stress-Rupture Testing

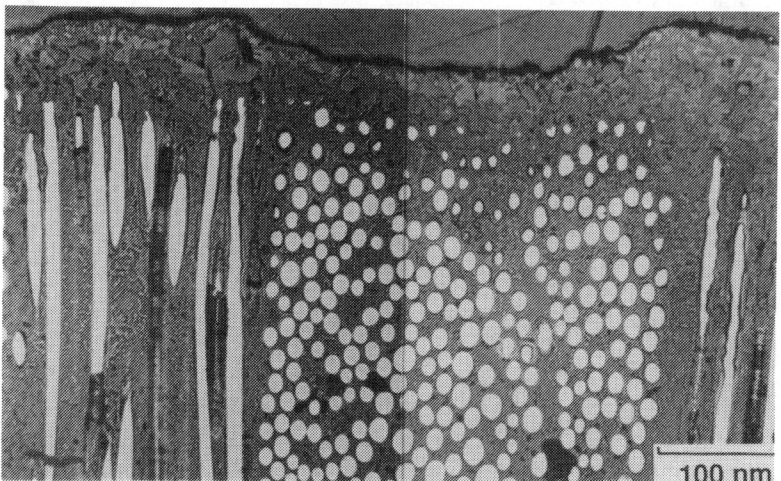

Fig. 11 - Microstructure of Composite Sample Edge Section After 1200°C, 69 MPa, 11,725 hr Tensile Stress-Rupture Testing

The second composite tested at 1100°C in long-time tensile stress-rupture consisted of a 0/90° ply layup configuration, and was loaded to a stress level of 138 MPa, which is significantly above the 1100°C fast fracture proportional limit stress of 88 MPa, as was shown in Table I. The composite sample remained at this stress level for 14,685 hrs (~1.7 yrs), when a failure of the gripping arrangement terminated the test. The sample was then reloaded at 1100°C and stressed to failure. The retained 1100°C tensile stress was 248 MPa, which is only slightly less than the normal fast fracture tensile strength of this composite configuration, as shown in Table I. SEM examination of the fracture surface revealed it to be quite fibrous, as shown in Fig. 12, with reactivity (oxidation) of the Nicalon fibers and the SiC/BN fiber coatings limited to a distance of less than 50μm from the composite surface.

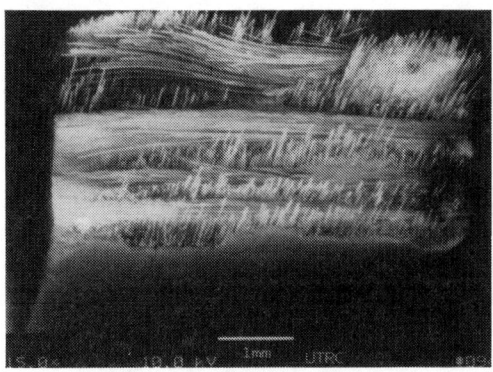 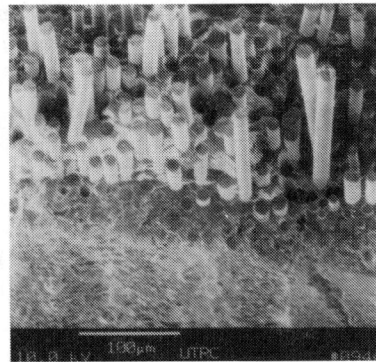

Fig. 12 - Fracture Surface of 0/90° Composite After 1100°C, 138 MPa, 14,685 hr Tensile Stress-Rupture Testing (Retained 1100°C UTS = 248 MPa).

From TEM and X-ray diffraction analysis of the above composites, it was found that the matrix regions near the composite surface had transformed from the barium osumilite BMAS phase to the celsian BAS phase ($BaAl_2Si_2O_8$). This transformation may be induced by the formation of excess silica through the oxidation of the Nicalon SiC fibers near the composite surface, along with the volatilization of MgO from the composite surface region during the long-time, high temperature exposures. Celsian is a highly refractory crystalline phase with a melting point over 1700°C, and a thermal expansion coefficient ($\sim 2.29 \times 10^{-6}/°C$) that is a good match to that of barium osumilite ($\sim 3.0 \times 10^{-6}/°C$). The formation of this celsian phase may play a role in limiting the diffusion of oxygen from the environment to the SiC/BN coated Nicalon fibers.

CONCLUSIONS

From the mechanical property data and microstructural information presented in this paper, it has been demonstrated that the ceramic composite system based on the reinforcement of a barium-magnesium aluminosilicate (BMAS) glass-ceramic matrix with CVD SiC over BN coated Nicalon SiC fibers exhibits the mechanical properties necessary for potential application as a high temperature, lightweight structural component. In particular, it has been shown that this composite system can withstand stress levels at elevated temperature for long periods of time and/or many cycles that are *higher* than the proportional limit, or matrix microcracking, stress of the composite. This indicates that the fiber/matrix interface consisting of CVD SiC over BN is performing as it should, i.e., a weakly bonded yet environmentally stable system that allows the deflection and diversion of matrix cracks such that the composite system behaves in a pseudo-ductile or "tough" manner, and when it fails does so in a graceful, non-catastrophic fracture mode.

ACKNOWLEDGEMENTS

The authors would like to thank Mr. Jim O'Kelly of 3M for the CVD SiC/BN coatings on the Nicalon fibers, Mr. Gerald McCarthy of UTRC for the TEM replica analyses, Mr. David Murphy of Pratt&Whitney for the composite mechanical testing, and Ms. Laura Austin of UTRC for the fabrication of the coated fiber composites, and to Dr. Alexander Pechenik of the Air Force Office of Scientific Research (AFOSR), Bolling AFB DC, for his sponsorship of this program.

REFERENCES

[1] K.M. Prewo, J.J. Brennan, and G.K. Layden, "Fiber Reinforced Glasses and Glass-Ceramics for High Performance Applications," *Am. Cer. Soc. Bull.*, 65 [2] 305-13, 322 (1986)

[2] J.J. Brennan, "Interfacial Characterization of Glass and Glass-Ceramic Matrix/Nicalon SiC Fiber Composites," pp. 549-60 in *Materials Science Research*, Vol. 20, Plenum Press, New York, (1986).

[3] R.F. Cooper and K. Chyung, "Structure and Chemistry of Fiber-Matrix Interfaces in SiC Fibre-Reinforced Glass-Ceramic Composites: An Electron Microscopy Study," *J. Mater. Sci.*, 22, 3148-60 (1987).

[4] C. Ponthieu, M. Lancin, J. Thibault-Desseaux, and S. Vignesoult, "Microstructure of Interfaces in SiC/Glass Composites of Different Tenacity, *Journal de Physique,* 51, C1-1021-26, (1990).

[5] R.W. Rice: "BN Coating of Ceramic Fibers for Ceramic Fiber Composites", US Patent 4,642,271, Feb. 10, 1987.

[6] R.N. Singh: "Fiber-Matrix Interfacial Characteristics in a Fiber-Reinforced Ceramic-Matrix Composite", *J. Am. Cer. Soc.*, 72 (9), 1764-67 (1989).

[7] J. Llorca and R.N. Singh: "Influence of Fiber and Interfacial Properties on Fracture Behavior of Fiber-Reinforced Ceramic Composites, *J. Am. Cer. Soc.*, 74 [11] 2882-90 (1991).

[8] R. Naslain, O. Dugne, and A. Guette: "Boron Nitride Interphase in Ceramic-Matrix Composites," *J. Am. Cer. Soc.*, [10] 2482-88 (1991).

[9] O. Dugne, S. Prouhet, A. Guette, R. Naslain, R, Fourmeaux, Y Khin, J. Sevely, J.P.Rocher, and J. Cotteret: "Interface characterization by TEM, AES, and SIMS in tough SiC (ex-PCS) fibers-SiC (CVI) matrix composites with a BN interphase," *J. Mat. Sci.*, 28, 3409-22 (1993).

[10] J. Brennan: "Interfacial Studies of Fiber Reinforced Glass-Ceramic Matrix Composites," High Temperature Ceramic Matrix Composites, HT-CMC-1, Editors: R. Naslain, J. Lamon, and D. Doumeingts, Woodhead Publishing, Ltd, Abington Cambridge, England, 269-84 (1993).

[11] E. Sun, S. Nutt, and J. Brennan: "Interfacial Microstructure and Chemistry of SiC/BN Dual Coated Nicalon Fiber Reinforced Glass-Ceramic Matrix Composites", *J. Am. Ceram. Soc.* 77, [5] 1329-39 (1994)

OXYGEN-FREE CERAMIC FIBERS FROM ORGANOSILICON PRECURSORS AND E-BEAM CURING

Hiroshi Ichikawa, Nippon Carbon Co.,Ltd., Shin-urashima-cho, Kanagawa-ku, Yokohama, 221, JAPAN
Kiyohito Okamura, University Osaka Prefecture, Sakai, Osaka,593, JAPAN
Tadao Seguchi, JAPAN Atomic Energy Research Institute, Watanuki, Takasaki,370-12, JAPAN

Oxygen free Si-C fibers"Hi-Nicalon" with C/Si ratio of 1.39 and "Hi-Nicalon S" with C/Si ratio of 1.05 were synthesized from polycarbosilane electron beam curing and pyrolysis with different processing conditions. Both fibers exhibited outstanding thermal stability up to 2073K as compared to Nicalon. Especially, the stoichiometric SiC fiber, Hi-Nicalon S, has high elastic modulus of 420GPa, good oxidation resistance at 1673K and excellent creep resistance at 1473K.

I INTRODUCTION

The Si-C-O fiber Nicalon synthesized from polycarbosilane(PCS) has high tensile strength, high tensile modulus and good thermal resistance in air at high temperatures.[1-4] Nicalon fiber has been produced industrially and applied widely to heat resistant materials and reinforcements for polymer, metal and ceramic matrix composites.[5,6] Especially, Nicalon is one of the best candidates for the reinforcement of ceramic matrix composites(CMC).[7]

Recently, CMC have attracted public attention as a the heat resistant structural material substituted for super heat resistant alloy. Especially, CMC is required with heat resistance at more than 1,500°C for the materials of the space plane and the high temperature gas-turbine. However, the ordinary Nicalon has been limited to 1,200°C in heat resistance. The Nicalon fiber is degraded on thermal exposure at over 1,300°C by evolution of carbon monoxide with a pyrolytic reaction.[8-10] This degradation of the fiber is caused by oxygen contained in the Nicalon fiber. The oxygen introduced in the oxidizing crosslinking reaction in the curing process.

We have developed a new curing process for polycarbosilane fiber using electron beam irradiation in an oxygen-free atmosphere to decrease the oxygen content of Nicalon fiber.[11-14] Consequently, we have developed the Si-C fiber "Hi-Nicalon" having less than 0.5wt% - oxygen content. Furthermore, we have synthesized the oxygen free Si-C fibers with various C/Si atomic ratio by E-beam curing and pyrolysis from PCS fiber. Then the stoichiometric SiC fiber "Hi-Nicalon S" has been developed.

In this study, the recent development results on these oxygen free Si-C fibers from organosilicon precursors and E-beam curing such as physical and mechanical properties, thermal stability and environmental resistance are reported.

II EXPERIMENTAL PROCEDURE

(1) Preparation of the Si-C fibers

Fig.1 shows preparation process of the Si-C fiber "Hi-Nicalon" using electron beam irradiation curing technique as compared to the oxygen curing which corresponds to the Si-C-O Nicalon fiber.

Polycarbosilane(PCS) was melt-spun into PCS fiber composed of 500 continuous filaments with a multi-holes spinning machine. The PCS fiber was cured by electron beam irradiation of 10~15 MGy in a helium gas.

Under electron beam irradiation, the Si-H group and the -CH_2- or CH_3- group of polycarbosilane were crosslinked and polymerized directly. The cured PCS fiber was pyrolyzed up to more than 1,500°C in an argon gas stream to obtain the Si-C fiber "Hi-Nicalon".

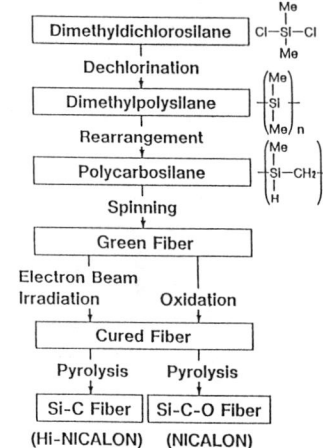

Fig.1 Preparation process of the SiC-based fibers (Hi-Nicalon and Nicalon)

The Si-C fibers with various C/Si atomic ratio were prepared from E-beam cured PCS fibers by pyrolysis with special conditions. The C/Si ratio of the fibers in this work ranged from 0.84 to 1.68.

The contents of all other elements, O, N, and H were<1wt%,<0.2wt%, and 0.1wt%, respectively. The stoichiometric SiC fiber "Hi-Nicalon S" was prepared from the fiber with C/Si atomic of 1.05 by heat-treating at high temperature.

(2) Characterization of the Si-C fibers

Chemical compositions of various types of SiC fibers were analyzed. Silicon analysis was done by the gravimetric method after alkali fusion, and carbon and oxygen analyses were done using carbon and oxygen analyzer (Leco company). The structure of these fibers was investigated by Xray diffraction and TEM. The tensile strength and modulus of the fibers were measured by the single filament method of JIS R 7601 with a gauge length of 25mm and a cross-head speed of 1mm/min, using Tensilon UTM- II (Orientec Co., Ltd.).

(3) Thermal stability, Oxidation and creep resistance at high temperature

Thermal stability was estimated by tensile properties and SEM observation after the thermal exposure test. Three types of SiC-based fibers (Nicalon, Hi-Nicalon, Hi-Nicalon S) were exposed for 10hours at 1773K to 2073K.

An oxidation test was performed by exposing the fibers for 10hours in dry air at 1673K. After the tests, the tensile strengths and moduli of each fibers were measured by single filament tensile method at room temperature.

Creep resistance was evaluated with bend stress relaxation test developed by Morscher and DiCarlo.[15] In this experiment, fibers were set in a graphite die and heat-treated for 1hour in argon at 1473K.

III RESULTS AND DISCUSSION

(1) Physical and mechanical properties of the Si-C fiber with various C/Si atomic ratios

Fig-2 shows density of the Si-C fibers pyrolyzed at 1573K as a function of C/Si atomic ratio. The density of the Si-C fibers undergoes a maximum when the C/Si ratio is close to the stoichiometric C/Si atomic ratio of 1.0. The stoichiometric composition gave the highest density, 3.02 g/cm^3. When the SiC fiber with near stoichiometric C/Si ratio (1,05) was pyrolyzed at 1773K, the density of the fiber reached 3.1 g/cm^3, which is close to theoretical value of SiC, 3.2 g/cm^3.

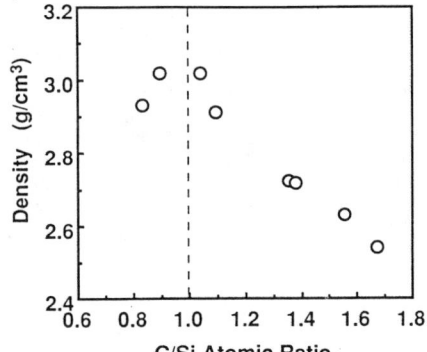

Fig.2 Density of the Si-C fibers as a function of C/Si atomic ratio[16]

Tensile strength and elastic modulus of the Si-C fibers are shown in Fig.3 and 4, respectively. The strength of each fiber is more than 2.5GPa. The tensile strength of the fiber seems to be independent of its chemical composition. On the other hand, the elastic modulus varies dramatically as a function of C/Si. It exhibits a maximum for a C/Si ratio close to 1. The near stoichiometric fiber exhibited the highest modulus of 400 GPa. The modulus appears to be related to density.

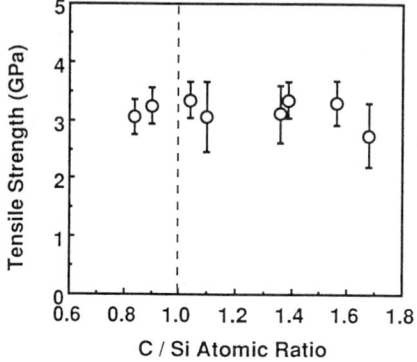 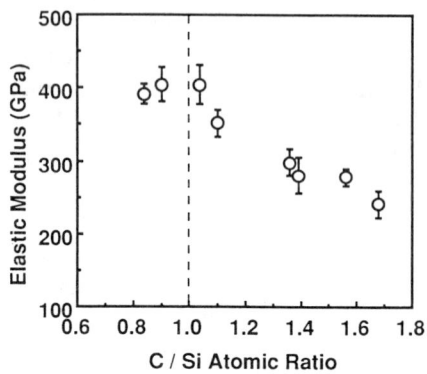

Fig.3 Tensile strength of the Si-C fibers as a function of C/Si atomic ratio[16]

Fig.4 Elastic modulus of the Si-C fibers as a function of C/Si atomic ratio[16]

(2) Structure and mechanical properties of the Si-C fibers, Hi-Nicalon and Hi-Nicalon S

Table 1 shows the typical properties of PCS-derived SiC-based fibers Nicalon, Hi-Nicalon and Hi-Nicalon S.

Table 1 Typical properties of PCS-derived SiC-based fibers
Each data indicate the average value.

Properties		Nicalon NL-200	Hi-Nicalon	Hi-Nicalon type S
Fiber diameter	(μm)	14	14	12
Number of filaments	(fil./yarn)	500	500	500
Tex	(g/1000m)	210	200	180
Tensile strength	(GPa)	3.0	2.8	2.6
Tensile modulus	(GPa)	220	270	420
Elongation	(%)	1.4	1.0	0.6
Density	(g/cm^3)	2.55	2.74	3.10
Specific resistivity	(ohm-cm)	10^3-10^4	1.4	0.1
Chemical composition	Si (wt.%)	56.6	62.4	68.9
	C	31.7	37.1	30.9
	O	11.7	0.5	0.2
	C/Si (atomic)	1.31	1.39	1.05

The oxygen content of Hi-Nicalon prepared by irradiation curing decreased remarkably to the minimum value of 0.5 wt.% as compared to that of Nicalon. The C/Si atomic ratio of Hi-Nicalon was 1.39. It is considered that the fiber consists of fine crystallite SiC and turbostratic carbon fringes. Fig. 5 shows AES depth profiles of Nicalon and Hi-Nicalon. Hi-Nicalon was covered with a carbon rich layer on the surface, though Nicalon was covered with SiO_2. In case of CMC reinforced with Nicalon fiber, the fiber surface of Nicalon is coated with a carbon layer for improved compatibility of the fiber and the matrix. On that point, the surface of Hi-Nicalon is suited to the reinforcement of ceramics.

The continuous Hi-Nicalon fiber had the same value of filament diameter and number of filaments per yarn as ordinary Nicalon. Hi-Nicalon had a slightly increased density and sharply increased tensile modulus as compared to Nicalon.

Based on the study of the various C/Si fibers, we have selected the fiber with C/Si atomic ratio of 1.05 as new grade of Hi-Nicalon. The fiber was pyrolyzed at high temperature with special conditions, and the crystallite size of the fiber grew without decreasing the strength. Then, we have developed the stoichiometric SiC fiber "Hi-Nicalon S" as a new grade of Hi-Nicalon.

Fiber diameter and the Tex of Hi-Nicalon S are smaller than the other two types of fibers. The tensile strength of each fiber is more than 2.5GPa, which is a sufficient value for the reinforcement of CMCs. Hi-Nicalon S shows the highest modulus of 420GPa and the highest density,$3.1 g/cm^3$. Specific resistivity of Hi-Nicalon S is much smaller. The reason may be explained by the existence of the carbon layer with 200 nm thickness on the outer skin structure. The oxygen content is less than that of Hi-Nicalon.

Fig.6 shows the XRD patterns of the three types of fibers. In the fibers, only β-SiC diffraction peaks were observed. The crystallite size of Hi-Nicalon S was 10.9nm and the largest of the three types of fibers.

The TEM micro graphs of SiC-based fibers are shown in Fig.7. The SiC grain size of the three types of the fibers was obviously different from each other, and increased in the order of Nicalon, Hi-Nicalon, Hi-Nicalon S.

The grain size of Hi-Nicalon and Hi-Nicalon S seemed to be larger than the crystallite size which was calculated from XRD patterns.

Fig.8 shows the fracture surfaces of Si-C fibers. Hi-Nicalon shows typical glassy feature. On the other hand, small grains are observed on the fracture surface of Hi-Nicalon "S", which is evidence that this fibers has higher crystallinity.

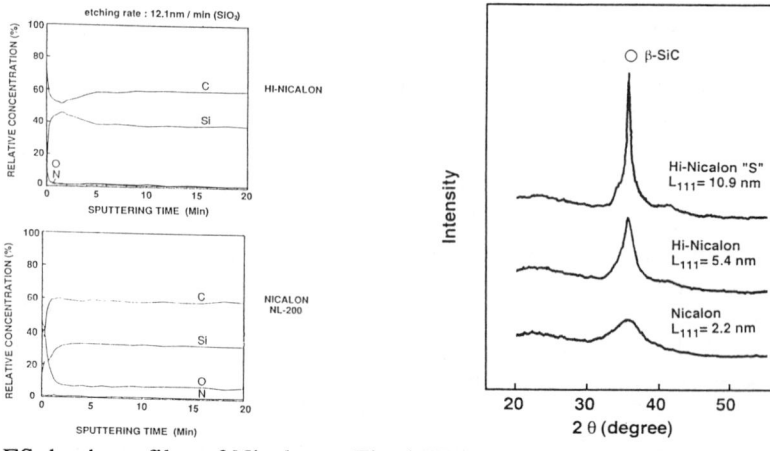

Fig.5 AES depth profiles of Nicalon and Hi-Nicalon

Fig.6 XRD patterns of the SiC-based fibers

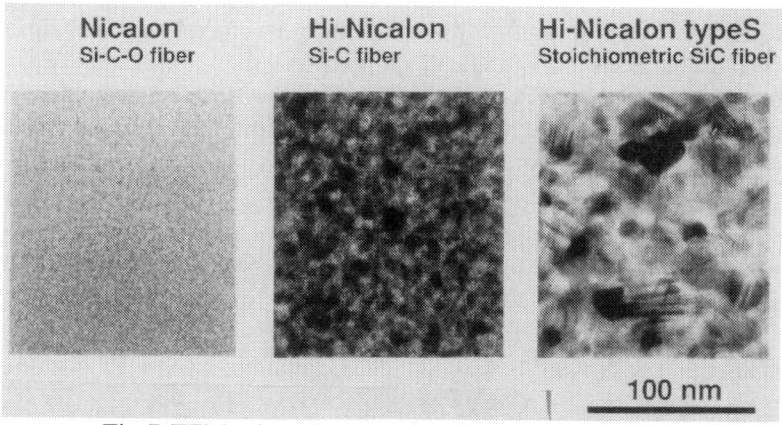

Fig.7 TEM microphotographs of the SiC-based fibers

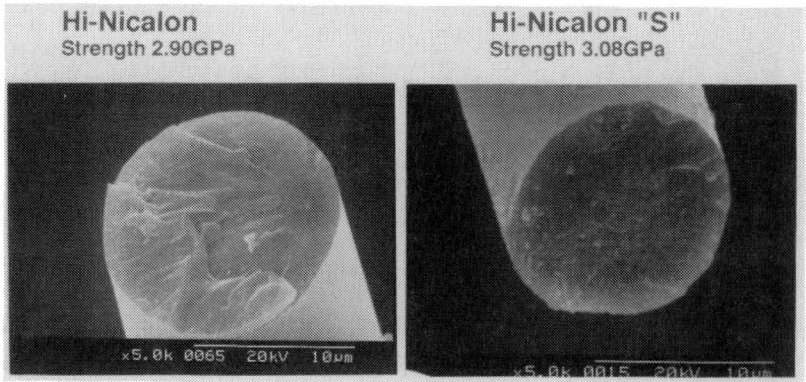

Fig.8 Fracture surfaces of the Si-C fibers

(4) Thermal stability

Fig.9 shows the tensile strength of the fibers after the thermal exposure test. The Nicalon fiber exhibited low strength after exposure at 1673K and no strength after the 1773K exposure. On the other hand, Hi-Nicalon and Hi-Nicalon"S" retained good strength even after the 1873K exposure. Hi-Nicalon"S" showed the highest strength,1.80GPa, after 10hours exposure in argon at 1873K.

Further tests at higher temperature were performed on Hi-Nicalon and Hi-Nicalon"S". After exposure for 1 hour at 2273K in an argon atmosphere, Hi-Nicalon fiber maintained the same appearance and flexibility as that before exposure and had moderate strength of 1.3GPa. Hi-Nicalon"S" after 10hours exposure in argon at 2073K was quite stable chemically, since no structural decomposition occured and it exhibited a good strength of 1.9GPa.

(5) Oxidation resistance

The tensile strength of the SiC-based fibers after exposure at 1673K for 10hours in dry air is shown in Fig.10. As a result, all the fibers were oxidized, and thickness of the oxide layer on the surface of the fibers was almost the same i.e. $0.3 \sim 0.4 \mu m$. All tested fibers undergo a strength decrease due to oxidation. However, the strength of the fiber after oxidation was different from each other. The stoichiometric SiC fiber Hi-Nicalon S showed the highest residual strength of 1.8GPa. Nicalon fiber is not only oxidized, but also thermally degraded. Hi-Nicalon had lower strength than Hi-Nicalon S after oxidation, because the free carbon phase in carbon rich Si-C fiber Hi-Nicalon, was oxidized with an evolution of carbon monoxide.

(6) Creep resistance

Fig.11 shows comparison of creep resistance in Si-based ceramic fibers,which is described by the stress relaxation ratio(m). Hi-Nicalon"S" exhibits improved creep resistance over Nicalon and Hi-Nicalon, which is at a similar level to that of Dow Corning and Carborundum polycrystalline fibers. It is well understood that larger crystallite size material shows less diffusional creep deformation. Grain boundary chemistry and structure are also important factors. Further analysis of this fibers is needed.

(7) Summary of properties of the SiC-based fibers

Table2 represents the properties of PCS-derived fibers, Nicalon,Hi-Nicalon and Hi-Nicalon S. Hi-Nicalon has much improved thermal stability over Nicalon, and also a higher elastic modulus and creep resistance. Hi-Nicalon S has the advantage of excellent thermal stability as does Hi-Nicalon. It has also the highest elastic modulus, best oxidation resistance and creep resistance of PCS-derived fibers.

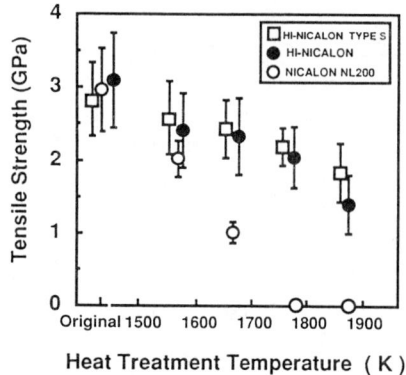

Fig.9 Tensile strength of the SiC-based fibers after thermal exposure test

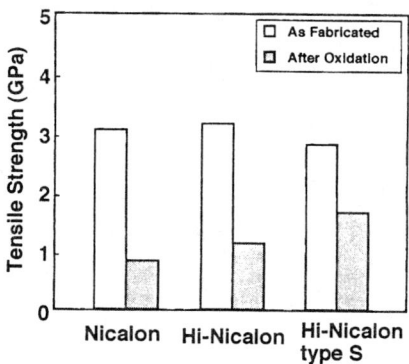

Fig.10 Tensile strength of the SiC-based fibers after exposure at 1673K for 10 hours in dry air

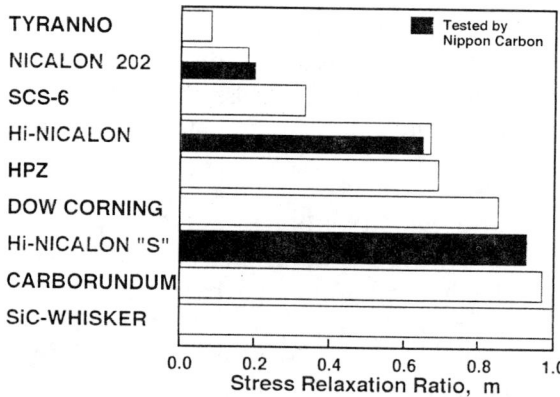

Fig.11 Comparison of creep resistance in Si-based ceramic fibers[15]

Commercial production of Hi-Nicalon has been started by Nippon Carbon Co., Ltd., and Hi-Nicalon samples such as yarns and clothes have been provided to customers.

Table 2 Comparison of properties on PCS-derived SiC fibers[16]

◎ excellent ○ good ▲ moderate		Nicalon → Si-C-O Fiber	Hi-Nicalon → Si-C Fiber	Hi-Nicalon type S Stoichiometric SiC Fiber
Chemical composition	C/Si O	1.31 12 wt%	1.39 0.5 wt%	1.05 0.2 wt%
Tensile strength		◎ 3.0 GPa	○ 2.8 GPa	○ 2.6 GPa
Elastic modulus		○ 200 GPa	○ 270 GPa	◎ 420 GPa
Thermal stability		○ 1473 K	◎ 2073 K	◎ 2073 K
Oxidation resistance		▲	▲	○
Creep resistance		▲	○	◎

IV CONCLUSIONS

Oxygen free Si-C fibers "Hi-Nicalon" were synthesized from poly-carbosilane by electron beam curing and pyrolysis. Hi-Nicalon had a high tensile strength of 2.8GPa and a high elastic modulus of 270GPa, and also exhibited outstanding thermal stability as compared to Nicalon. The Si-C fibers with various C/SiC atomic ratio have been prepared from E-beam cured PCS fibers with different processing conditions, and then stoichiometric, crystalline SiC fiber "Hi-Nicalon S" has been developed.

Hi-Nicalon S has excellent thermal stability up to 2073K as does Hi-Nicalon. It has also higher elastic modulus of 420GPa, higher creep resistance at 1473K and better oxidation resistance at 1673K than other PCS-derived Si-C fiber.

Hi-Nicalon S should be the best candidate for the reinforcement of ceramic matrix composites.

REFERENCES

(1) S.Yajima, J.Hayashi, M.Omori, K.Okamura: "Development of a SiC Fiber with High Tensile Strength," Nature,261,683-685(1976).
(2) T.Ishikawa and T.Nagaoki: *Recent Carbon Technology*, JEC Press Inc., Cleveland, OH,348-350(1983).
(3) T.Ishikawa, H.Ichikawa and H. Teranishi: "Strength and structure of SiC fiber

after exposure to high temperature," Proc.Electrochem. Soc. 88-5:Proc. Symp. High Temp. Mater. Chem.4,,205-217(1987).
(4) D.J. Pysher, K.C.Goretta, R.S.Hodder Jr., and R.E Tressler: "Strengths of Ceramic fibers at Elevated Temperatures," J.Am. Ceram. Soc.,72(2), ,284-288(1989).
(5) H.Ichikawa, S.Mitsuno, Y.Imai,and T.Ishikawa: "Mechanical and Electrical properties of SiC fiber(Nicalon) and their composites," Proc. 1st Japan International SAMPE Symposium, Nov.28 Dec1,,923-928(1989).
(6) Y.Imai, H.Ichikawa, and T.Ishikawa: "Multi-filament, continuous SiC fiber "NICALON" reinforced aluminum composites wire," Proceedings of 20th International SAMPE Technical. Conference., 1-12(1988).
(7) J.R.Strife, J.J Brennan, and K.M.Prewo: "Status of Continuous Fiber-Reinforced Ceramic Matrix Composite Processing Technology," Ceram. Eng. Sci. proc.,11(7-8) 871-919(1990).
(8) T.J.Clark, R.M.Arons, J.B.Stamatoff, and J.Rabe: "Thermal Degradation of Nicalon SiC Fibers," Ceram. Eng. Sci. Proc., 6(7-8), 576-588(1985).
(9) P.L.Coustumer, M.Monthioux, and A.Oberlin: "Thermal Degradation Mechanisms of a Nicalon Fiber as Deduced from TEM Observation," Int'1. Symp. Carbon,Tsukuba, Japan, 1,182-185(1990).
(10) O.Delverdier, M.Monthioux, and A.Oberlin: "Thermal Evolution of a Cured Fiber and Uncured PCS-Based Ceramic Micro textural Aspect," ibid, 1, 190-193(1990).
(11) K.Okamura, T.Seguchi, and S.Kawanishi: "High-Temperature Strength Improvement of Si-C-O fiber by the Reduction Oxygen content," Proceedings of. 1st Japan. International. SAMPE Symposium.,929-934(1989).
(12) M.Takeda, Y.Imai, H.Ichikawa, T.Ishikawa, T.Seguchi, and K.Okamura, "Properties of the Low Oxygen Content SiC Fiber on High Temperature Heat Treatment," Ceram. Eng. Sci. Proc.,12(7-8),1007-1018(1991).
(13) M.Takeda, Y.Imai, H.Ichikawa, T.Ishikawa, T.Seguchi, and K.Okamura: "Thermal Stability of the Low Oxygen Silicon Carbide Fibers Derived from polycarbosilane," Ceram. Eng. Sci. Proc.,13(7-8),209-217 (1992).
(14) M.Takeda, Y.Imai, H.Ichikawa, T.Ishikawa, N.Kasai, T.Seguchi, and K.Okamura, "Thermomechanical Analysis of the Low Oxygen Silicon Carbide Fibers Derived from Polycarbosilane," Ceram. Eng. Sci. Proc., 14(9-10),540-547(1993).
(15) G.N.Morscher, J.A.DiCarlo, and T.Wagner,"Fiber Creep Evaluation by Stress Relaxation Measurements," Ceram. Eng. Sci. Proc., 12(7-8), 1032-1038(1991).
(16) M.Takeda, J.Sakamoto, Y.Imai, H.Ichikawa and T.Ishikawa, "Properties of stoichiometric silicon carbide fiber derived from polycarbosilane", Ceram. Eng. Proc.,Vol.15, No.4, 133-141(1994).

Si-B-(N,C) A NEW CERAMIC MATERIAL FOR HIGH PERFORMANCE APPLICATIONS

H.-P. Baldus, G. Passing, D. Sporn*, and A. Thierauf*
Bayer AG, Zentrale Forschung, 51368 Leverkusen, FRG
*Fraunhofer-Institut für Silicatforschung, 97082 Würzburg, FRG

ABSTRACT

A new multinary, single phase (carbo-)nitride ceramic with outstanding properties has been developed. The "single-source" precursor $Cl_3Si-NH-BCl_2$ reacts with alkylamines to form a preceramic polymer which can be a liquid or a fusible or infusible solid depending on reaction conditions. Polymer pyrolysis in ammonia or nitrogen leads to amorphous $Si_3B_3N_7$ and $SiBN_3C$ ceramics, respectively. The present paper reports on the physical and chemical properties of $SiBN_3C$ ceramics, with emphasis on its outstanding oxidation resistance. The resistance towards crystallization up to 1800 °C enables the development of considerably improved high-performance inorganic fibers.

INTRODUCTION

The development of a new generation of gas turbines or jet propulsion engines for higher combustion temperatures requires new materials which withstand temperatures of up to 1500 °C for several thousand hours. Unfortunately, materials such as SiC, Si_3N_4 or fiber-reinforced composites suffer from several drawbacks, e. g. low thermal shock resistance (SiC), low oxidation resistance (sintered Si_3N_4) or, in case of fibers, the detrimental oxidation behavior of carbon or the loss of fiber-strength due to crystallization (Si_3N_4, SiC or SiCN). To overcome these drawbacks, novel multinary non-oxide materials were synthesized in the system Si-B-(N,C), starting from the "single-source" precursor molecule $Cl_3-Si-NH-B-Cl_2$, which has the stoichiometry and basic atom structure desired in the final ceramic

Figure 1, Chemical structure proposed for borosilazane II

material. We report on the properties of the borosilazane polymer intermediate and the final ceramics derived with particular emphasis on thermal stability, crystallization behavior and oxidation properties of the ceramic materials. Furthermore applications such as ceramic fibers, coatings, infiltrations and structural parts are discussed.

POLYMER SYNTHESIS AND PROPERTIES

The single-source precursor Cl_3-Si-NH-B-Cl_2[1] reacts with NH_3 or aliphatic amines, R-NH_2. The reaction with NH_3 yields a carbon-free, porous, insoluble and unmeltable inorganic polymer referred to in this text as I, whereas the reaction of the precursor with H_3C-NH_2 yields a thermosetting oligomer referred to as II, which is soluble in n-hexane, o-xylene or methylene chloride. The oligomer II can be polymerized in a controlled fashion to a polyborosilazane called PBS-Me, whose proposed chemical structure is shown in Fig. 1.[1] The degree of crosslinking of the oligomer can be tuned by variation of temperature and reaction duration. Fig. 2 shows the time dependence of the viscosity of the oligomer II (R = CH_3) at 150 and 160 °C, indicating the onset of polymerization near 160 °C. Thus the ceramic precursor PBS-Me can be supplied to customers as a low viscosity liquid, a meltable polymer or an infusible solid, offering a wide range of processing pathways.

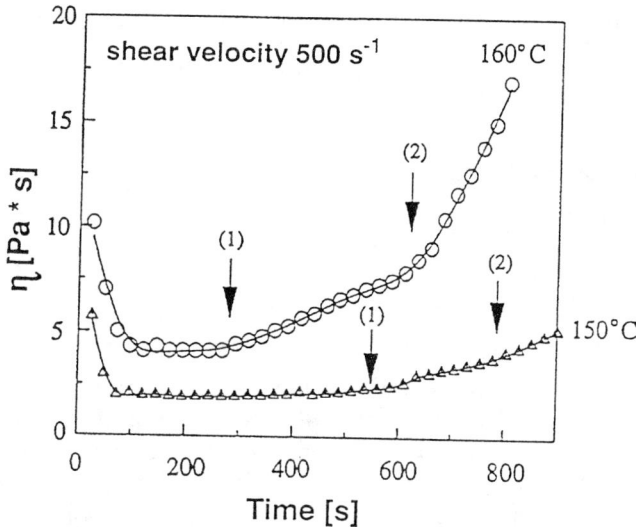

Figure 2, Time dependence of the viscosity of oligomer II (R = CH$_3$) at 150 and 160 °C

PHYSICAL AND CHEMICAL PROPERTIES OF Si-B-(N,C)

Pyrolysis of I in N$_2$ at 1000 °C leads to a white, amorphous single-phase ceramic with the approximate stoichiometry Si$_3$B$_3$N$_7$. The specific surface area of 60 m^2/g is due to the intrinsic high surface area of polymer I. Heating of PBS-Me in NH$_3$ to 1000 °C (1 °C/min) and additional annealing in N$_2$ at 1450 °C yields a similar product, but with a specific surface area of only 0.5 m^2/g. The elemental analysis has been described elsewhere.[1,2] Pyrolysis of PBS-Me in N$_2$ or Ar at 1450 °C yields a black, single phase ceramic.[2] Electron diffraction showed no crystalline phases. The elemental analysis suggests an empirical stoichiometry of SiBN$_3$C: Si = 32.1 wt.-%, B = 11.7 wt.-%, N = 40.5 wt.-%, C = 12.7 wt.-%, O < 0.5 wt.-% and Cl = 0.02 wt.-%. The physical properties of the materials are summarized in Table 1. All described amorphous materials have excellent thermal stabilities superior to Si$_3$N$_4$; for example: Si$_3$B$_3$N$_7$ decomposes into Si, BN and N$_2$ at 1700 °C in He-atmosphere, while Si$_3$N$_4$ is only stable up to 1450 °C. In N$_2$-atmosphere α- and β-Si$_3$N$_4$ are formed at 1750 °C within the Si$_3$B$_3$N$_7$-matrix. BN or other phases are not detected at this temperature by XRD. The thermal stability of the carbonitride SiBN$_3$C is even higher than that of

Table 1. Properties of pyrolized SiBN$_3$C and Si$_3$B$_3$N$_7$-ceramics

Starting material	PBS-Me	PBS-Me	Polymer I
Atmosphere	Ar, N$_2$	NH$_3$	N$_2$
Approximate stoichiometry	SiBN$_3$C	Si$_3$B$_3$N$_7$	Si$_3$B$_3$N$_7$
Ceramic yield [%]	66	63	95
Density [kg/m^3]	1790	1740	1860
Specific surface area [m^2/g]	0.19	0.5	60

Si$_3$B$_3$N$_7$. Crystallization of SiBN$_3$C starts at temperatures above 1900 °C in N$_2$. After 6 h at 1940 °C in N$_2$ (0.5 MPa, N$_2$) weak XRD patterns of α- and β-Si$_3$N$_4$ are detected, no BN or B$_4$C are present. Although the material stays amorphous at 1800 °C (6 h, N$_2$), it begins to crystallize at the same temperature in Ar-atmosphere (Fig. 3). This crystallization is accompanied by a loss of nitrogen of 7 wt.-%.

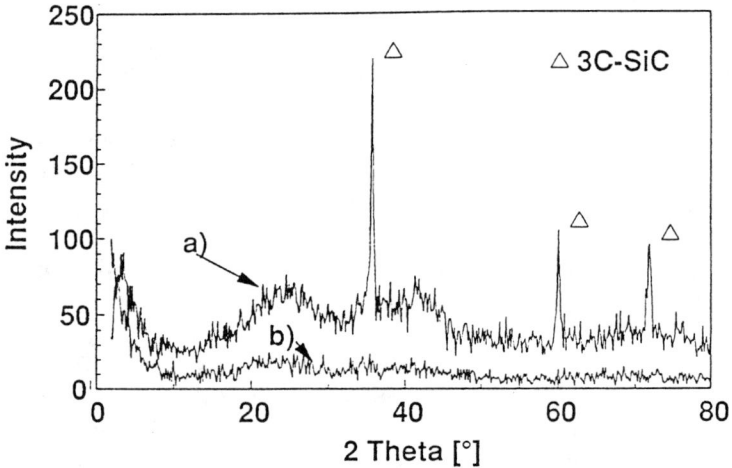

Figure 3, XRD-pattern of SiBN$_3$C annealed at 1800 °C for 6 h in a) Ar, b) N$_2$

The stability of SiBN$_3$C in liquid Si was tested at 1600 °C for 1 h. The carbonitride was not attacked and only a small reaction layer of about 150 nm thickness was formed in the melt.

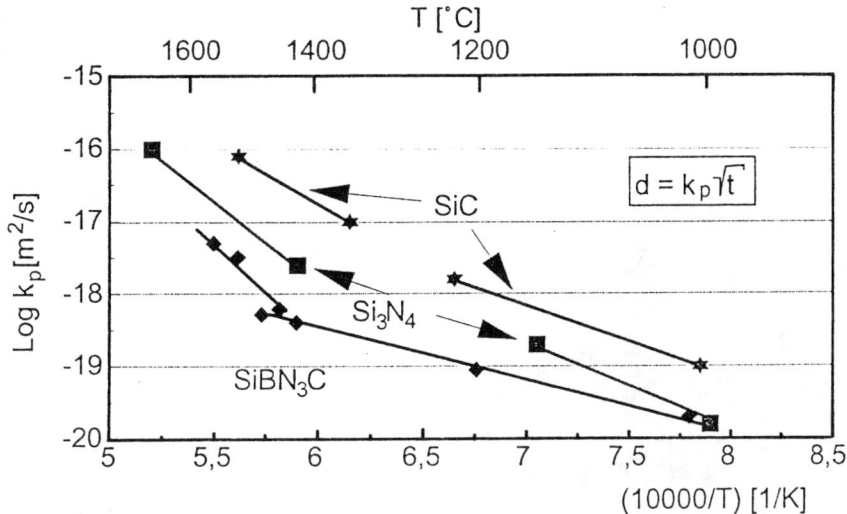

Figure 4, Oxidation rate k_p vs. 1/T. Comparison of oxidation rates of SiBN$_3$C, CVD-Si$_3$N$_4$ and SiC

The oxidation behavior of SiBN$_3$C was established between 1200 °C and 1600 °C in air, using coarse powder particles about 0.5 mm in diameter. The oxidation reaction obeys a parabolic time law. The k_p-values shown in Fig. 4 were determined from the thickness of the oxide layers. The k_p-values from SiBN$_3$C are lower than those of CVD-Si$_3$N$_4$ or SiC. The excellent oxidation resistance is probably due to the formation of an intermediate BNO-layer. In Fig. 5. the polished cross section of a sample is shown that preveously has been oxidized at 1600 °C in air for 48 h. The dark intermediate BNO-layer is visible in the middle of the SEM-picture (BSE-mode, Fig. 5 a) between the SiBN$_3$C-bulk on the left and the silica layer on the right. The EDAX-analysis indicates that N and B have evaporated out of the silica layer as N$_2$ and B$_2$O$_3$ respectively (Fig. 5 b) and C (not shown) burned off. Inside the intermediate layer B has been remarkably enriched by diffusion out of the oxide layer. Also N and small amounts of O are present in this zone. This BNO layer is stable up to 1600 °C in contrast to Si$_2$N$_2$O which protects Si$_3$N$_4$ only up to 1450 °C. Since no O can be detected in the SiBN$_3$C-bulk, the BNO layer acts as an efficient diffusion barrier, protecting the ceramic from further oxidation.

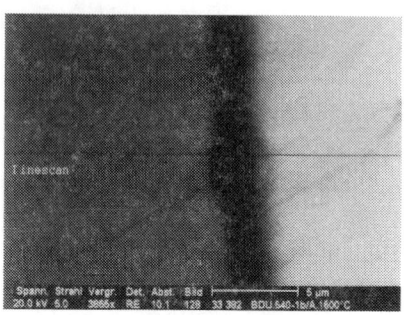 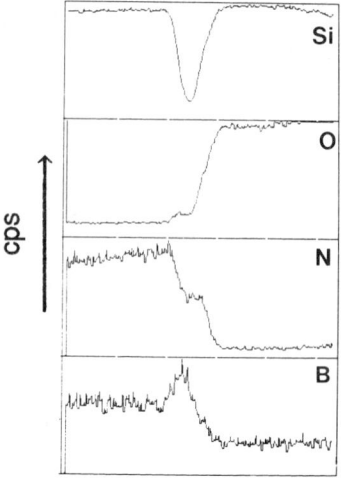

Figure 5, left: SEM-micrograph (BSE-mode) of oxidized SiBN$_3$C, right: EDAX line-scan analysis

We compared oxidation resistance and onset of crystallization of SiBN$_3$C with an other single-source precursor derived Si-B-(N,C) ceramic, which was prepared according to a Seyferth's synthesis route[3]. Elemental analysis for this material (referred to as SiBNC-S) yields: Si = 43.0 wt.-%, B = 9.7 wt.-%, N = 27.6 wt.-%, C = 16.6 wt.-% and O = 1.3 wt.-%. Ceramic yield is 85 % and the specific surface area 0.13 m^2/g. Oxidation of both Si-B-(N,C)-ceramics was performed at 1550 °C for 48 h in air and the k_p-values were calculated from the measured O-contents and the specific surface areas determined before testing. SiBN$_3$C exhibits k_p-value of $1 \cdot 10^{-18}$ m^2/s slightly lower than that of SiBNC-S ($k_p = 4 \cdot 10^{-18}$ m^2/s). More apparent is the difference regarding the crystallization behavior. SiBNC-S begins to crystallize at 1650 °C (N$_2$, 6 h). After 6 h at 1800 °C (N$_2$) the XRD-diagram in Fig. 6 shows the presence of α-, β-Si$_3$N$_4$ and α-SiC, whereas SiBN$_3$C still is amorphous to X-rays (Fig. 3b). It is presumed that the different behavior of SiBN$_3$C and SiBNC-S has its origin in the different structure of the preceramic polymers. PBS-Me contains only Si-N-, B-N- and N-C-bonds. In contrast to our material, the Seyferth polymer, however shows also Si-C-bonds and borazine like ring structures.[3] It has been previously reported,[1] that the atom bonds present in PBS-Me are still preserved after the conversion to the ceramic material. The structure of SiBN$_3$C-ceramic is therefore different from SiBNC-S, which may explain the unique properties of PBS-Me generated SiBN$_3$C.

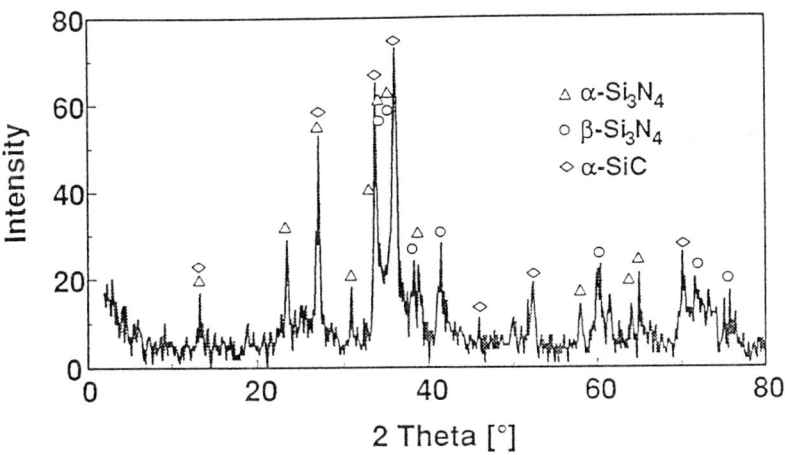

Figure 6, XRD-pattern of SiBNC-S, annealed at 1800 °C for 6 h in N_2

APPLICATIONS

The unique crystallization behavior and the excellent oxidation resistance make $SiBN_3C$ an ideal material for the production of ceramic fibers. Depending on the degree of crosslinking within the polymer by melt spinning at 100 - 180 °C or by a dry spinning process from solution. Due to extensive thermal crosslinking of PBS-Me during the melt spinning procedure both methods do not require any additional curing step to get an infusible green fiber. Pyrolysis at 1500 °C yields black, smooth $SiBN_3C$- or colorless $Si_3B_3N_7$-fibers, depending on the pyrolysis atmosphere (N_2 or NH_3) and heating conditions. In crystalline materials, diffusion at grain boundaries and grain growth lead to an increase in creep and a decrease in bending strength. These phenomena do not occur in $SiBN_3C$-fibers, since they remain amorphous up to 1900 °C in N_2. In air, the fibers can be used up to 1600 °C due to the high oxidation resistance of $SiBN_3C$. The low reactivity of $SiBN_3C$ with molten Si offers the opportunity to manufacture fiber-reinforced composites with a RBSN-matrix.

Coating and/or infiltration of woven carbon fibers, carbon fiber composites (CFC) or porous ceramics is an other interesting application. CFC-samples were vacuum-infiltrated with a 40 % solution of PBS-Me in o-xylene and subsequently heated to 750 °C (10 °C/h). This process was

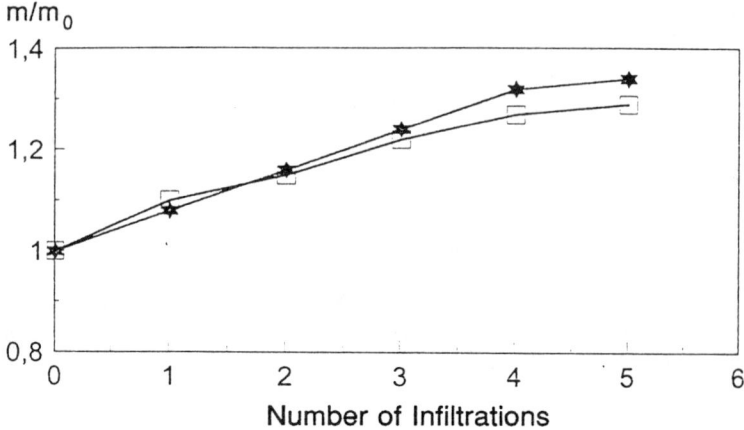

Figure 7, Mass-uptake of CFC, infiltrated with a 40 % solution of PBS-Me, pyrolyzed at 750 °C

repeated 5 times. Fig. 7 correlates the mass uptake of the CFC samples with the number of infiltrations with PBS-Me solution. The mass-uptake levels off between the 4th and the 5th infiltration which indicates that the open porosity is filled. The density of $SiBN_3C$ (1790 kg/m^3) is very low in comparison with e. g. $Si_{1.73}C_{1.0}N_{1.56}$ (2280 kg/m^3).[4,5] Therefore fewer problems occur due to volume shrinkage during pyrolysis in comparison with precursors which yield ceramic materials with higher densities. The volume change can be calculated by the following formula:

$$V_{ceramic} = (V_{polymer} \cdot \rho_{polymer} \cdot \text{ceramic yield}) / \rho_{ceramic}.$$

Starting from 1 cm^3 polymer with a density of 1100 kg/m^3, 0.41 cm^3 $SiBN_3C$- or 0.35 cm^3 $Si_{1.73}C_{1.0}N_{1.56}$-ceramic material is formed, indicating the advantage of low specific density ceramic materials derived from polymers. Also, the low coefficent of thermal expansion of $1.7 \cdot 10^{-6}$ K^{-1} is similar to that of amorphous carbon.

An additional application of preceramic polymers is the preparation of monolithic ceramic bodies from pressed polymer powders.[5] Starting from an unmeltable PBS-Me-powder, the process leads to amorphous bodies, exhibiting a high hardness of H_V = 15 GPa and the above quoted coefficient of thermal expansion.

CONCLUSIONS

The synthesis of a novel polyborosilazane (PBS-Me) has been accomplished by tailoring a "single-source" molecular precursor to the elemental stoichiometry of the desired material. The properties of this preceramic polymer (viscosity, melting point or fusibility) can be varied by crosslinking, with respect to the desired processing technology.

Pyrolysis of the precursor material yields $SiBN_3C$ or $Si_3B_3N_7$ in N_2- and NH_3-atmosphere, respectively. The unique structure of these materials (presence of Si-N-B-bonds) and the absence of Si-C- and B-C-bonds is responsible for their outstanding phase stability. Both ceramic materials are thermally stable, remain amorphous up to very high temperatures (1750 °C for $Si_3B_3N_7$ and 1900 °C for $SiBN_3C$) and exhibit a superior oxidation resistance.

ACKNOWLEDGEMENT

We thank Mr. Sahabi, KFA Jülich, FRG for conducting corrosion experiments with $SiBN_3C$ in liquid Si.

REFERENCES

1) H.-P.BALDUS, O.WAGNER and M.JANSEN, "Synthesis of advanced ceramics in the systems Si-B-N and Si-B-N-C employing novel precursor compounds", pp. 821-826, in: "Better Ceramics Through Chemistry V", Mat. Res. Soc. Symp. Proc. Vol. 271, eds. M.J.Hampden-Smith, W.G.Klemperer and C.J.Brinker, Materials Research Society, Pittsburgh, PA, USA, 1992.

2) H.-P.Baldus, M.Jansen, and O.Wagner, "New materials in the system Si-(N,C)-B and their characterization", pp. 75-80, in: "Key Engineering Materials" Vol. 89-91, eds. M.J.Hoffmann, P.F.Becher and G.Petzow, Trans Tech Publications, Aedermannsdorf, Switzerland, 1994.

3) D.SEYFERTH, H.PLENIO, "Borasilazane polymeric precursors for borosilicon nitride", J. Am. Ceram. Soc., 73, 7, 2131-2133, 1990.

4) M.FRIESS, J.BILL, F.ALDINGER, D.V.SZABÓ and R.RIEDEL, "Crystallization of amorphous silicon carbonitride investigated by

transmission electron microscopy (TEM)", pp. 95-99, in: "Key Engineering Materials" Vol. 89-91, eds. M.J.Hoffmann, P.F.Becher and G.Petzow, Trans Tech Publications, Aedermannsdorf, Switzerland, 1994.

5) G.PASSING, H.SCHÖNFELDER, R.RIEDEL and R.J.BROOK, "Monolithic crack-free Si-C-N ceramic bodies derived from polysilazane forms", Br. Ceram. Trans., **92**, 1, 21-22, 1993

STRUCTURAL AND MECHANICAL CHARACTERISATION OF SOME ALUMINA AND SiC BASED FIBRES

A.R. BUNSELL, M.H. BERGER and N.HOCHET
Centre des Materiaux de l'Ecole des Mines de Paris
BP 87. 91003 Evry cedex. France

ABSTRACT

Small diameter ceramic fibres are used to reinforce metals and ceramics for high temperature use. These fibres are limited by grain growth, chemical degradation and creep. In some fibres, but not all, a reduction in oxygen content leads to improved properties.

1. INTRODUCTION

Ceramic fibres have been developed in the last two decades which can be used at high temperature. These fibres fall into two classes consisting of oxide fibres which began to appear in the 1970s and fibres generally based on silicon carbide made from organosilicide precursors, which were first produced commercially in the early 1980s. These latter fibres are degraded by internal oxidation reactions occurring at high temperature. Interest has therefore been renewed in alpha alumina based fibres because of the high chemical stability and resistance to oxidation of alumina coupled with its high strength and stiffness. A review of those fibres which have been made available commercially is given in this paper.

2. THE OXIDE FIBRES

Oxide fibres are generally based on alumina in one of its phases, often combined with silica, sometimes in the form of mullite. In this paper three oxide fibres will be considered. These are an almost pure and dense alpha alumina fibre, the FP fibre and a zirconia reinforced alpha alumina fibre, the PRD 166 fibre, both developed by DuPont but not commercially produced and a porous alpha alumina fibre, the Almax fibre, from Mitsui Mining.

The FP fibre was composed of more than 99% of alpha alumina, which when completely densified had a density of 3.92 g/cm^3 and a diameter of 20 μm [1]. It had high stiffness but a low elongation to failure. This brittleness made it unsuitable for weaving. The PRD 166 fibre consisted of alpha alumina to which 20% wt of zirconia had been added to increase the elongation to failure [2]. Zirconia is known to improve strength by phase transformation toughening and to limit grain growth when added to bulk alumina. The fibre has a density of 4.2g/cm^3. The Almax fibre from Mitsui Mining is composed of almost pure alpha alumina. It has a diameter of 10μm, half the diameter of the FP and PRD 166 fibres, which allows Mitsui Mining to produce woven cloth from the fibre. The fibre has a lower density of 3.60 g/cm^3 [3] compared to the other two fibres tested.

3. SILICON CARBIDE FIBRES FROM PRECURSORS

Four types of small diameter ceramic fibre made from organosilicide precursors will be considered. Two of these fibres are produced by Nippon Carbon. These are the Nicalon 202 fibre, which represents the present generation of reinforcing fibres for ceramic matrices. This fibre consists of 58%(wt) Si, 30% C and 12% O [4]. The other is the Hi-Nicalon which has a much reduced oxygen content, comprising 63.7%Si, 35.8% C and 0.5% O. Two other fibres produced by Ube Industries will also be considered. These are the Tyranno LOX-M fibre containing 13% O, 2% Ti, 28% C and 57% Si and a new generation fibre the Tyranno LOX-E consisting of 5% O, 2% Ti, 38% C and 55% Si.

4. FIBRE PRODUCTION

The three oxide fibres considered here were made by blending the alumina in powder form with water and other compounds to form a mixture which could be spun. In the case of the PRD-166 fibre a solution of zirconium acetate was added to the mixture. The mixture was then extruded through spinnerets, drawn and fired at a temperature at or above 1300°C.

The work of Yajima and his colleagues in Japan was first published in the mid-1970s and has given rise to the first and so far the most successful series of fine ceramic fibres based on silicon carbide [5]. This fibre is produced by Nippon Carbon under the name of Nicalon and by Ube Industries under the name Tyranno. The manufacture of the Nicalon fibres involves the production of a polycarbosilane precursor fibre. The Tyranno fibres are produced from polytitanocarbosilane precursor. The titanium is introduced into the precursor in the form of tetra-2 ethyl hexyltitanate.

The Nicalon 202 and Tyranno LOX-M fibres are produced by subjecting the precursor fibres to heating in air at about 300°C to produce crosslinking of the structure by oxidation. This oxidation makes the fibre infusible but introduces oxygen into the structure which remains after pyrolysis. The ceramic fibre is obtained by a slow increase in temperature in an inert atmosphere up to 1300°C. The fibre which is obtained contains a majority of SiC but also significant amounts of free carbon and excess silicon combined with oxygen and some carbon in an amorphous phase. The crosslinking process for the Hi-Nicalon and Tyranno LOX-E fibres is achieved in the absence of oxygen by electron beam radiation. These latter fibres as a consequence have a reduced oxygen content although the latter still contains a significant amount as it is introduced with the titanium.

5. FIBRE CHARACTERISATION

A single fibre testing machine, [6], was used for these tests and an electrical resistance oven allowed the fibres to be tested up to their mechanical limits as a function of temperature. The fibres a have been observed with a PHILIPS 501 scanning electron microscope and the microstructures of the fibres have been determined by TEM using a Philips EM 430 TEM-STEM with an acceleration voltage of 300 kV and a maximum punctual resolution of 0.23 nm. Thin foils were obtained by glancing ionised argon sputtering by a method detailed elsewhere [7]. The granulometry of the oxide fibres has been quantified by image analysis of TEM images.

6. RESULTS

The as received FP fibre was circular and extremely granular, Figure 1 with a diameter of 18 ± 1.4 μm and composed of one population of grains with a mean size of 0.50 ± 0.25 μm. The grains, which can be seen in Figure. 2, showed no marked preferential elongation and they enclosed a small amount of small spherical porosity. A small amount of triangular porosity was found at triple points. No obvious second phase was detected [8].

The PRD 166 fibre was seen to have a diameter of 17.6 ± 1.6 μm and to be composed of three populations of grains: alpha alumina, intra and intergranular zirconia, see Figure 3. The alpha alumina grains had a mean size of 0.34 ± 0.18 μm with a very slight elongation parallel to the fibre axis. The alumina grains contained a small amount of porosity and zirconia particles. Their size varied from a few tens of nanometers to 0.2 μm.

The Almax fibre was circular in cross section with a diameter of 9.5 ± 0.8 μm and consisted of one population of grains of around half a micron. The fibre exhibited a large amount of porosity which was essentially transgranular, as can be seen from Figure 4. The diffraction contrast varied inside the same grain indicating a curvature of the grain.

The Nicalon 202 fibre, as received, consisted of β–SiC grains of around 2 nm, free carbon aggregates of about one nanometer in size and an amorphous phase consisting of silicon, carbon and oxygen. The free carbon is described as being composed of dicoronene-like molecules similar to the basic structural units seen in carbon fibres [9]. The Hi-Nicalon fibre consisted of randomly oriented β–SiC grains, as can be seen in Figure 5 the biggest of which were 20 nm but 5 to 7.5 nm was more usual. Such grains consist of 20 to 30 (111) SiC lattice fringes. Figure 6 reveals that the free carbon is in the form of turbostratic carbon consisting of 5 to 10 distorted layers with an interfringe distance of 0.34 nm. The Tyranno fibres possessed similar microstructures to one another consisting of some β–SiC grains up to 2nm in size, surrounded by free carbon aggregates made up of several turbostratic carbon layers and all embedded in an amorphous phase of Si-O-C-Ti.

All the fibres exhibited linear elastic behaviour with brittle failure at room temperature and up to around 1000°C. The failure stresses σ_R and failure strains ε_R of the fibres are presented in Table I for a displacement speed of 0.75 mm/mn and a gauge length of 150 mm, chosen as this enabled high temperature tests to be conducted using the tubular furnace and maintaining cold grips. Test at various gauge length have been conducted at room temperature and will be published elsewhere [8].

TABLE 1 Mechanical characteristics in traction of the fibres at room temperature

	σ_R (GPa)	ε_R (%)	E (GPa)
FP	1.23 ± 0.37	0.29 ± 0.07	414
PRD 166	1.46 ± 0.30	0.40 ± 0.08	366
Almax	1.02 ± 0.39	0.30 ± 0.11	344
Nicalon 202	2.2 ± 0.7	1.2 ± 0.09	185
HI-Nicalon	2.6 ± 0.6	0.9 ± 0.03	263
Tyranno LOX-M	2.5 ± 0.8	1.2 ± 0.14	180
Tyranno LOX-E	2.9 ± 1.06	1.2 ± 0.13	200

The FP and PRD 166 failure surfaces were flat without any large apparent defects initiating failure and the propagation mode of the crack was both inter- and intra-granular, with a predominance of the intergranular mode. The Almax failure surfaces revealed large cavities of up to 2 or 3 μm and a more pronounced intragranular propagation mode than that was seen with the other fibres. The Nicalon and Tyranno fibres showed classic brittle failure surface normal to the fibre axis direction consisting of a mirror zone, a mist zone and a hackle zone.

All three oxide fibres exhibited linear elastic behaviour up to 1000°C for the FP and Almax fibres and up to 1100°C for the PRD 166 fibre. Cross head speed was 0.75 mm/mn. Beyond 1000°C the mechanical characteristics decreased rapidly, especially for the Almax fibre, Figure 7. The Nicalon and Tyranno fibres showed essentially linear behaviour to 1400°C however strength was observed to fall for each of the four fibres beyond a certain threshold, Figure 8. This decrease was seen from 800°C for the Tyranno LOX-M fibres whilst the decrease occurred at 1000°C for the Tyranno LOX-E fibres. The latter fibres showed higher mechanical properties than the former fibres up to 1300°C. The Tyranno fibres were seen to possess slightly lower properties than the Nicalon fibres which showed no change in their properties to above 1000°C.

No significant creep occurred below 1000°C for the FP and Almax fibres and below 1100°C for PRD 166 fibre. Greater time to failure and lower creep strain rates were obtained for the PRD 166 compared to the FP and also for the FP compared to the Almax fibres. Similar results have been published elsewhere for the first two fibres [10]. Logarithmic plots of the strain rate as function of the stress applied at the beginning of secondary creep are presented in Figure 9 [11]. Creep was not detected with any of the SiC based fibres at temperatures below 900°C and then at the highest of stresses. The creep threshold temperatures as a function of applied stresses are shown in Figure 10 The Hi-Nicalon fibre can be seen to possess the greatest resistance to creep whilst the Tyranno fibres show very similar behaviour.

After 24 hours at 1300°C in air, with no applied load, the microstructures of the FP and PRD 166 fibres were seen to be almost stable whereas the Almax fibre showed an isotropic grain growth of 40%. Similar heat treatment under load produced grain growth in the two former fibres and this was more marked in the PRD-166 fibre. Failure of the FP and PRD-166 fibres became mostly intergranular after heat treatment. Tensile failure of the Almax fibre under load at 1250°C revealed isotropic grain growth up to 55%. The intergranular porosity and dislocations were still present, without any relaxation of the internal stresses and failures surfaces showed the presence of large cavities.

The Nicalon 202 fibres were seen to show grain growth from around 1100°C [12] with growth to around 3nm occurring after twenty hours at this temperature. Grain growth was observed from around 1200°C, or possibly lower, in the Hi-Nicalon leading to a doubling in average grain size after 24 hours at 1300°C and tripling, compared to the original state, to around 15 nm, at 1400°C. Heating in air produces a silica layer at the surface of these fibres. At 1200°C the silica can be seen to a discontinuous double layer consisting of crystalline cristobalite, identified by electron diffraction and amorphous silica. At 1400°C the layer was completely crystalline and bubbles could be seen to at the cristobalite/SiC interface. The thickness of the crystalline layer was 0.5nm at 1200°C and 2nm at 1400°C. Grain growth in the Tyranno fibres was inhibited as the average grain size after 24 hours at 1400°C was only 8nm although some larger grains could be seen.

7. DISCUSSION

The Young's modulus of a polycrystalline alumina depends strongly on its density, and its strength on the granulometry and the presence of flaws in the ceramic. The elastic modulus of the FP fibre is typical of dense alumina but its strength is close to twice that of the strength of a bulk material having a comparable microstructure [13]. The small cross section which leads to the high surface/volume ratio of the fibre form lowers the probability of the presence of large volume flaws compared to the bulk ceramic. The high porosity of the Almax, around 8 %, leads to a reduced Young's modulus compared to the FP fibre and the presence in the Almax fibre of large cavities and porosity also reduces strength due to the smaller load supporting cross section. The PRD-166 fibre contains 20%wt of zirconia which has a lower Young modulus of 200 GPa, than that of alumina resulting in an elastic modulus 366 GPa. The observed increase in strength compared to the FP fibre may be due to the toughening effect of the tetragonal zirconia particles, which will transform around the crack tip to the monoclinic symmetry by a martensitic reaction absorbing energy, and inducing a resulting strain in the matrix[14]. Creep in the FP fibre is due to grain boundary sliding due to intergranular movement of dislocations and accommodated by interfacial diffusion mechanisms. The PRD-166 fibre has improved resistance to creep compared to the two other oxide fibres as the dispersed zirconia particles limits the mobility of intergranular dislocations, at least up to 1200°C above which the effect is lost. The Almax fibres do not exhibit classical creep behaviour due to the movement and growth of the intergranular porosity which leads to higher rates of strain and more rapidly induced failure.

It has been widely reported, for example see refs.9 and 15, that the Nicalon 202 fibres suffer, above around 1100°C, from grain growth and when heated in air a layer of silica is formed at their surface. Above 1250°C the fiber suffers from internal degradation. Similar behaviour is seen with the Tyranno fibres although grain growth is inhibited. The absence of an amorphous phase in the Hi-Nicalon fibre accounts for the higher mechanical properties than those of the other SiC based fibres but this also allows earlier grain growth. Growth of silica at the surface of these fibres occurs at temperatures above around 1100°C. These fibres creep at high temperatures, as has been shown. The mechanism in the Nicalon and Tyranno fibres is grain boundary sliding whilst this is clearly limited in the Hi-Nicalon fibres which have the greatest resistance to creep.

8. CONCLUSION

Fine fibres have been developed which have prompted the development of high temperature composites. The oxide systems which have been examined clearly resist oxidation at high temperatures however grain growth and intergranular sliding limits those fibres which have been examined, to temperatures lower than 1100°C. The fibres based on SiC based systems also suffer above this temperature due to grain growth and for those which contain oxygen from internal oxidation processes. In addition oxidation at the surface both reduces fibre properties and provides a silica interface with any matrix, which is likely to be detrimental to composite properties. The Hi-Nicalon possess higher mechanical properties than the earlier versions of this type of fibre but grain growth is a problem at high temperature as is the formation of silica at the surface. Those fibres presented in this review article represent the presently available commercial fibres. These fibres have limited properties at high temperature but an understanding of the processes restricting their use may enable fibres of greater stability to be produced. Many laboratories are working to produce such fibres but until they become generally available for evaluation claims of, particularly improved creep, behaviour must await verification.

REFERENCES

1. A.K. DHINGRA, *Phil. Trans. R. Soc. Lond.* **A 294** (1980) 411.
2. J.C. ROMINE, *Cer. Eng. & Sci. Proc.* **8** (1987) 755.
3. ALMAX *Commercial Literature of Mitsui Mining*.
4. TAKEDA, M., IMAI, Y., ICHIKAWA, H., KASAI, N., SEGUCHI, T. and OKAMURA, K. Cer.Eng. & Sc. Proc. Jul.Aug.(1993)540
5. YAJIMA,S., YASEGAWA,Y., HAYASHI, J. & I. IMURA, M. J.Mat.Sci., 13(1978)2569
6. A.R. BUNSELL, J.W.S. HEARLE and R.D. HUNTER, *J. Phys. E Sci. Instru.* **4** (1971) 868.
7. M.H. BERGER and A.R. BUNSELL, *J Mater. Sci. Let.* **12** (1993) 825.
8. V.LAVASTE, M.H.BERGER, J.BESSON and A.R.BUNSELL to be published in J.Mat.Sci.
9. Le COUSTUMER P., MONTHIOUX, M. and OBERLIN A. J.Europ.Cer.Soc. 11(1993)95
10. D.J. PYSHER and R.E. TRESSLER. J. Mat. Sci. 27, 2 (1992) 423.
11. V. LAVASTE, J. BESSON, M.H. BERGER, A.R. BUNSELL accepted in *J. of Amer. Cer. Soc.*
12. R.M. CANNON, in Advances in Ceramics vol 10, "Structure and properties of MgO and Al_2O_3 ceramics" edited by W.D. Kingery (The American Ceramic Society, 1984) p. 818.
13. R.W. DAVIDGE, R.C. PILLER, A. BRIGGS, et al. in "Technical Ceramics" edited by H. Nosbusch and I.V. Mitchell (Elsevier, 1988) p. 163.
14. F.F. LANGE, *J. Mat. Sci.* **17** (1982) 247.
15. G.SIMON and A.R.BUNSELL,J.Mat.Sci. 19 (1984) 3649.

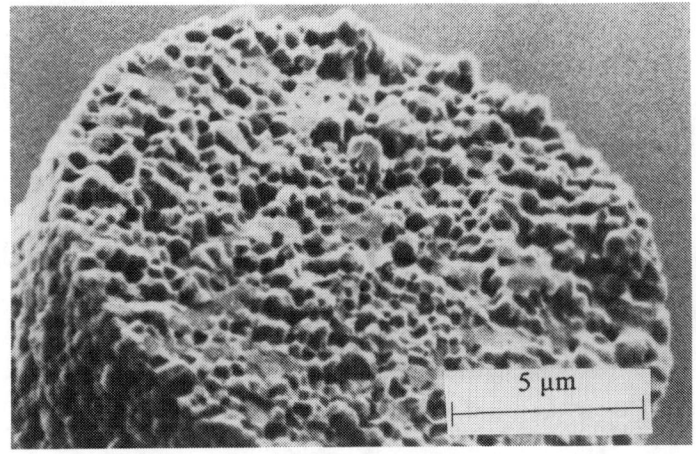

Fig. 1 : SEM image of the FP fibre failed at room temperature

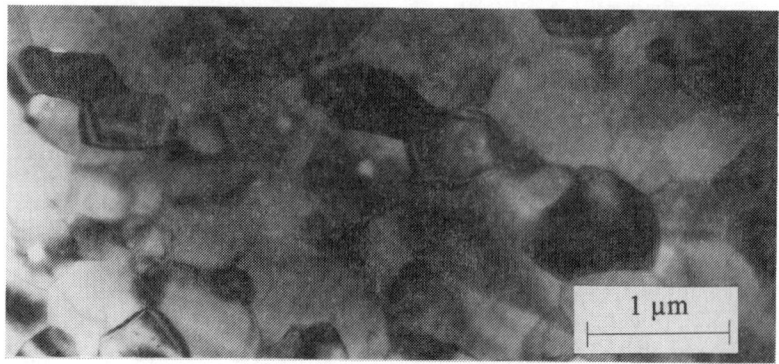

Fig. 2 : TEM image of the microstructure of the as received FP fibre.

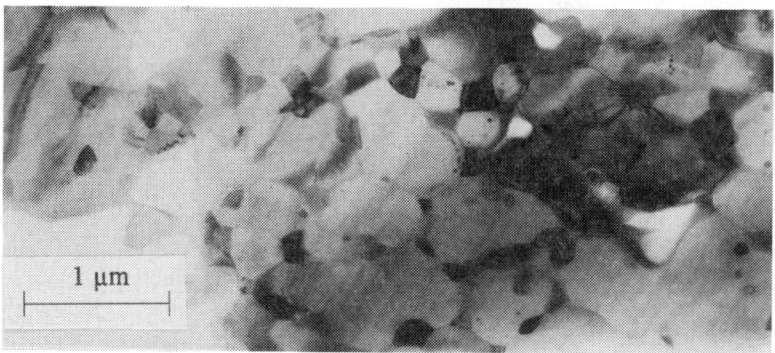

Fig. 3 : TEM image of the microstructure of the PRD 166 fibre

Fig.4 : TEM image of the microstructure of the Almax fibre.

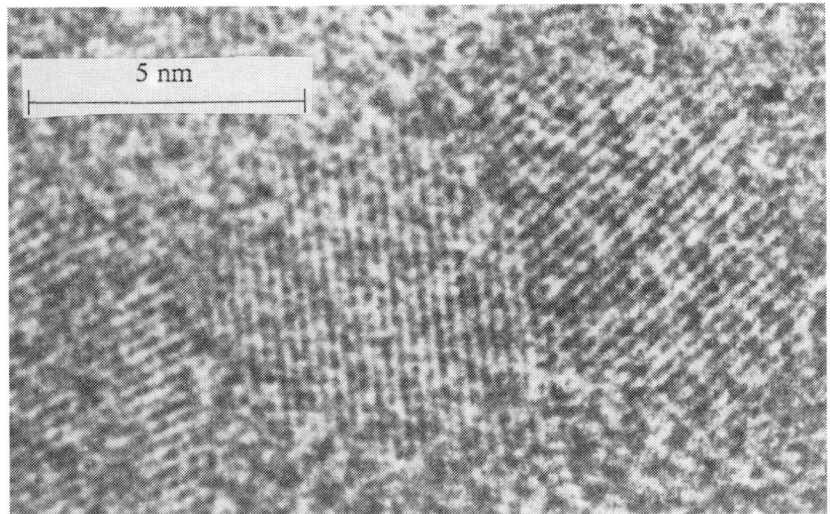

Fig.5 : Lattice fringes image showing a continuum of SiC grains.

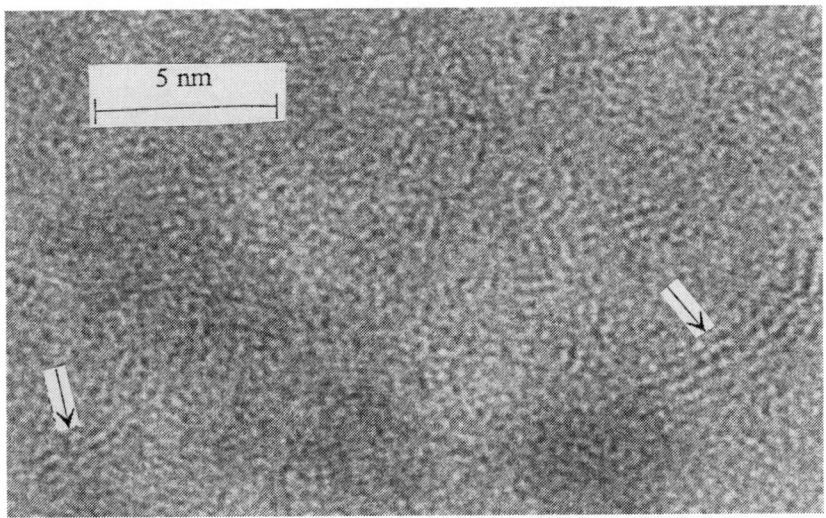

Fig.6 : Lattice fringes showing ~ 0.34 nm fringes characteristic of turbostratic carbon.

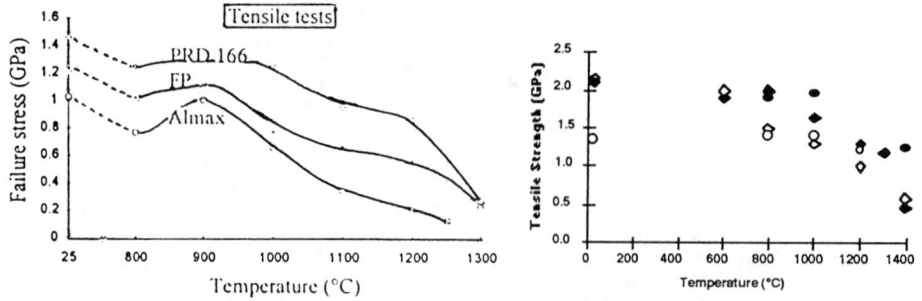

Fig.7: Failure stresses of the oxide fibres as a function of temperature.

Fig.8 : Failure stresses of the Nicalon NLM-202 (o), Hi-Nicalon(•), Tyranno Lox-M (◊) and Tyranno Lox-E (♦) fibres.

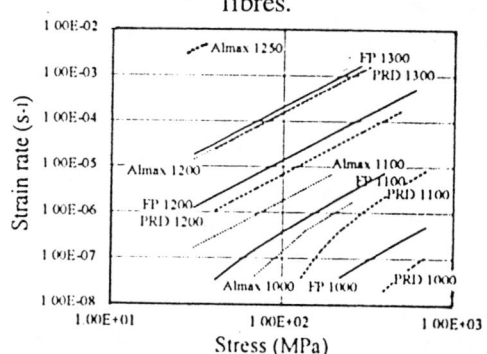

Fig. 9 : Logarithmic plots of the creep strain rates as a function of applied stress at different temperatures for the oxide fibres.

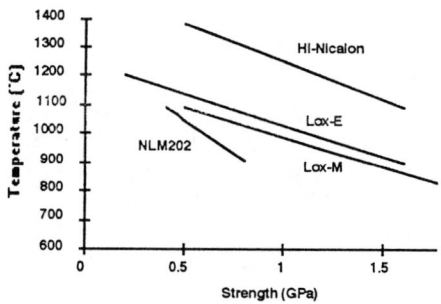

Fig.10 : Creep threshold levels for the SiC based fibres as a function of temperature and applied load.

DESIGN AND PROCESSING OF ALL-OXIDE COMPOSITES

Robert Lundberg, Lena Eckerbom

Volvo Aero Corporation
Trollhättan, Sweden

ABSTRACT

All-oxide ceramic composites as a material with potential for long life-time applications at temperatures in the 1400-1600°C range in combustion environments were studied. The properties of available polycrystalline and single crystal oxide fibres were summarised. The literature on stable weak interfaces in all-oxide composites was reviewed. Composites with single crystal fibres, a polycrystalline matrix of the same material as the fibres, and a compatible high temperature stable weak oxide interphase was suggested to be the most promising approach. Processing of all-oxide composites was performed. ZrO_2-coated sapphire fibres in reaction bonded alumina and in hot pressed alumina showed crack deflection and fibre pull-out. In reaction bonded mullite crack deflection and pull-out was observed even for un-coated sapphire fibres. This was attributed to thermal expansion mismatch. A recently started European project aiming at development, scale-up and property evaluation of all-oxide composites is briefly outlined.

INTRODUCTION

Oxide fibre reinforced oxide composites are studied today by aeroengine manufacturers all over the world. One driving force for the development of oxide composites is the potential for use at extremely high temperatures for several thousand hours in a combustion environment. For temperatures in the 1400 - 1600°C range composites based on single crystal oxide fibres and polycrystalline oxide matrices and oxide interphases look most promising. In this paper the design (i.e. fibre/interface/matrix selection) and processing of oxide composite systems are discussed. Some successful examples of recently developed oxide composites are also briefly described.

To the extent authorized under the laws of the United States of America, all copyright interests in this publication are the property of The American Ceramic Society. Any duplication, reproduction, or republication of this publication or any part thereof, without the express written consent of The American Ceramic Society or fee paid to the Copyright Clearance Center, is prohibited.

FIBRE/INTERFACE/MATRIX SELECTION

A methodology for fibre/interphase/matrix selection based on thermodynamic stability calculations followed by experiments with particulate composites and laminate composites has been previously presented[1]. First oxides that are stable at high temperatures were screened. The screening was made looking at both the high temperature stability and to the potential possibility of producing fibres of the selected oxide material. Similar assessments of potential high temperature ceramics have been made by other authors. Complex oxides in particular have attracted a lot of interest. A summary of some oxides of interest is given in Table 1.

Table 1. High temperature stability for some oxides[2,5,6]

Oxide	$T_{max.use}$ [°C]	$T_{m.p.}$ [°C]	remarks
Al_2O_3	1200-1800	2054	
ZrO_2	2300 (950)	2600-2765	
$SrZrO_3$	1700		
$SrHfO_3$	1800		
$La_2Hf_2O_7$		2300	moisture sensitive
CaO		2825-2927	high volatility
SrO		2454	
MgO	2637	2832	
$CaZrO_3$		1995	
mullite, $Al_6Si_2O_{12}$		1750	
YAG, $Y_3Al_5O_{12}$		1940	
$CaZrO_3$		2325-2350	grain growth
$SrZrO_3$		2650-2750	segregates in O_2 gradient
Cr_2O_3		2270-2330	
olivine, Ca_2SiO_4		2130	
spinel, $MgAl_2O_4$		1995-2135	low strength at high temperature

It is well known that most polycrystalline oxides usually have very poor creep properties. This has stimulated the development of single crystal fibres. Another drawback of polycrystalline oxides is that they are prone to grain growth. In polycrystalline oxide fibres grain growth has been shown to cause severe strength degradation. Polycrystalline Al_2O_3-fibres before and after ageing in air at 1400°C are shown in Figure 1. Extensive grain growth can be seen. The main advantage of polycrystalline fibres is that they are available as small diameter fibres and can easily be woven into fabrics or 3D fibre preforms. Some available polycrystalline fibres are summarised in Table 2.

Although single crystal fibres do not suffer from grain growth, there are other defects that develop and grow at high temperatures. These defects are pores, bubbles and also step-like features on the fibre surface. Typical defects on the surface of single crystal sapphire (Al_2O_3) fibres appearing after heat treatment at 1400°C are shown in figure 2. Sapphire is today the only commercially available continuous single crystal fibre. Several other single crystal oxides, especially complex oxides, with even better creep resistance have also been reported in the literature, either as fibres or as potential fibres. The literature on single crystal fibres is summarised in Table 3. In this table the creep resistance of single crystal oxides is also indicated[19].

Table 2. Polycrystalline oxide ceramic fibres (from ref:(5)(7-12)(14)

Fibre	Manufact.	Comp.	E, [GPa]	Ø [μm]	Density, [g/cm³]
Fibre FP	Du Pont	α-Al_2O_3	380-400	~20	3,9
PRD-166	Du Pont	80% Al_2O_3, 20% ZrO_2	360-390	~20	4,2
Almax	Mitsui Mining	α-Al_2O_3	320-340	10	~3,6
Nextel 312	3M	24%SiO_2, 14%B_2O_3, 62%Al_2O_3	150	10-12	2,7-2,9
Nextel 440	3M	28%SiO_2, 2%B_2O_3, 70%Al_2O_3	220	10-12	3,05
Nextel 480	3M	28%SiO_2, 2%B_2O_3, 70%Al_2O_3	220	10-12	3,05
Nextel 550	3M	27%SiO_2, 73%Al_2O_3	193	10-12	3,03
Nextel 610	3M	Al_2O_3	373	10-12	3,75
Nextel Z-11	3M	32%ZrO_2, 68%Al_2O_3	76	14	3,7
Altex	Sumitomo	15%SiO_2, 85%Al_2O_3	200-230	9-17	~3,2

When a fibre has been selected, a compatible matrix has to be chosen. In order to avoid residual thermal stresses it is often a good idea to use the same material for matrix and fibre. An example of this is single crystal sapphire in a polycrystalline Al_2O_3 matrix. To obtain the desired tough composite fracture behaviour a weak interface is then needed. Several approaches have been studied[1,3,4,6,22,23]. One is fugitive interphases like carbon[4,23], molybdenum or tungsten[4]. Another approach is weak layered oxides that promote crack deflection in the interphase[22]. To ensure a thermodynamically stable system the interface material most likely to succeed is probably an oxide with good high temperature stability that does not form intermediate phases, exhibit solid solubility or have low temperature eutectics with the selected fibre or matrix oxide[1]. Having the same material as fibre and matrix is again helpful, since the number of possible reactions with the interphase is reduced. Some oxides that are compatible with Al_2O_3 (and sapphire) have been previously suggested[1]. A summary of compatibility studies from the literature is presented in Table 4.

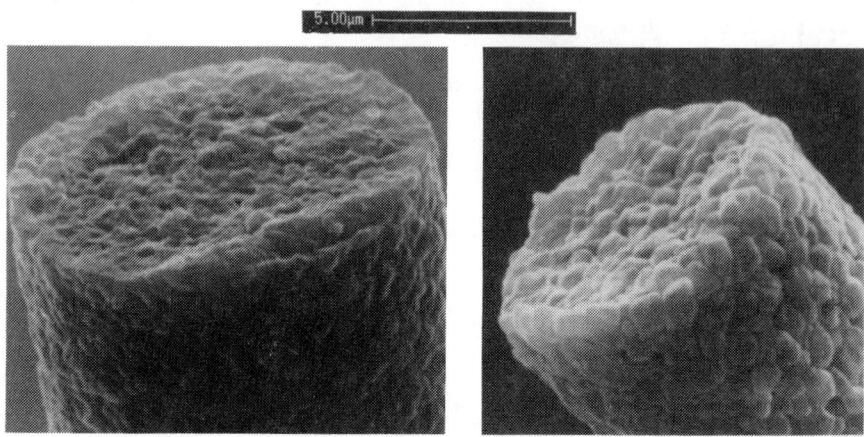

Figure 1. Polycrystalline Al_2O_3 fibres (Almax). **a.** as-received fibres, **b.** aged at 1400°C, 100 h in air. (bar = 5 µm)

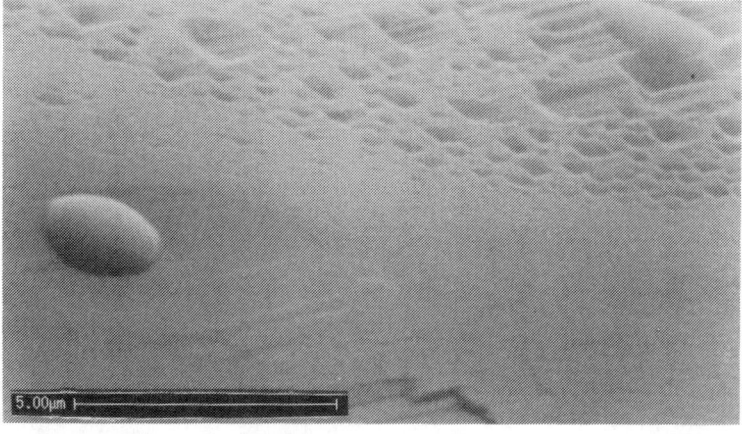

Figure 2. Surface defects found on sapphire fibres (Saphikon) aged at 1400°C, 100 h in air. (bar = 5 µm)

Table 3. Properties of some single crystalline oxides.

	Structure	$T_{max.}$ [°C]	$T_{m.p.}$ [°C]	E, [GPa]	Tensile strength [GPa]	Creep resistance	Ref.
sapphire, Al_2O_3	hexagonal	~1500	2050	350-450	2.4-3.4	good	13,15,16 17,19
spinel, $MgAl_2O_4$	cubic		>1927		2.2	very good	15,18,19
YAG, $Y_3Al_5O_{12}$	complex bcc		1940	334	1.0	very good	2,15,16 19,21
chrysoberyl, $BeAl_2O_4$	ortho-rhombic	>1820	1870	~470		extremely good	19,20
ThO_2	fluorite					good	19
MgO			2800			poor	19
BeO	hexagonal					good	19
mullite, $Al_6Si_2O_{12}$	cubic		1850			very good	16,19
Y_2O_3	C-type R_2O_3		2410			poor	19
ZrO_2,	cubic		2800			poor	19

PROCESSING EXAMPLES

Oxide composites could be produced in a variety of ways[27]. Green forming using sol/gel or slurry routes such as slurry (or sol) infiltration of fibre preforms, or stacking and thermo pressing of green fibre/matrix prepregs is possible[27]. Densification could be achieved through uniaxial hot pressing, reaction bonding (metal powder→oxide)[25,28,30,31], or pyrolysis and repeated infiltration (sol-gel). Another possible processing route is fibre preform infiltration by directed melt oxidation (DIMOX™)[27]. Three examples of oxide composites with single crystal Al_2O_3 fibres will be discussed here.

Reaction Bonded Alumina Composites

Small samples of reaction bonded alumina with a few sapphire fibres (Saphikon Inc., USA) were produced. The reaction bonding and slurry processing used in these experiments has been previously reported[25]. It has been shown that by using reaction bonding the problem of shrinkage during densification could be avoided[25,30]. Reaction bonding therefore is an attractive and rapid way of producing fibre reinforced ceramics in complex shapes. The sapphire fibres were coated with ZrO_2 using a proprietary technique. Both coated and uncoated fibres were incorporated in the matrix. After reaction bonding, cutting and polishing, crack deflection was studied. Cracks were generated in the matrix by diamond indentation. Crack deflection was observed only at the ZrO_2-coated fibres. This is shown in figure 3, where a coated and uncoated fibre are compared. Fracture surfaces were also created and fibre pull-out detected for the ZrO_2-coated fibres.

Table 4. Compatibility data for some material combinations.

material I (matrix)	material II (or fibre)	Interface material (or fibre coating)	Remarks	Ref.
Al_2O_3	Al_2O_3	Mo, W, Cr, porous Al_2O_3,	fibre debonding	4
		γ-TiAl	Al_2O_3 dissolves	4
		Nb, NiAl	No debonding	4
		SnO_2	diffusion barrier weak bonding	26 1
		fugitive C		4
		ZrO_2, YAG	variable debonding	4
		HfO_2, ZrO_2, Al_2TiO_5	weak interface	1
		$KAl_{11}O_{17}$	weak interface	22
	single cryst. Al_2O_3		very promising	6
	$CaZrO_3/CaHfO_3$		reaction above ~1750°C	6
	mullite ($Al_6Si_2O_{12}$)		different CTE cause multiple cracking	6
	YAG ($Y_3Al_5O_{12}$)		stable to 1700°C	6
	ZrO_2, HfO_2		liquid phase at ~1865°C	6
YAG	single cryst. YAG		very promising	6
	single cryst. Al_2O_3	fugitive carbon		23
ZrO_2	single cryst. Al_2O_3			24
$MgAl_2O_4$	Al_2O_3	$KMg_2Al_{15}O_{25}$	weak interface	22
Ca-ZrO_2	Al_2O_3	$CaAl_{12}O_{19}$	weak interface	22
$GdAlO_3$ or $Gd_3Al_5O_{12}$	Al_2O_3	$GdMgAl_{12}O_{19}$	weak interface	22

Reaction bonded mullite composites

Reaction bonding of mullite from Al and SiC has been reported by other authors[31]. An alternative process starting from an Al:Si alloy has recently been developed[28]. This process was used to produce Sapphire/reaction bonded mullite composites. Details on the processing of these composites has been previously presented[28].

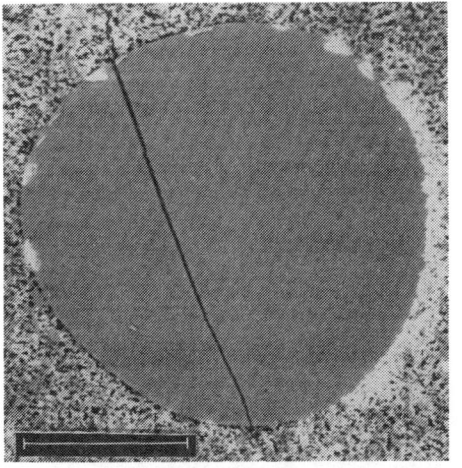 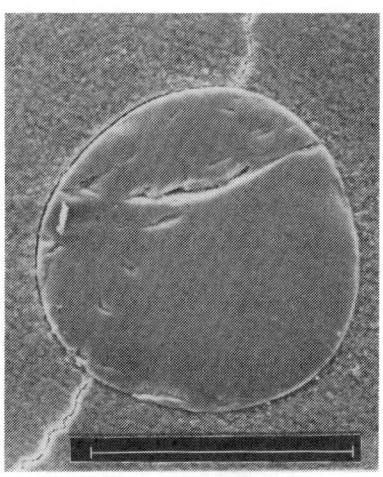

a. **b.**

Figure 3. Reaction bonded alumina with sapphire fibres. **a.** No crack deflection for un-coated fibres. BEI image (bar = 50 µm) **b.** Crack deflection at interface for ZrO_2-coated fibres. SEI image (bar = 100 µm)

Figure 4. Fibre pull-out detected in reaction bonded mullite[28] with un-coated sapphire fibres (Saphikon). (bar = 50 µm)

The fibres were uncoated. Due to the thermal expansion mismatch between mullite and sapphire a gap at the interface formed after cooling from the reaction bonding temperature. This gap resulted in a weak interface at room temperature and fibre pull-out was detected, as shown in figure 4.

Hot Pressed and Hot Isostatically Pressed (HIPed) Alumina Composites

Zirconia-coated sapphire fibres (the same type of coating and fibre as in the reaction bonded alumina samples) were placed in alumina powder and hot pressed at 1600°C, 0.5 h and 8 MPa. Crack deflection experiments showed crack deflection both at the sapphire/ZrO_2 interface, and within the ZrO_2 interphase (see figure 5).

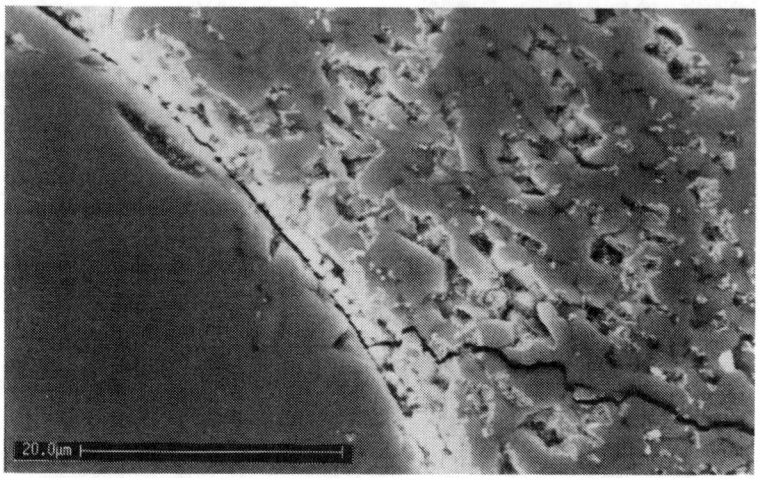

Figure 5. Crack deflection detected in hot pressed alumina containing ZrO_2-coated sapphire fibres (Saphikon). (BEI image, bar = 20 μm)

In a recently started European project[29] green forming, hot pressing and HIPing of oxide composites will be developed. The best fibre/interphase/matrix combination will be scaled up and a full mechanical characterisation will be made. A model component (combustor tile) will be designed and rig tested at realistic aeroengine conditions. The partners in the project are Rolls Royce plc (UK), Snecma and Onera (France), Volvo Aero Corporation and AC Cerama AB (Sweden).

CONCLUSIONS

The literature on available oxide fibres and studied all-oxide fibre/interphase/matrix combinations has been reviewed. Single crystal oxide fibre reinforced oxide composites with stable oxide interfaces seem to be the most promising material for long lifetime applications at very high temperatures. Zirconia-coated sapphire fibres exhibit crack deflection both in reaction bonded alumina and in hot pressed alumina. Uncoated sapphire fibres show pull-out in a reaction bonded mullite matrix. The weak bond in this composite was attributed to thermal expansion mismatch. For the future it is believed that complex oxide composites such as YAG/YAG or mullite/mullite with a compatible weak interphase will be even more interesting than the sapphire reinforced oxides presented here. Oxide composites are currently being developed in a European project addressing scale up and extensive property evaluation.

ACKNOWLEDGEMENTS

Mr. N. Ahlen, Stockholm University is thanked for the hot pressing experiments. Miss A. Kristoffersson and Mr. J. Brandt, Swedish Ceramic Institute are sincerely thanked a for fruitful reaction bonding collaboration. Prof. R. Warren at Luleå University and Prof. M. Nygren, Stockholm University are also thanked for sharing their thoughts on creep behaviour of oxides and oxide/oxide compatibility.

REFERENCES

1. Lundberg, R., Pejryd, L., Butler, E., Ekelund, M., Nygren, M., "Development of Oxide Composites" *Proc. Int. Conf. HTCMC-1, Ed. Naslain, Lamon, Doumeingts, Woodhead Publ. UK*, 167-174 (1993)
2. Courtright, E.L.,"Engineering Property Limitations of Structural Ceramics and Ceramic Composites Above 1600°C.",*Cer. Eng. Sci. Proc.*, **12** [9-10] 1725-44 (1991)
3. Kerans, "Control of Fiber-Matrix Interface Properties in Ceramic Composites", Proc. Int. Conf. HT-CMC-1, see ref 1, 301-312 (1993)
4. Davis, J. B.; Löfvander, J. P. A. ; Evans, A. G. ; Bischoff, E.; Emiliani, M. L.,"Fiber Coating Concepts for Brittle-Matrix Composites.", *J. Am. Ceram. Soc.*, **76** [5] 1249-1257 (1993)
5. Raj, Rishi,"Fundamental Research in Structural Ceramics for Service Near 2000°C.", *J. Am. Ceram. Soc.*, **76**,[9], 2147-2174 (1993)
6. Courtright, E.L.; Graham, H.C.; Katz, A.P.; Kerans, R.J.,"Ultrahigh Temperature Assessment Study: Ceramic Matrix Composites/Final report", *NASA STAR Technical Report, AD A262740* , (1990)
7. Berger, M.H., Lavaste, V., Besson, J., Bunsell, A.R.,"Microstructure and Characteristics of Alpha Alumina Based Fibres.", *Proc. Int. Conf. ECCM-6*, p.169-186, (1993)
8. Cooke, T.F.,"Inorganic Fibers - A Literature Review.", *J. Am. Cer. Soc.*, **74** [12] 2959-2978, 1991
9. Pysher, D.J.; Goretta, K.C.; Hodder, Robert S, J, "Strengths of Ceramic Fibers at Elevated Temperatures.", *J. Am. Ceram. Soc.*, **72** [2] 284-288 (1989)
10. DiCarlo, James A.,"High Temperature Structural Fibres- Status and Needs", *NASA TM-105174*, (1991)
11. Yun H. M., Goldsby J. C., DiCarlo J. A.,"Stress-Rupture Behaviour of Small Diameter Polycrystalline Alumina Fibres.", *NASA Technical Memorandum 106256*, (1993)
12. Klein, A.J.,"Specialty Reinforcing Fibers.", *Adv. Compos.* **3** [3] pp32-34, 38, 42, 44 (1988)
13. Haggerty, J.S.; Wills, K.C.; Sheehan, J.E.,"Growth and Properties of Single Crystal Oxide Fibers.", *Cer. Eng. Sci. Proc.*, **12** [9-10] 1785-1801 (1991)

14. Johnson, D. D.; Holtz, A. R.; Grether, M. F.,"Properties of Nextel 480 Ceramic Fibers.", *Cer. Eng. Sci. Proc.*, **8** [7-8] 744-754 (1987)
15. Corman, G.S.," High-Temperature Creep of Some Single Crystal Oxides.", *Cer. Eng. Sci. Proc.* **12** [9-10] 1745-66 (1991)
16. Jones, L.E.; Tressler, R.E.,"The High Temperature Creep Behaviour of Oxides and Oxide Fibers.", *NASA Contract Report 187060*, (1991)
17. Sayir, H.; Sayir, A.; Lagerlöf, K. P. D.,"Temperature-Dependent Brittle Fracture on Undoped and Impurity-Doped Sapphire Fibers.", *Cer. Eng. Sci. Proc.*, **14** [7-8] 581-589 (1993)
18. Sheehan, J.E.; Sigalovsky, J.; Haggerty, J.S.; Porter, J.R.,"Mechanical Properties of $MgAl_2O_4$ Single Crystal Fibers.", *Cer. Eng. Sci. Proc.* **14** [7-8] 660-670 (1993)
19. Deng, S., Warren, R., "Creep Properties of Single Crystal Oxides Evaluated by a Larson-Miller Procedure", to be published in J. Eur. Ceram. Soc. (1995)
20. Whalen, P.J., Narasimhan, D., Gasdaska, C.G., O'Dell, E.W., Morris, G.C.,"New High-Temperature Oxide Composite Reinforcement Material: Chrysoberyl.", *Cer. Eng. Sci. Proc.*, **12** [9-10] 1774-1784 (1991)
21. Corman, G.S.,"Creep of Yttrium Aluminum Garnet Single Crystals.", *J.Mater. Sci. Letters* **12** [6] 379-382 (1993)
22. Morgan, P.E.; Marshall, D.B.,"High-Temperature Ceramic Composites.", *US Pat.US-5137852*, (1992)
23. Mah, T.; Keller, K.; Parthasarathy, T.A.; Guth, J.,"Fugitive Interface Coating in Oxide-Oxide Composites: a Viability Study.", *Cer. Eng. Sci. Proc.*, **12** [9-10] 1802-1815 (1991)
24. Brownie, P.M.; Ponton, C.B.; Marquis, P.M.,"Coating of Single Crystal Alumina Fibers with Zirconia Sols.",*Br. Ceram. Proc.*, [50] 121-130 (1993)
25. Kristoffersson, A., Warren, A., Brandt, J., Lundberg, "Reaction Bonded Oxide Composites" , *Proc. Int. Conf. HTCMC-1, see ref.1*, 151-158 (1993)
26. Patankar, S.N., Venkatesh, R., Chawla, K.K., "Effect of Tin Oxide Coating on Tensile Strength of Alumina Fibres" *Scripta Metallurg. Mater.* **25**, 361-366 (1991)
27. Warren, R., Lundberg, R., "Principles of Preparation of Ceramic Composites" in *Ceramic Matrix Composites, Ed. R. Warren, Blackie Publ. UK*, 35-63 (1992)
28. Brandt, J., Lundberg, R., "Processing of Mullite-Based Long-Fibre Composites via Slurry Routes and by Oxidation of an Al:Si Alloy Powder", Proc. Int. Workshop Mullite '94, Irsee, Germany 7-9 Sept 1994. To be published in J. Eur. Ceram. Soc. (1995)
29. "Novel Oxide Ceramic Composites" , *Brite/Euram-Project BE-7125 Contract BRE2-CT94-0537*
30. Claussen, N., Wu, S., Holz, D., "Reaction Bonding of Aluminum Oxide (RBAO) Composites: Processing, Reaction Mechanisms and Properties", J. Eur. Ceram. Soc. 14, 97-109 (1994)
31. Wu, S., Claussen, N., "Reaction Bonding and Mechanical Properties of Mullite/Silicon Carbide Composites", J. Am. Ceram. Soc., **77** [11] 2898-2904 (1994)

DYNAMIC ANALYSIS OF PRESSURE INFILTRATION PROCESSES

Dhiman K. Biswas, Jorge E. Gatica and Surendra N. Tewari
Department of Chemical Engineering - 455 Stilwell Hall - 1960 E. 24th Street,
Cleveland State University, Cleveland, OH 44115 (FAX: 216 / 687-9220).

ABSTRACT

Unidirectional pressure infiltration of porous preforms by molten metal/alloys is investigated in this study. The dynamics of the process is analyzed via the numerical solution of a mathematical model. Comparison against classical asymptotic analyses shows that, for realistic samples, end effects may become important and render asymptotic results unreliable. A comparison with experiments proves the model to be an efficient predictive tool in the analysis of infiltration processes for different preform/melt systems.

INTRODUCTION

For the pressure infiltration process, porous preforms are normally prepared by packing paricles/whiskers or as microporous structures (cf. Fig. 1). A superheated molten metal/alloy is then infiltrated into the preform by applying an external pressure gradient. The pressure gradient, velocity of propagation, and temperature of the preform and superheated melt; are critical variables determining the infiltration kinetics and microstructure of the final composite. Before solidification is observed, the preform permeability is constant and the infiltration velocity is only a weak function of the infiltration length. As solidification starts, a two-phase region emerges with a time-space variable solid fraction and the infiltration dynamics becomes strongly dependent on the infiltration length and solid fraction. This two-phase zone is confined between two sharp fronts; a <u>remelting</u> front at the point where the superheated metal enters the two-phase zone and an <u>infiltration</u> front. These two fronts have independent dynamics resulting in a two-phase zone which expands with the infiltration time. Because of its considerable engineering relevance to MMC fabrication, the infiltration process has been studied by several authors, from both theoretical and experimental

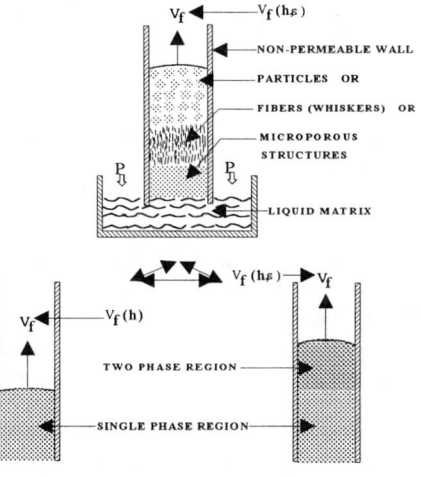

Figure 1: Schematic of the Pressure Infiltration

standpoints. Nagata and Matsuda (1982) investigated the infiltration of beds of particles ranging in size from 37 to 3400 µm. These authors proposed the existence of a critical preheating temperature, based upon physical constants of the metal and particles, above which the particles must be heated in order for infiltration to occur. Mortensen et al. (1989) investigated the critical pressure necessary for melt infiltration, and the effect of infiltration pressure on fiber perform deformation. Girot et al. (1990) performed a numerical analysis of the infiltration of liquid alloys into fibrous preforms, and proposed that flow ceases when the metal cools to its liquidus temperature and, therefore, did not account for the release of latent heat of solidification by the metal in their calculations. In this study, all these variables are investigated separately.

MODEL EQUATIONS AND ASSUMPTIONS

For analysis of fluid flow and heat transfer, a volumetric element ΔV is considered. The flow velocity through ΔV is assumed within the range of validity of Darcy's law. The dimensionless form of the momentum equation is:

$$V + \frac{\partial P}{\partial X} + Ra\theta \check{z} = 0 \qquad \text{with,} \quad \frac{\partial P}{\partial X} \approx -\frac{\Delta P}{X_f} \qquad [1]$$

where, $Ra = \frac{K}{\mu}\left(\frac{\rho_c c_{p_c}}{k_c}\right) L \rho_{mo} g(-\beta)\frac{\Delta H_m}{c_{p_c}}$ is the Rayleigh number, $\beta = -\frac{1}{\rho_{mo}}\left.\frac{\partial \rho}{\partial T}\right|_{T_m}$ is the thermal expansion parameter, V is the dimensionless velocity, L is the total preform length, X_f is dimensionless infiltration front position, ΔP is the pressure drop in melt, and K is the preform permeability. In developing the above equation, stationary solid and fiber phases were assumed stationary, and the difference between solid and liquid metal densities was assumed negligible, i.e. the momentum transfer due to phase change is negligible. The pressure drop was assumed independent of the infiltration front velocity, and the applied pressure is considered high enough for the flow to be a slug-type flow.

The dimensionless forms of energy balance equations in the melt/composite and preform are:

$$\frac{\partial \Phi_c}{\partial t} + V\frac{\partial \Phi_m}{\partial X} = \hat{\rho}_c \frac{\partial}{\partial X}\left(\frac{\partial \theta}{\partial X}\right) + S \qquad [2]$$

$$\frac{\partial \Phi_p}{\partial t} + V\frac{\partial \Phi_g}{\partial X} = \hat{\rho}_p \frac{\partial}{\partial X}\left(\frac{\partial \theta}{\partial X}\right) + S \qquad [3]$$

where $\hat{\rho} = \frac{\rho}{\rho_m}$, ρ is density, Φ is the dimensionless enthalpy, θ is the dimensionless temperature and S is a source term (or energy generation in the system); the subscripts "c", "m", "p" and "g" refer to composite, matrix, preform and gas respectively. The time for heat exchange between the fibers and the metal is very short when compared to the infiltration characteristic time. Viscous heat dissipation, compression work, and volumetric energy and species sources are neglected. The interface between solid and liquid metal is assumed at thermal equilibrium, and the effect of curvature on the melting point of metal, T_m, is considered negligible.

To solve equations [1] and [2] (or [3]) simultaneously, one initial condition and three (or four) boundary conditions are necessary. The temperature in the melt and preform zone is assumed a continuous function; therefore to avoid a temperature discontinuity, the initial temperature distribution in the preform zone is found from the analytical solution of the energy balance equation in the preform zone, i.e.

$$\theta_p(t,X) = (\theta_o - \theta_p)\,erfc\left[\frac{X}{C\Delta X}\right] + \theta_p\,; \quad @\ t=0 \qquad [4]$$

where θ_p and θ_o are the preform and melt initial temperatures, respectively.

Two of the boundary conditions can be formulated at the two interfaces. The *infiltration front interface*, which separates uninfiltrated preform from infiltrated two phase zone, and at the *remelting front interface*, which separates two phase zone from remelted single phase zone. Thermal equilibrium at the infiltration front interface establishes:

$$\vec{n_f}\vec{v_f}[C_{pf}V_f\rho_f(T_m-T_f)+(-\Delta H_m)(1-V_f)\varepsilon_s\rho_s] = \underbrace{\vec{n_f}k_p\vec{\nabla}T_p}_{\text{very small}} - \overbrace{\vec{n_f}k_m\vec{\nabla}T_m}^{\text{zero(pure metal)}} \qquad [5]$$

where, T_m is the melting temperature of the metal, v_f is front velocity, V_f is the volume fraction of fiber, ε_s is the fraction of the melt that is solidified, ΔH_m is heat of fusion of the melt, and n is the surface normal unit vector. The equilibrium condition at the remelting front interface is:

$$\overbrace{\vec{n_b}\left(-k_m\vec{\nabla}T_m\right)}^{\text{zero(pure metal)}} + \vec{n_b}\left(k_c\vec{\nabla}T_c\right) = (1-V_f)\varepsilon_s(\Delta H_m)\rho_s\vec{n_b}\vec{V_b} + \underbrace{c_{pf}v_f\rho_f\left(\widetilde{T}_f-T_m\right)\vec{n_b}\vec{V_b}}_{\text{very small}} \qquad [6]$$

where, $\widetilde{T}_f$ is the fiber temperature after remelting all solid, and V_b is the bottom, remelting, front velocity. The remaining two boundary conditions, inlet and exit conditions, are assumed as suggested by Danckwerts (1953) for flow systems.

For flow perpendicular to the fiber axes, the permeability based on the numerical calculations of Sangani and Acrivos(1982) is given as:

$$K = \frac{2\sqrt{2}\,r_{sf}^2}{9V_{sf}}\left[1-\sqrt{\frac{4V_{sf}}{\pi}}\right]^{\frac{5}{2}} \qquad [7]$$

where, r_{sf} is the radius of the solid material and V_{sf} is its volume fraction. Eq. [7] is valid for solid volume fractions, V_{sf}, ranging between 0.2 and $\pi/4$.

SOLUTION PROCEDURE

To solve the system of differential equations [1] [2] and [3], the system is discretized, with the maximum number of cells limited by the hypothesis of continuum. The governing equations are applied for each cell, with upstream discretization for first order derivatives and central difference for second order derivatives being used. The results for temperature and enthalpy are updated after each iteration, while the solid fraction and infiltration velocity are updated only after filling each computational cell. The numerical stability is ensured by selecting the proper number of steps to "fill" a cell. The discretized equations are used directly both in the preform and melt section for each cell, except for the cell where the infiltration front is located. Incoming and outgoing fluid for the convection term for the cell where the front is, are melt and gas respectively. The heat conduction takes place on one side of this cell through the composite (fiber and melt) and on other side through the preform (fiber and gas). Therefore, the rate of energy exchange by conduction at the "front" cell is obtained by one-sided differences, which are used to find the thermal gradient for each side. Different thermal conductivities are used for each side.

RESULTS AND DISCUSSION

Typical results are presented in Figure 2, this figure shows a time sequence of thermal profiles for the infiltration of pure aluminum in an array of fibers preheated to 200 K below the metal melting temperature. It can be seen how the two-phase zone develops as soon as the melt reaches the melting temperature. Two regions can be clearly distinguished ahead and behind the infiltration front (cf. Fig. 2a), which is indicated by the steep temperature drop at the front-preform interface. As time progresses the two-phase region expands with the two fronts (remelting and infiltration, cf. Fig. 2b) exhibiting independent dynamics. At the beginning of the process the infiltration is very fast. As solidification takes place the two-phase zone permeability decreases and the infiltration is slowed down significantly. It is interesting to note that the infiltration rate determining zone is neither at the remelting or infiltration fronts, but in the bulk of the two-phase zone where the solid fraction reaches a maximum (cf. Fig 2c).

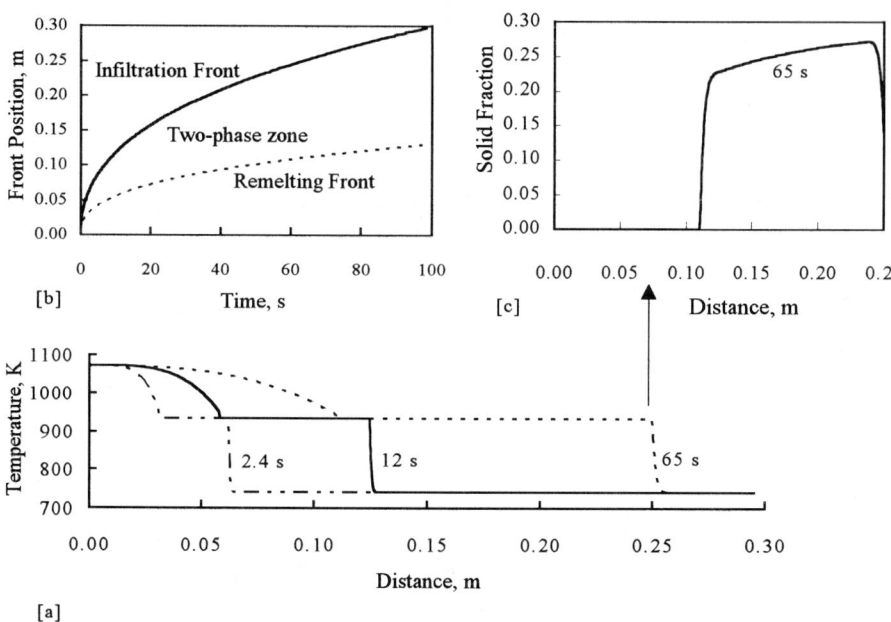

Figure 2: Infiltration dynamics for Aluminum infiltration of a fibrous preform.

When the initial fiber temperature, T_f, is sufficiently high (or when infiltration velocity is high), no solidification of the matrix occurs. The permeability in the composite is constant everywhere. This simple case is considered first to compare with the analytical model presented by Mortensen et al. (1989). The analytical and numerical results of the infiltration length as a function of time are compared (cf. Fig 3a). Satisfactory agreement between the numerical and analytical prediction validates the numerical model.

Numerical predictions can also be compared with experimental and analytical results for unidirectional adiabatic infiltration with solidification. When the initial fiber preform temperature, T_f, and the initial metal temperature (or the infiltration velocity) are sufficiently low to avoid solidification, solid metal forms at the infiltration front as a coating surrounding the packing fibers. The analytical treatment of the governing equations (Mortensen et al., 1989) solves the problem in terms of the similarity variable, $\Psi = L/\sqrt{t}$. For instance, for the conditions of Fig. 3b, the similarity variable, Ψ, for the analytical model was found to be 0.0299 m/s$^{1/2}$. For the same conditions, this parameter was found experimentally as 0.0304. This indicates a satisfactory agreement between the analytical model and the experimental results. When the analytical results are compared with the numerical model (cf. Fig 3b) some discrepancies are noticeable. These discrepancies can be explained as follows: to find the parameter, Ψ, Mortensen et al. (1989) measured the slopes of plots of the experimental infiltration length as a function of square root of the infiltration time. Since the plots were non-linear, the parameter Ψ was determined from a selected section of the plot (arbitrarily defined as that corresponding to "before sufficient solidification occurred"). The agreement between the analytical predictions and experiments, therefore, corresponds only to a portion of the experimental results, not to the entire infiltration process. A similar plot is produced in Figure 3c with the results from the numerical simulation; the numerical results are clearly consistent with the experimental observations. Indeed, the plot in Figure 3c is not linear, which is the basis for similarity solution and the main reason for disagreement between analytical and numerical results.

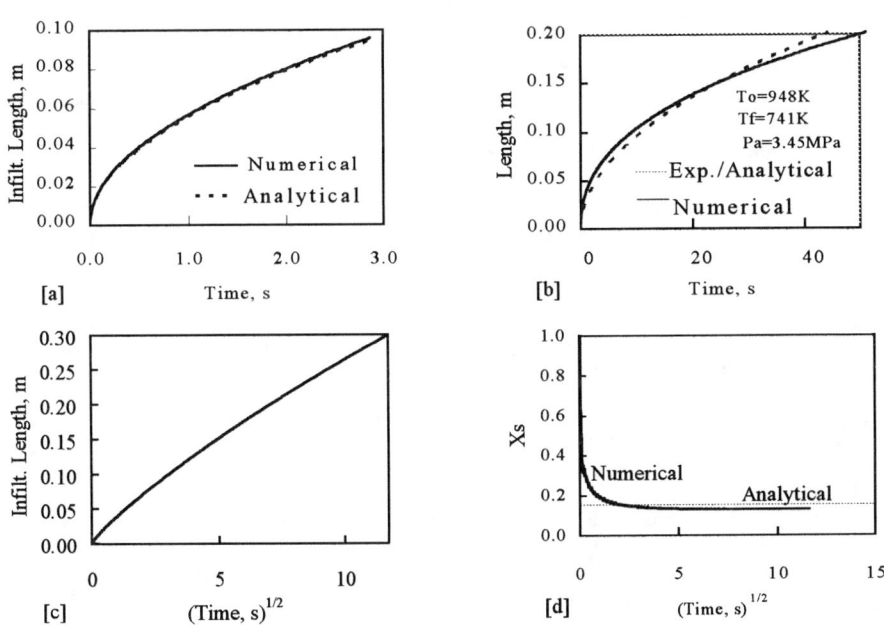

Figure 3: Comparison between numerical and experimental/analytical results.

In the analytical model, Xs, the ratio of the remelting front position to the infiltration front position, is assumed constant for given process conditions. This assumption assumes a constant-pattern propagation during the infiltration process. The parameter Xs was investigated numerically for different preform temperatures. Results for the experimental conditions reported by Masur et al. (1989) are shown in Figure 3d. The plot shows that, at beginning, the value of Xs is 1.0, i.e., the remelting and the infiltration fronts are at the same position, (i.e. the initial condition). As the melt infiltrates and solidification starts, a two-phase zone emerges and starts spreading, i.e., the value of Xs decreases and approaches an asymptotic value. The plot shows that, the value of Xs will coincide with the analytical value at about 4 seconds into the infiltration process, before this the two-phase zone is shorter and after this it is longer than the analytical value. For these experimental conditions, the analytically predicted Xs is 0.163, while numerically it was found that Xs approaches a constant value of about 0.14 after 25 seconds. Therefore, using a constant value of 0.163 for Xs, the value of the overall permeability is underestimated for the first 4 seconds of infiltration. While, after 25 seconds, the overall permeability is overestimated in the analytical model. Therefore, the prediction of the infiltration rate by the analytical model will be underestimated at the beginning, and overestimated at the end of the infiltration process. The impact on the predictive ability will depend, therefore, on the dimensions of the preform being infiltrated. For instance, Mortensen et al. (1989) presented comparisons for infiltration lengths in the range of 4-5 cm which would occur in the first 2-3 seconds of infiltration.

CONCLUSIONS

The numerical analysis presented above leads to the following conclusions :
- Actual infiltration processes never show constant pattern propagation (initial and final dynamics cannot be neglected.
- Conduction in the preform zone cannot be neglected. Although it might be negligible in determining the temperature profile.
- The two phase zone dynamics is very sensitive to the initial preform temperature.
- Asymptotic analyses can only be used as an indicative tool; not as a design tool.

ACKNOWLEDGMENTS

Support provided by the Processing Science and Technology Branch at the NASA Lewis Research Center (Cleveland, OH) is gratefully acknowledged.

REFERENCES

Danckwerts, P.V. *Chem. Eng. Sci.*, 2(1), pp. 1-18 (1953).
Girot, F.A., Fedon, R., Quenisset, J.M., and Naslain, R., *Journal of Reinforced Plastics and Composites.*, 9, pp. 456-69 (1990).
Masur, L.J., Mortensen, A., Cornie, J.A., and Flemings, M.C., *Metall. Trans. A*, 20A, pp. 2549-57 (1989).
Mortensen, A., Masur, L.J., Cornie, J.A., and Flemings, M.C. *Metall. Trans. A*, 20A, pp. 2535-47 (1989).
Nagata, S., and Matsuda, K., *Trans. Jpn. Foundarymen's Soc.*, 2, pp.616-20 (1982).

FABRICATION OF SiC MATRIX COMPOSITES USING A LIQUID POLYCARBOSILANE AS THE MATRIX SOURCE

L.V. Interrante*, C.W. Whitmarsh[@], and W. Sherwood[@], Department of Chemistry, Rensselaer Polytechnic Institute, Troy, NY 12180-3590* and Starfire Systems, Inc., 877 25th St, Watervliet, NY 12189[@]

ABSTRACT

A polymer precursor-based route to SiC-matrix composites is described. This route employs a proprietary hydridopolycarbosilane (HPCS) as a liquid phase source of SiC. The polycarbosilane is prepared from methyltrichlorosilane, a low-cost starting material, via a process which is amenable to large scale production. The resulting liquid polymer can be handled in air and has physical and chemical properties appropriate to SiC-matrix formation by liquid phase infiltration of fiber, whisker or particulate preforms followed by thermal crosslinking and pyrolysis. This process has been used to prepare Nicalon SiC-fiber and SiC-powder/whisker reinforced SiC/SiC composites which show improved uniformity, overall density and flexure/tensile strengths compared to comparably reinforced CVI-SiC matrix composites.

INTRODUCTION

The use of organometallic compounds and polymers as pyrolytic precursors to ceramics offers several potential advantages over existing ceramic processing methodologies [1]. Chief among these is the prospect of employing the properties commonly found for molecular systems (e.g., volatility, solubility, fluidity, etc.) to advantage in the fabrication of ceramic coatings, fibers, matrices for composites, etc., final forms that are often difficult, if not impossible, to achieve by more conventional processing methods. In the particular case of SiC/SiC composites, polymer precursors have been used primarily as a source of the continuous SiC fiber that is generally used as the reinforcement phase in such composites [2]. Such precursors have been less commonly employed as the source of the SiC matrix in these composites, where Chemical Vapor Infiltration and liquid Si infiltration into C-filled preforms are more commonly employed. Both of these methods have inherent difficulties that have limited their large scale application; in the case of CVI it is largely a question of an intrinsically high processing cost, brought upon by the need for expensive vacuum equipment and the typically long time period required to bring a part to maximum final density.

To the extent authorized under the laws of the United States of America, all copyright interests in this publication are the property of The American Ceramic Society. Any duplication, reproduction, or republication of this publication or any part thereof, without the express written consent of The American Ceramic Society or fee paid to the Copyright Clearance Center, is prohibited.

Moreover, under typical CVI processing conditions, the closing off of surface access channels to internal porosity leads to severe limitations in terms of the ability to fabricate thick parts, as well as inhomogeneous matrix filling, particularly for complex part shapes, that can lead to low strengths. In the case of Si metal infiltration, the chief problems include the intrinsically high processing temperatures required (>1400 °C), the exposure of the reinforcement phase (and interphase, if present) to highly corrosive liquid silicon, and the incomplete reaction of the Si with the free carbon present, all of which lead to limitations in the mechanical properties and the use temperature for the final part. In theory, most or all of these problems could be overcome through the use of a suitable polymeric precursor to SiC as the matrix source, where the well-developed technology of vacuum impregnation of fiber-filled preforms could potentially provide a relatively low cost alternative to CVI, while allowing a much wider range of final part shape and dimensions.

Reports relating to the use of a polymer precursor as the SiC matrix source have been quite limited [3, 4], although rather more attention appears to have been directed towards the use of siloxanes as silicon oxycarbide ("black glass") matrix sources [5]. In the latter case, the limited thermal stability of the matrix material set corresponding limits on the effective use temperature of the final composite (well below 1000 °C under certain circumstances; *vide infra*). As far as SiC precursors are concerned, limited work has been carried out by using both the Yajima "polycarbosilane" [3] and a vinylic polysilane [4] as the matrix source; however, the results have been fairly disappointing in terms of the reported strengths of the final composites. Moreover, these precursors are known to produce "SiC" with a large amount of excess carbon, with expected detrimental effects in terms of oxidative stability and mechanical properties.

The use of a polymer precursor as a SiC matrix source requires a certain combination of physical and chemical properties that must be built into the polymer precursor structure. Thus the precursor must be sufficiently fluid to penetrate the interstices of the fiber preform, it must wet the surface of the fibers, it should be air stable, and it should have the requisite chemistry for crosslinking and subsequent high-yield, pyrolytic conversion to stoichiometric SiC. In addition to these specific requirements of precursor properties, the precursor must be readily obtainable at a sufficiently low cost to justify its use for this application.

Since the original reports by Yajima of the successful application of a "polycarbosilane" derived from thermal isomerization of polydimethylsilane as a "processable" precursor to SiC [6], considerable attention has been directed to the synthesis of polymeric precursors to Si-based ceramics and their application to ceramic processing [1,2], with particular attention toward the use of such precursors to prepare continuous, weavable, ceramic fiber for use as the reinforcement phase in both ceramic and metal matrix composites [2]. With the exception of "polymethylsilane" [7], which has a 1:1 Si:C ratio and has been reported to yield SiC with a 1.1:1 Si:C ratio, virtually all of these precursors contain excess carbon over the requisite 1:1 Si:C ratio and on pyrolysis yield SiC with considerable amounts of excess carbon. Moreover, few of these precursors have the requisite physical properties for use as SiC matrix sources, and the char yields are rarely above the at least 80% that would be needed to minimize the number of infiltration cycles required to reach full final density. Moreover, most of them are easily oxidized on exposure to air, and require the use of inert atmosphere handling methods in order to avoid the introduction of appreciable amounts of oxygen into the final "SiC" phase. This paper describes the use of a liquid, pre-ceramic polymer (XHPCS) that pyrolyzes to near stoichiometric silicon carbide in >80 % yield and is amenable to simple resin transfer molding (RTM) of fabric plies or braided preforms. Moreover, this polymer can be handled in air and is amenable to large scale production from a relatively low cost starting material (MTS).

PREPARATION OF HPCS AND ITS OLEFINIC DERIVATIVES

HPCS is a hydridopolycarbosilane which is prepared in high yield by Mg coupling of $ClCH_2SiCl_3$ followed by reduction with $LiAlH_4$ [8]. The $ClCH_2SiCl_3$ is obtained from MTS via a similarly high-yield chlorination process. The overall process yields a hydridopolycarbosilane of the approximate composition "SiH_2CH_2". Due to the trifunctional nature of the Si atom in this monomer unit, the resulting polymer has a highly branched structure with a distribution of $(-CH_2)_n SiH_{4-n}$ sites ranging from n = 1-4.

This polymer is an oily liquid that is readily miscible with ether and hydrocarbon solvents. It undergoes crosslinking on heating under an inert atmosphere between 200 and 400 °C, or in the presence of a dehydrocoupling catalyst at 100 °C, to yield a hard, clear solid [9]. After crosslinking it can be pyrolyzed under N_2 to produce a near-stoichiometric "SiC" at 1000 °C in 70-90% yield. The SiC is largely amorphous at 1000 °C but crystallizes to a nanocrystalline β-SiC ceramic between 1400 and 1600 °C. One of the unique advantages of the HPCS system is its apparent stability to air exposure, at least at room temperature over a period of days. Samples exposed in a TGA apparatus to flowing air at room temperature do not change weight over several hours and show no changes in IR or NMR spectra.

At the intermediate "chloropolycarbosilane" stage of the HPCS preparation, the polymer can be substituted with a small amount (typically 2.5-10%) of olefinic functionality (such as vinyl or allyl) prior to reduction of the remaining Si-Cl groups to Si-H, so as to modify the viscosity and crosslinking characteristics [9]. These olefin-modified HPCS polymers [XHPCS; where X = A (allyl) or V (vinyl)] thermoset via a thermally-induced hydrosilation process at 150-400 °C, yielding a hard, clear, glass with little loss in weight. The resultant crosslinked polymer converts to "SiC" in 80-90% yield on pyrolysis to 1000 °C. The "SiC' so obtained is amorphous and contains a slight excess of carbon over the stoichiometric 1:1 formula (ca. 4-8 mole % for 5-10 mole % substitution of A or

V) [10]. This amorphous SiC can be converted to a polycrystalline β-SiC with little further loss in weight and with retention of the shape of the amorphous SiC preform on further heating under Ar to 1600 °C. In this manner, pellets of polycrystalline SiC having densities (by hexane immersion) of 2.7-3.0 g/cm^3 were prepared from XHPCS by the following procedure: AHPCS-derived SiC powder obtained at 1000 °C was heated to 1600 - 1700 °C, cold pressed with more polymer at 1000 psi and then crosslinked by heating in the press to 250 °C. The resultant cured right circular cylinders were then pyrolyzed to 1000 °C, cut into pellets with a diamond saw, and then subjected to further RTMP cycles, followed by a final heating to 1600 - 1700 °C. These pellets were gray-green in color, quite hard and showed β-SiC as the only crystalline phase by XRD.

The long term thermal stability of the 1000 °C AHPCS SiC in air was tested in comparison with a sample of silicon oxycarbide that was derived from a commercial precursor source. Coarsely ground (ca. 6-8 mm pieces) samples of both materials were placed in a furnace held at 600 °C in air and periodically weighed after cooling to room temperature. Whereas the "black glass" sample was found to slowly and continually decrease in weight over a period of several weeks, amounting to an average weight loss of about 1%/week, the AHPCS-SiC remained constant in weight and in physical appearance. After 300 h at this temperature the black glass had lost about 2.5% of its original weight and parts of the sample had acquired a white surface coating. These results suggest that even at such relatively low temperatures as 600 °C, significant changes in microstructure that are accompanied by measurable weight losses occur in this material and that such changes are not experienced in the case of the AHPCS-derived SiC material. Tests at higher temperatures, along with a detailed microstructural examination of both materials after long-term thermal treatment, are in progress.

FABRICATION OF SiC/SiC COMPOSITES BY USING AHPCS

A partially allyl-substituted HPCS (AHPCS) was used to prepare SiC/SiC composites via a Resin Transfer Molding and Pyrolysis (RTMP) process [11]. Preforms consisting of fiber cloth lay-ups or dry-pressed powders/whiskers held in a simple press were employed in this process. The fiber reinforcement (Nicalon™; 8-harness satin, 1800 denier, carbon coated) was cut to size and the desired number of plies (6-30) were stacked in a molybdenum boat. Introduction of the SiC matrix was accomplished via a RTMP process which employed an autoclave for the polymer curing step. The cloth layers are compressed by a steel fixture and infiltrated with the AHPCS resin. After curing at 400°C and 800 psi of Ar, the plate was removed from the mold and pyrolyzed to 1000°C to convert the cured resin into amorphous silicon carbide (SiC). The free standing plates were then placed into the molybdenum boat and reinfiltrated, cured, and pyrolyzed. The final composite plates ranged from 0.25 cm to 1.20 cm in thickness.

A similar infiltration process has been used in connection with particulate reinforcements, such as chopped SiC fiber and whiskers as well as both SiC and B_4C powder, to prepare cylinders, plates and other shaped objects. A plate or other component typically undergoes 8-10 RTMP cycles to achieve maximum density. After the first RTMP cycle is complete, the composite is sufficiently bonded that it can be handled and subjected to further RTMP cycles without any need for clamping. After partial densification, the composite is strong enough, yet sufficiently porous, to allow further shaping by cutting, grinding or drilling, which can be accomplished by using simple hardened steel or carbide tooling. In this manner, complex shapes such as rods, disks, tubes, and simulated turbine shroud segments have been fabricated and brought to full density through further infiltration and pyrolysis cycles.

CHARACTERIZATION OF THE RESULTANT COMPOSITES

The fiber volume fractions in the final processed plates were around 34-37% for the Nicalon fiber plates. The final densities, as measured by hexane immersion after 6-8 VPIP cycles, were found to be independent of the thickness of the plate and in the range of 90-95% of the theoretical density. These densities were about 2.16-2.33 g/cm^3 for the Nicalon-reinforced plates. The open porosity, as measured by the amount of hexane absorbed on immersion, was less than 5% and generally below 2%.

The rate of densification of these plates also appears to be virtually independent of the thickness of the plate (determined by the number of cloth plies used), at least up to ca. 1.25 cm (30 cloth plies), reaching ca. 30% of theoretical density after the first RTMP cycle and >80% after 4 cycles. Further densification proceeds more slowly, requiring about 3 more RTMP cycles to exceed 90% of the theoretical density. Typically, the plates were subjected to 8-10 RTMP cycles in order to reach the final densities of 90-95% of theoretical. There was also very little difference between the calculated (geometric) and measured (Archimedes) densities of the plates of different thickness, again suggesting very little accessible (open) porosity. Three plates, comprised of 10, 20 and 30 cloth plies, respectively, were sectioned, polished and examined by SEM in cross section. Oblate pores ranging up to a few 0.1 mm long were observed between the cloth plies; however, no significant variations in porosity were detected through the thickness of the plates, even for the 30 layer plate. Recently, plates up to 1.5 cm thick, employing 30 layers of Nicalon cloth along with added crystalline SiC (600 grit powder) as a particulate filler have been fabricated which appear to have slightly higher final densities and a lower degree of macroporosity relative to the plates prepared without the SiC filler.

Several of the composite plates were cut into bend bars and tested in flexure in a 4-point bend fixture (as per Mil STD1942). Tests on ten bend bars cut from three different Nicalon-reinforced plates (without added SiC filler) showed an average fracture stress of 378 MPa (54.8 ksi), with a range of 358 - 419 MPa (51.2 - 60.8 ksi). This is almost 80 MPa higher than the values measured (316 MPa) on comparable CVI SiC matrix/Nicalon composite specimens [12]. Preliminary flexure studies on samples obtained from one of the plates to which SiC powder had been added indicated a slightly higher average flexure strength (383 MPa) and improved uniformity (372 - 391MPa) compared to those without the filler.

The flexure specimens exhibited "graceful failure" with incomplete separation of the broken sections and a saw-tooth load vs. displacement curve. Moreover, a considerable amount of fiber pullout was evident in the fracture surface of the broken Nicalon/AHPCS-SiC plate. These observations are suggestive of appreciable toughness for the composites prepared by this RTMP process.

This conclusion is supported by the results of some preliminary tensile tests carried out on the 10 (Nicalon cloth) layer composites at room temperature[12]. The results are presented in Figure 1 for two different samples obtained from the same 10-layer plate. The data indicate a tensile strength of ca. 250 MPa and a modulus of ca. 100 GPa. These tensile strength values are comparable to [13], or even higher [14] than, those previously reported for CVI-SiC matrix composites reinforced by Nicalon cloth; however, the modulus values are apparently somewhat lower than those exhibited by the CVI composites [13]. The relatively large strain-to-failure (0.48-0.55%) is also indicative of considerable toughness, as is the appearance of the fracture surface which shows extensive fiber pullout.

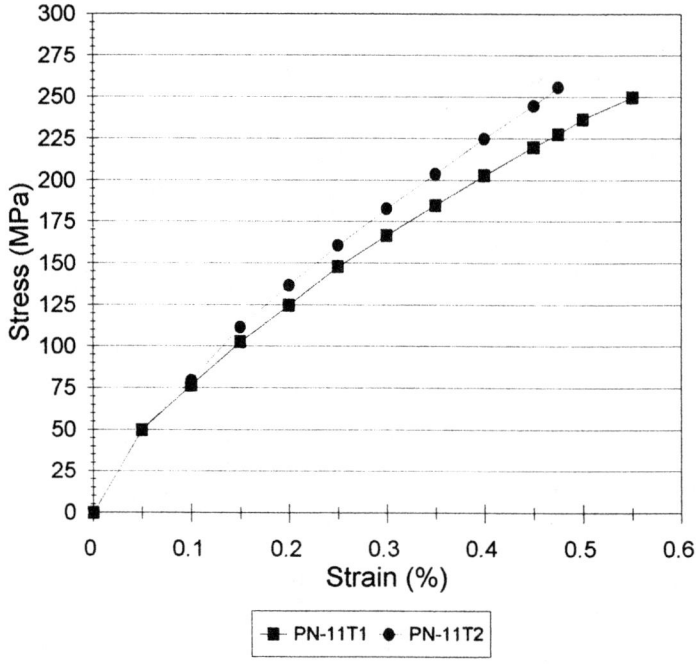

Figure 1. Results of tensile tests on 10 layer Nicalon/AHPCS-SiC 2-D composite

OPTIMIZATION OF XHPCS SYNTHESIS

A new, proprietary, procedure has been developed recently for the preparation of the XHPCS family of SiC precursors which has resulted in a higher yield of polymer per batch while decreasing the time required to complete the polymerization. This has rendered the polymer amenable to large scale production and discussions are currently underway with a chemical company to produce it commercially under license to Starfire Systems, Inc. Moreover, the product obtained via this new procedure provides an improved yield of amorphous SiC, through the crosslinking and pyrolysis stages of the RTMP process, and gives a product that is even closer to the ideal SiC formula, relative to the polymer obtained by the earlier procedure [8,10]. In particular, the SiC derived from the new 5% AHPCS prepared by this procedure was found to have 32 wt% C) and only 2 wt % O indicating, by application of the rule of mixtures, an approximate composition of 92 wt % SiC, 4.5 wt % C, and 3.75 wt % SiO_2. As a result of this lower O and C content, the weight loss on heating the 1000 °C product to 1600 °C, is only 3%, giving a overall ceramic yield of ca. 80% from the initial polymer to the final polycrystalline SiC product (and 83% from the polymer to the amorphous SiC (1000 °C) stage).

The new AHPCS product is being employed in all of our current composite fabrication efforts and preliminary results indicate an overall reduction in the time and cost of the RTMP processing operation (through reduction in the number of RTMP cycles required to reach full density). The final products are expected to have higher densities, due to reduced macroporosity, relative to those described above that were prepared with the prior AHPCS formulation, suggesting further improvements in mechanical properties. Efforts to test these assumptions are currently in progress, as well as efforts to examine in detail the effects of fillers (B_4C and SiC) on the mechanical properties of both fiber-reinforced and particulate-reinforced composites prepared using XHPCS.

ACKNOWLEDGMENT

This work was supported by the U.S Army under Phase I Small Business Innovation Research (SBIR) contract number DAALO1-93-C-4012, Phase II SBIR contract No. DAAL01-94-C-4042 (both to to Starfire Systems, Inc., C. Whitmarsh, P.I.), and New York State Energy Authority Agreement No. 4015-IABR-IA-95 (with Starfire Systems and RPI).

REFERENCES

[1]. K.J. Wynne and R.W. Rice, "Ceramics Via Polymer Pyrolysis", Ann. Rev. Mater. Sci., 14, 297-333 (1984); R.M. Laine and F. Babonneau, "Preceramic Polymer Routes to Silicon Carbide", Chem. Mater., 5, 260-79 (1993); D. Seyferth, "Preceramic Polymers: Past, Present, and Future", in "Materials Chemistry - An Emerging Discipline", L.V. Interrante, L.A. Caspar, and A.B. Ellis, eds., Advances in Chemistry Series, Vol. 245, pp. 131-60 (1995).

[2]. K. Okamura, "Ceramic Fiber and Whisker Requirements to Advanced Structural Inorganic Composite", in "Advanced Structural Composites", P. Vicenzini, ed., Elsevier Science Publ. B.V., Amsterdam, pp. 19-34, (1991); J. Lipowitz, "Polymer-derived Ceramic Fibers", Am. Ceram. Soc. Bull. 70[12], 1888-94 (1991); (c) T.F., Cooke, "Inorganic Fibers - A Literature Review", J. Amer. Ceram. Soc., 74[12] 2959-78 (1991).

[3]. W. Fohey, M. Battison, J. Halada, and T. Nielson, U.S. Government Report No. WL-TR-92-4019, July 1992.

[4]. J.R. Strife, J.P. Wesson and H.H. Streckert, "A Study of the Critical Factors Controlling the Synthesis of Ceramic Matrix Composites from Preceramic Polymers", US Government Report No. AD-A23 686, December 1990; R.P. Boisvert, "Ceramic Matrix Composites via Organometallic Precursors", M.S. Thesis, Rensselaer Polytechnic Institute, 214 pages (1988); B.C. Mutsuddy, "Use of Organometallic Polymer for Making Ceramic Parts by Plastic Forming Techniques", Ceramics International, 13, 41-53 (1987).

[5]. F.I. Hurwitz, L. Hyatt, J. Gorecki and L. D'Amore, "Silsesquioxanes as Precursors to Ceramic Composites", Ceram. Eng. Sci. Proc. 8, 732-43 (1987); H. Zhang and C.G. Pantano, in "Ultrastructure Processing of Advanced Materials", D.R. Uhlmann, D.R. Ulrich (eds.), J. Wiley &Sons, p. 223 (1992).

[6]. S. Yajima, K. Okamura, J. Hayashi, and M. Imura, "Synthesis of Continuous SiC Fibers with High Tensile Strength", J. Amer. Ceram. Soc. 59[7-8] 324-327 (1976).

[7]. Z-F. Zhang, F. Babonneau, R.M. Laine, Y. Mu, J.F. Harrod and J.A. Rahn, "Poly(methylsilane) - A High Ceramic Yield Precursor to Silicon Carbide", J. Amer. Ceram. Soc. 74, 670-73 (1991).

[8]. C.K. Whitmarsh, L.V. Interrante, "Synthesis and Characterization of a Highly Branched Polycarbosilane Derived From Chloromethyltrichlorosilane", Organometallics, 10, 1336 (1991); C.K. Whitmarsh, L.V. Interrante, "Carbosilane Polymer Precursors to Silicon Carbide Ceramics", US Patent No. 5,153,295, October 6, 1992.

[9]. C-Y. Yang and L.V. Interrante, "Thermally and Chemically Induced Crosslinking of a Hydridopolycarbosilane for Optimization of Silicon Carbide Yield", Polymer Preprints (ACS Polymer Divis.), 33 [2], 152-153 (1992).

[10]. Elemental analysis of the 1000 °C HPCS- and APCS-SiC gave the following results: HPCS: %C, 27.9; %H, 0.7; %Si, 63.1; %O, 3.5; 5% AHPCS: %C, 34.8; %O, 2.8; %Si, 62.4 (by difference).

[11]. Preliminary results were reported at the 1993 Amer. Ceram. Soc. PACRIM Conference [Ceramic Trans.,Vol 42, pp. 57-69 (1994)] and at the Fall 1994 MRS Meeting [MRS Sympos. Proc., Vol 365 (1995), in press].

[12]. Tensile tests carried out by Dr. John Holmes at the Univ. of Michigan (monotonic tension, edge-loaded tensile grips, 20 °C, air, 25 MPa/s).

[13]. DuPont/Lanxide Product Data Sheet.

[14]. D.A. Woodford, D.R. Van Steele, J.A. Brehm, L.A. Timms and J.E. Palko, "Testing the Tensile Properties of Ceramic-Matrix Composites", JOM May pp. 57-63 (1993).

DEVELOPMENT OF A TWO-STEP, FORCED CHEMICAL VAPOR INFILTRATION PROCESS

W.M. Matlin, D.P. Stinton, and T.M. Besmann
Metals and Ceramics Division
Oak Ridge National Laboratory
Oak Ridge, TN 37831-6063
USA

ABSTRACT

A two-step forced chemical vapor infiltration process was developed that reduced infiltration times for 4.45 cm dia. by 1.27 cm thick Nicalon™ fiber preforms by two thirds while maintaining final densities near 90 %. In the first stage of the process, micro-voids within fiber bundles in the cloth were uniformly infiltrated throughout the preform. In the second stage, the deposition rate was increased to more rapidly fill the macro-voids between bundles within the cloth and between layers of cloth. By varying the thermal gradient across the preform uniform infiltration rates were maintained and high final densities achieved.

INTRODUCTION

In spite of the significant advantages of forced-flow, thermal-gradient chemical vapor infiltration (FCVI),[1] the composites produced may still be too expensive to be commercially viable. Furnace operation during preform infiltration make up the largest fraction of the composite's cost.[2] Therefore, reducing the time to infiltrate the preforms would have the greatest impact on the economics of the process.

Sheldon[3] has approached optimization of the isothermal, isobaric chemical vapor infiltration (ICVI) process by dividing the process into two steps. In the first step, intrabundle infiltration would be optimized. In the second step, the much larger interbundle voids would be more efficiently filled.[4] Since the transport phenomena for ICVI, diffusion, is different than for FCVI, which relies on forced convection, the conclusions from this work could not be directly applied to FCVI optimization. However, the general two-step approach should be applicable.

To the extent authorized under the laws of the United States of America, all copyright interests in this publication are the property of The American Ceramic Society. Any duplication, reproduction, or republication of this publication or any part thereof, without the express written consent of The American Ceramic Society or fee paid to the Copyright Clearance Center, is prohibited.

In previous experimental FCVI optimization efforts, reductions in processing time have always corresponded to reductions in final density.[5] If this trend was to be reversed a better understanding of time dependent relationship between process variables would be required. The Georgia Tech Chemical Vapor Infiltration Model (GTCVI),[6] a three dimensional finite volume program for modeling chemical vapor infiltration processes, was used to gain a phenomenological understanding of the process and as a guide in identifying the relative importance of each of the involved variables.

Several sets of modeling runs were performed to identify the optimum conditions for a two step process to infiltrate a 4.45 cm dia. by 1.27 cm thick preform, comprised of 52 layers of Nicalon™ cloth (Nippon Carbon, Tokyo, Japan), with SiC deposited from CH_3SiCl_3 in hydrogen. The object of these runs was to develop a first step optimized to fill the micro-voids within the fiber bundles, and a second step optimized to fill the macro-voids between bundles in the cloth and between layers of fabric. This process scheme should result in much shorter infiltration times.

TWO STEP MODELING

A two-step process consisting of the optimized conditions for each of the stages was modeled and compared to the original one-step process. In the first set of computations one-step conditions were used: 1200°C hot-side temperature, 750°C cool-side temperature, 500 cm³/min H_2, 45 cm³/min CH_3SiCl_3. The simulation ran until the pressure drop across the preform reached 70 kPa. The second run combined the thermal gradient optimization of the first stage with the increased CH_3SiCl_3 concentration of the second stage. The first five hours of the run used a 1200°C hot-side temperature 950°C initial cool-side temperature, 500 cm³/min H_2, and 45 cm³/min CH_3SiCl_3. After the first five hours the cool-side temperature was decreased to 850°C and the CH_3SiCl_3 flow was increased from 45 cm³/min to 83 cm³/min. This simulation ran until a back-pressure of 170 kPa was reached.

The two-step simulation predicted a processing time 25% shorter than the one-step process, with more uniform final density (Figs. 1 and 2).

EXPERIMENTAL VERIFICATION

Two experimental runs were preformed and compared to the model predictions. In the first run (no. 907) the preform was infiltrated for five hours using a 1200°C

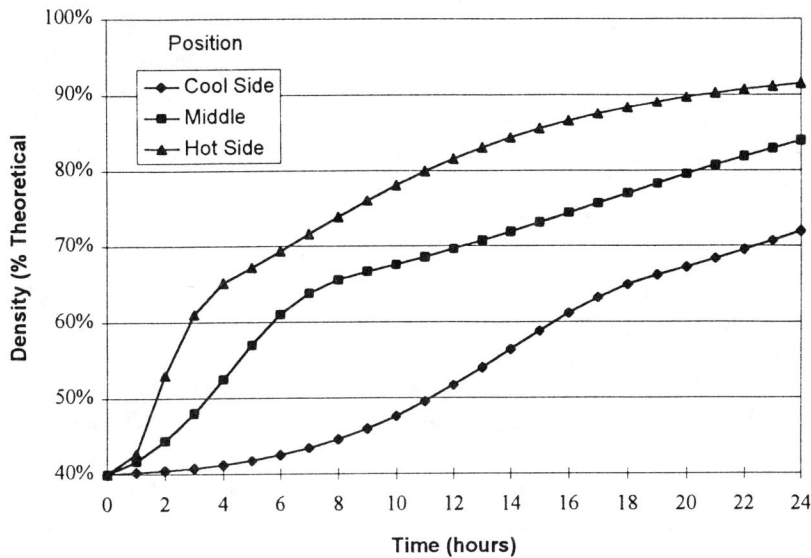

Figure 1. Density vs. time for 750 °C cool-side, one-step GTCVI simulation.

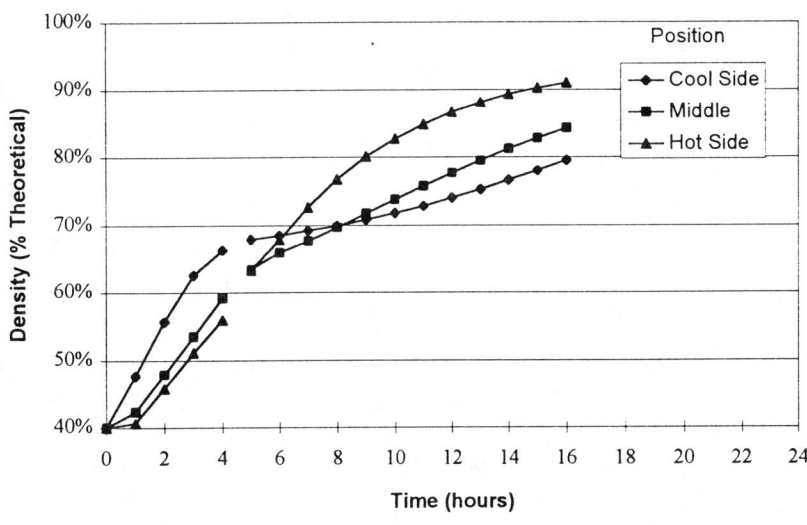

Figure 2. Density vs. time for 950 °C cool-side, two-step GTCVI simulation.

hot-side temperature, 920°C cool-side temperature, 500 cm³/min H_2, and 45 cm³/min CH_3SiCl_3. Process conditions were then changed to 1200°C hot-side temperature, 825°C cool-side temperature, 500 cm³/min H_2, and 83 cm³/min CH_3SiCl_3. The process automatically shutdown when a back-pressure of 170 kPa was reached.

The second preform (no. 906) was infiltrated using standard conditions: 1200°C hot-side temperature, 750°C cool-side temperature, 500 cm³/min H_2, and 45 cm³/min CH_3SiCl_3. The process, again, automatically shutdown when a back-pressure of 170 kPa was reached.

The two step process required ten hours for infiltration and displayed a relatively uniform density averaging 89.8% (Table 1). In contrast, the run using standard conditions took 4.4 hours longer and had an average density 6.6% lower (Table 1). Additionally, the density gradients were significantly larger. Image analysis of the specimens was used to characterize the relative percentage of micro-porosity and macro-porosity. The sample fabricated using standard conditions had a total porosity of 15.4%, consisting of 8.7% macro-porosity and 6.8% micro-porosity. The sample fabricated using the two-step process had a total porosity of 14.5%, consisting of 9.4% macro-porosity and 5.1% micro-porosity.

Table 1. Experimental verification runs.

Process Number	Process Time (h)	Cool side Temp. (°C)	CH_3SiCl_3 Flow (cm³/min)		Density (% Theoretical) Radial Position				
					Center				Edge
One-Step (906)	14.4	750	45	Hot side	89.8	89.7	89.5	88.9	86.6
				Middle	80.6	87.3	86.8	84.1	84.4
				Cool side	75.1	75.6	75.9	77	77.1
Two-Step (907)	5 + 5 = 10	920 + 750	45 + 83	Hot side	90.4	92.7	92	94.1	93.2
				Middle	92.8	92.2	92.5	92	91.6
				Cool side	85.6	86.2	86.3	85	80.9

SUMMARY

The two-step process significantly reduced processing time without reducing final density or uniformity. The key to this success was focusing on maintaining uniform infiltration rates, which prevent density gradients from forming within the preform. Gradients are caused by closure of macro-voids within one section of the preform while other sections remain relatively porous.

The utility of the GTCVI model was also demonstrated. Based on the accuracy of the GTCVI simulations for the 4.45 cm diameter system, GTCVI is expected to be instrumental in developing similarly optimized processing conditions for larger-scale facilities.

REFERENCES

1. T. M. Besmann, B. W. Sheldon, R. A. Lowden, and D. P. Stinton, "Vapor phase Fabrication and Properties of Continuous-Filament Ceramic Composites", Science, Vol. 253, pp 1104-9 (1991).

2. Y.G. Roman, D. P. Stinton, "The Preparation and Economics of Silicon Carbide Matrix Composites by Chemical Vapor Infiltration"; in Symposium Proceedings, Ceramic Matrix Composites (Boston, MA, 1994). Edited by R.A. Lowden and J. R. Hellman, Materials Research Society, Pittsburg, PA, 1994 (in press).

3. H.C. Chang, T.F. Morse, B.W.Sheldon, "Minimizing Infiltration Times During the Initial Stage of Isothermal Chemical Vapor Infiltration", Journal of Materials Processing & Manufacturing Science, Vol. 2, pp 437-455 (1994).

4. B.W. Sheldon, H.C. Chang, "Minimizing Densification Times During the Final Stage of Isothermal Chemical Vapor Infiltration", Silicon-Based Structural Ceramics, B.W. Sheldon, S.C. Danforth (eds.), Ceramic Transactions, Vol. 42, American Ceramic Society, (1994).

5. R.A. Lowden, A.J. Caputo, D.P. Stinton, T.M. Besmann, and M.D. Morris, "Effect of Infiltration Conditions on the Properties of SiC/Nicalon Composites", Oak Ridge National Laboratory Report ORNL/TM-10403, (1987).

6. T.L. Starr, "Model for Rapid CVI of Ceramic Composites" Proc. 10th Int. Conf. CVD, (The Electrochemical Society, Inc., Pennington, NJ, 1987), pp 1147-1155, (1987).

7. D. P. Stinton, O.J. Schwarz, and J. C. McLaughlin, "Fabrication of Fiber-Reinforced Composites by Chemical Vapor Infiltration", Proceedings of the Eight Annual Conference on Fossil Energy Materials, Oak Ridge National Laboratory Report ORNL/FMP-94/1, pp 1-9, (1994).

CARBON FIBRE-REINFORCED SILICON NITRIDE COMPOSITES BY SLURRY INFILTRATION

C. Grenet, L. Plunkett, J.B. Veyret and E. Bullock
Joint Research Centre, Institute of Advanced Materials,
Commission of the European Communities, P.O. Box 2, 1755 ZG Petten, Netherlands.

ABSTRACT:

The present paper reports on the fabrication of long-carbon fibre reinforced silicon nitride matrix composites by liquid infiltration of an aqueous Si_3N_4 slurry followed by hot-pressing. A methodology for the maximum volume and uniform infiltration of preforms has been developed by optimising slurry rheology and fibre wetting conditions. Fully infiltrated green forms of 55% theoretical density are achieved with some 40% volume fraction of fibres. The quality of the composites has been assessed by microstructural analysis and mechanical characterization.

INTRODUCTION:

Development of effective methods of processing CMC's has been a subject of intense research in the last 15 years and several manufacturing processes have been identified[1]. The primary difficulty appears to be the achievement of adequate toughness while retaining the near-theoretical composite density required for high material strength. The process described in this work involves the low cost green forming process by infiltration/consolidation of long carbon fibre preforms with submicron Si_3N_4 powder slurries, followed by hot-pressing.
The major difficulty of slurry infiltration is the filling of the interfibre space uniformly and densely with the matrix phase implying a minimum of entrapped voids [2]. The variables of particular importance are:
- stable and well-dispersed suspensions (high repulsive forces between particles) with maximum solid content compatible with low viscosity,
- excellent wetting of the fibres by the slurry,
- small powder particle size compare to interfibre spacing to allow easy access through pre-form channels,
- a controlled consolidation technique to optimise uniform microstructure and high packing density in green compacts.

The present work studies some of these variables and their effect on the structure and properties of the final composite.

EXPERIMENTAL PROCEDURE:

A deionized water based slurry was used consisting of a mixture of 65wt% solids, of which 90wt% was submicron α-phase silicon nitride powder, with 6.9wt% Y_2O_3 and 3.1wt% Al_2O_3 sinter additives (see Table 1). The slurry was attrition milled with 2mm diameter silicon nitride balls for 1.5 hour in water to break up agglomerates and pH adjusted to 10.5 with tetramethylamonium hydroxide (TMAH) solution. An ammonium polymethacrylate (Darvan C) was used to assist dispersion and the surfactant Aerosol OT, to enhance fibre wetting and the penetration of the slurry into the fibre preform. Rheological measurements were made using a concentric cylinder viscometer (Contraves Rheomat 115). Flow curves (shear stress, t , versus shear rate, D) were measured with shear rates up to 130 sec^{-1} applied over a 200 second ramp. Viscosity, η, was obtained from the flow curves using the relation: $\eta = t/D$.

Table 1: Powder characteristics

Powder	Producer	Specific surface area (m^2/g)	Mean particle size d$_{50}$ (μm)
Si$_3$N$_4$ (SN-E10)	UBE	10-14	0.55
Y$_2$O$_3$ (grade C)	H.C. Starck	10-16	<1
Al$_2$O$_3$ (RA207LS)	Alcan Chemicals	7	0.5

For the fabrication of the composite, unidirectional high modulus (650 Gpa) pitch-based carbon-fibre (Tonen FT 600) preforms were individually infiltrated with the ceramic slurry in a bath (Figure 1). Infiltrated mats were stacked and consolidated by gas pressure infiltration (0.5MPa). The fibre volume fraction content of the composite was determined by measuring the compaction during pressure consolidation. Green bodies were dried in a dessicator, heat-treated at 400°C in air to remove organic additives, and hot-pressed at 1800°C/27 MPa.
Room-temperature mechanical properties were measured by four-point bending, with an outer to inner span ratio of 30/10 and a crosshead speed of 0.5 mm.min^{-1} on 45 x 4 x 3 mm^3 bars. The morphology of fracture surfaces was examined by SEM.

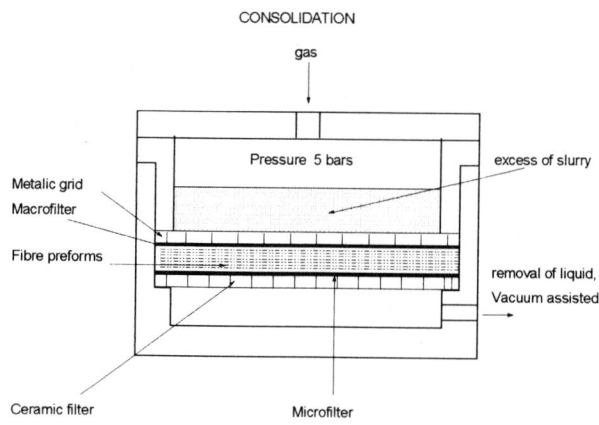

Figure 1. Slurry infiltration process.

RESULTS AND DISCUSSION:

(1) Slip Rheology:
The influence of pH on slip rheology is controlled by the acid/base reaction of Si$_3$N$_4$ (inevitably with a silica surface layer) in water, which is dominated by the hydrolysis of SiO$_2$. The effective surface potential arises from the pH dependent dissociation of the weakly acidic silanol group on the particle surface according to [3]:

$$[-Si(OH)_2^+] \xleftarrow{H^+} [-Si(OH)] \xrightarrow{OH^-} [-SiO^-] + H_2O.$$

Silica is an acidic oxide, so that the right hand reaction is dominant over a large range of pH. Microelectrophoresis measurements show that particles develop an increasingly negative charge as the pH increases over the range of 4 to 10.7 (0 at pH 4 and -60 mV at pH 10.7) [4]. At pH 10.5 the surface charge on the Al_2O_3 and Y_2O_3 is also highly negative with zeta potential values of -15 mV[5] and -20 mV[6] respectively. A pH of 10.5 was selected for further study.

For dispersant effects, Figure 2 shows the rheological behavior (shear rate versus shear stress) at several concentrations of deflocculant (Darvan C) for 65 wt% solid content suspensions adjusted to pH 10.5. At 0.2 wt% Darvan C the suspension is relatively Newtonian, the shear stress was almost linearly proportional to the shear rate. For suspensions with below and above 0.2 wt% Darvan C shear thinning was observed by the decrease in the instantaneous slope of the shear rate versus shear stress curves. Shear thinning [7] is characteristic of a flocculated slip since under low shear rate conditions the liquid is immobilized in the inter-particle voids of the flocs and floc networks. When the shear rate increases the breakdown of the floc networks releases the entrapped liquid. Figure 3 shows the shear rate versus viscosity curves derived from Figure 2. The minimum viscosity conditions (15 cps) at 0.2 wt% Darvan C represents the best dispersion conditions. The viscosity increase for dispersant concentration less than 0.2 wt% is probably due to particle agglomeration while the viscosity increase for concentrations above 0.2 wt% may be attributed to an increase of pore fluid viscosity contributed by the dispersant [8].

Addition of the wetting agent Aerosol OT (0.25 wt%) lowered the surface tension of the water from 73 to 27 dynes/cm [9] without changing significantly the flow behavior (shear rate versus shear stress) of the slurry (Figure 4). The powder remained well dispersed with a viscosity value of 16 cps (Figure 5).

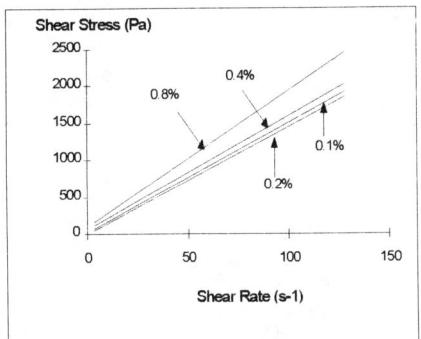

Figure 2. Shear rate versus shear stress for suspensions (65wt% solids content) at pH 10.5 with Darvan C concentrations indicated

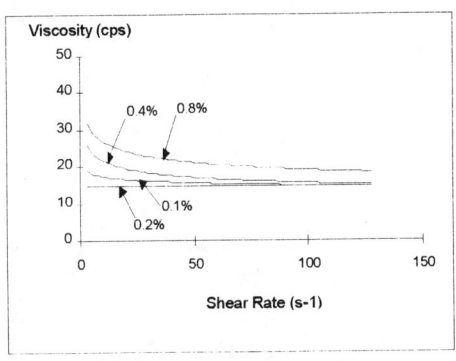

Figure 3. Shear rate versus viscosity for sus--pensions (65wt% solid content) at pH 10.5 with Darvan C concentrations indicated

(2) Consolidation and Hot-pressing:
The pressure filtration technique of Jamet et al. [10] has been adapted for the present work. In a trial experiment monolithic green compacts were fabricated by pressure filtration, packed to a relative density of 0.62. The rate of compaction proceeds with parabolic kinetics, showing uniform packing and allowing ready calibration of the consolidation of the green body. A series of trials produced optimum conditions of applying 0.5 MPa gas pressure for 20 min, to obtain fully infiltrated preforms of 0.55 relative density with 37 vol% fibres.

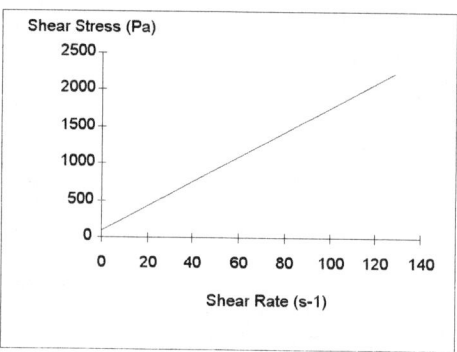 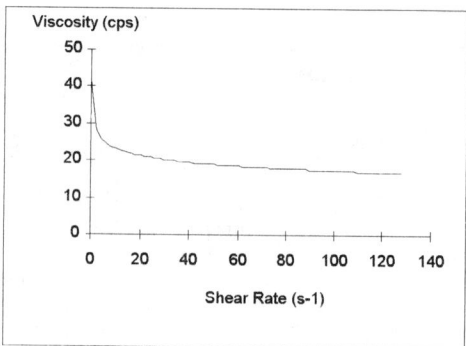

Figure 4. Shear rate versus shear stress for suspensions (65wt% solid content) at pH 10.5 with 0.2wt%Darvan C and 0.25wt% Aerosol OT

Figure 5. Viscosity versus shear rate for suspensions(65wt% solid content) at pH 10.5 with 0.2wt%Darvan C and 0.25wt% Aerosol OT

Hot-pressing of a C-fibre/Si_3N_4 composite above 1750°C is inevitably a compromise between the achievable final density and the risk of chemical reaction between the species with evolution of gaseous N_2. In order to prevent the chemical reaction of carbon fibres with Si_3N_4, Guo et al. [11] decreased the sintering temperature to 1450°C by using lower melting sinter phases (LiF-MgO-SiO_2). Using Y_2O_3 as additive in this study increases the intergranular phase softening temperature but requires a higher sintering temperature (1800°C). However, the chemical reaction of carbon fibres with Si_3N_4,/Y_2O_3/Al_2O_3 combination was significantly reduced by applying pressure to close porosity as soon as secondary grain boundary phases start softening (1300°C), but late enough to avoid damaging of the fibres. The early application of pressure to close porosity inhibits N_2 evolution and escape, inhibiting the decomposition of the Si_3N_4 phase and interphase reaction[12]. Resulting composites reveal a fully infiltrated material with uniform distribution of carbon fibres in a dense Si_3N_4 matrix (Figure 6).

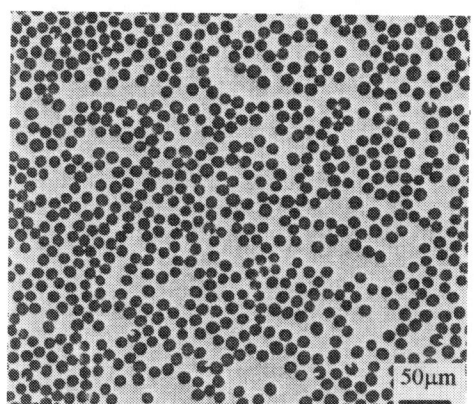 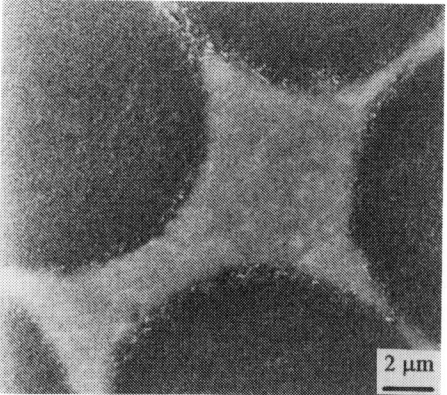

Figure 6. Macrostructure (A) and microstructure (B) showing good infiltration and uniform distribution of carbon fibres in the C/Si_3N_4 composite

4) Mechanical properties:

The quality of the composites prepared by slurry infiltration was assessed from the mechanical properties of typical compositions. Simple composite plate samples were fabricated in two compositions, a simple C/Si_3N_4 and a composite in which the fibres have been precoated with a 30 nm thick B_4C coating (Reactive Chemical Vapor Deposition, University of Villeurbanne, Lyon I, France) in order to evaluate the influence of the coating on composite fracture behaviour, in particular with respect to fracture toughness, energy dissipative mechanisms (fibre debonding, crack deflection and fibre pull-out).

Typical room temperature strength deformation curves are shown Figure 7. The composite with uncoated fibres shows elastic strain followed by a sawtooth extension, characteristic of periodic matrix microcracking with strain and the progressive transfer of the load to the fibres. Finally, the composite fails catastrophically with the simultaneous fracture of the fibres. For the coated fibre composite the load increases linearly up to 240 MPa until the first matrix microcracks are formed. Above this, the curve displays sawtooth extension behavior with progressive microcracking, but at the maximum fracture stress (370 MPa) the compliance changes associated with partial debonding at the interface. Catastrophic fracture is resisted by fibre sliding friction. The failure was graceful and the fracture surfaces (Figure 8) exhibited high amount of pull-out.

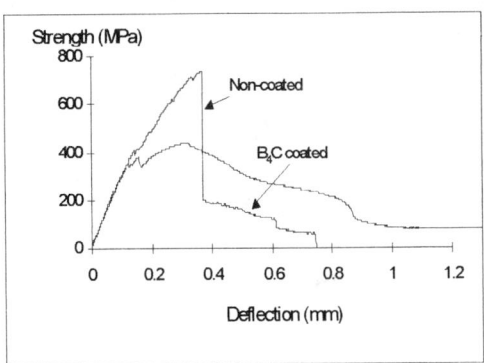

Figure 7. Flexure load-displacement curves of non-coated carbon fibre and B_4C-coated carbon fibre reinforced Si_3N_4

Figure 8. SEM fracture surface of B_4C-coated carbon fibre reinforced Si_3N_4

A 30 nm RCVD B_4C coating on the fibres appears to degrade the fibres themselves, promoting graceful failure of the composite by fibre-matrix debonding and early fibre fracture, with high pulll-out fibre lengths, but to the detriment of composite strength.
CONCLUSION

A technique for the fabrication of long-fibre reinforced carbon/ Si_3N_4 composites by low cost slurry infiltration, followed by Hot pressing has been optimised to produce fully infiltrated composites with uniform distribution of ca. 37% volume fraction carbon fibres in a dense Si_3N_4 matrix. The importance of tailoring the procedure for green forming has been highlighted. Composites fabricated by the technique show excellent room temperature mechanical properties and confirm the value of the technique for experimental research into fracture toughening of fibre composites.

REFERENCES

1. R. Naslain and F. Langlais, "CVD processing of ceramic-ceramic composite materials", Mat. Sci. Res., Vol. 20, Plenum Press New-York, p. 145-64 (1985).
2. F.F. Lange, D.C.C. Lam, O.Sudre, B.D. Flinn, C. Folsom, B.V. Velamakani, F.W. Zok and A.G. Evans, "Powder Processing of Ceramic Matrix Composite," Mat.Sci.Engi. A144 143-52 (1991).
3. J. Lyklema, "Interfacial Electrochemistry of disperse systems," in Emergent Process Methods for High-Technology Ceramics, ed. R.F. Davis, H. Palmour III and R.L. Porter, Plenum Press, New York, p. 1-24, 1984.
4. P. Greil, A. Nagel and H. Stadelmann, "Review: Colloidal Processing of Silicon Nitride Ceramics", Proc. 3rd Int. Symp. On Ceramics in Engines, Las Vegas, Amer. Ceram. Soc., p. 319-29, 1988.
5. P. McFadyen and D. Fairhurst, "Zeta Potentials of Nanoceramic Materials-Measurement and Interpretation" in Nanoceramics, British Ceramic Proceedings No. 51, Ed. by R. Freer, p. 175-86, 1992.
6. J.P. Pollinger "Dispersion and Stabilization of Multicomponent Silicon Nitride Based Aqueous Suspensions", Proc. 3rd Int. Symp. On Ceramics in Engines, Las Vegas, Amer. Ceram. Soc., p. 369-79, 1988.
7. M.D. Sacks, "Properties of Silicon Suspensions and Cast Bodies," Ceram.Bull., 63 [12] 1510-15 (1984).
8. B.E. Novich and D.H. Pyatt, "Consolidation Behaviour of High-Performance Ceramic Suspensions," J. Am. Ceram. Soc., 73 [2] 207-12 (1990).
9. L.I. Osipow, Surface Chemistry, Reihold Publishing Corporation, New York 1964.
10. J. Jamet, D. Damange and J. Loubeau, Fr. Patent 2, 526, 785, November 18, 1983.
11. J.K. Guo, Z.Q. Mao, C.D. Bao, R.H. Wang and D.S. Yan, "Carbon Fibre-Reinforced Silicon Nitride Composites," J. Mat. Sci., 17 3611-16 (1982).
12. K. Luthra and H.O. Park, "Chemical Considerations in Carbon-Fibre/Oxide-Matrix Composites," J. Am. Ceram. Soc., 75 [7] 1889-98 (1992).
13. R.A. Lowden, "Fibre Coatings and Mechanical Properties of Fibre-Reinforced Ceramic Composites," in Whisker and Fibre-Toughened Ceramic, ed. R.A. Bradley, D.E.

REACTION-BONDED Si_3N_4 REINFORCED WITH CONTINUOUS SiC FIBERS: PROCESSING AND INTERFACE CHARACTERISTICS

G. H. Wroblewska and G. Ziegler
Institute for Materials Research (IMA), University of Bayreuth, 95440 Bayreuth, Germany

ABSTRACT

The thermal stability of various SiC fibers was investigated under nitridation conditions of RBSN by thermogravimetry and X-ray diffraction. Powder processing of RBSN was optimized by powder design and nitridation conditions. The influence of various material parameters (particle size, additives, impurities) and H_2 content in nitridation atmosphere on shear strength of the fiber/matrix interface was studied with microcomposites.

INTRODUCTION

Continuous fiber-reinforced composites can be produced by chemical vapor infiltration (CVI), by liquid ceramic precursor technique following a pyrolysis procedure (polymer pyrolysis), by liquid silicon impregnation of C/C preforms (SiC-C/C) and by the powder route. The powder route implies many problems due to the limited fiber content and especially due to thermal degradation and pressure-induced damage of the fibers during sintering, hot-pressing or hot-isostatic pressing. To avoid these problems, reaction bonding was chosen as a promising process because of the relatively low nitridation temperature and the low shrinkage of the reaction-bonded Si_3N_4 matrix. Three topics will be discussed in this paper:
- the thermal and chemical stability of various SiC fibers under nitridation conditions, including the newly developed Hi-Nicalon fibers
- the improvement of powder processing and nitridation process of RBSN reinforced with SiC-fiber bundles, and
- the variation of shear strength of the interface fiber/matrix based on changes of matrix microstructure and nitridation atmosphere.

EXPERIMENTAL PROCEDURE

Thermal stability of various SiC fibers was investigated under nitridation conditions: NL 607, Hi-Nicalon, C-coated (20-30 nm C coating) Hi-Nicalon (Nippon Carbon Co. Japan) and SCS-6 (Avco Specialty Materials, Textron Inc., Lowell

MA). The examinations were performed with a thermal balance (STA 409, Netzsch) in flowing N_2 (<3 vpm O_2, < 5 vpm H_2) or N_2+5 vol% H_2 (<20 vpm O_2, < 20 vpm H_2O) with a flow rate of 125 cm³/min. The fibers were heated up to 1420 °C with a heating rate of 2 °C/min (exposure time at 1420 °C/2 h, cooling rate 10 °C/min). Moreover, the fibers were heat-treated under thermal conditions of the nitridation process (20 °C-800 °C: vacuum; 800-1200 °C: N_2+5% H_2; 1200-1380°C: N_2; 1380 °C: 60 h, N_2 atmosphere). The X-ray analyses of fibers in the as-received and in the heat-treated state were performed by using the X-ray diffractometer Siemens D 5000 with capillary goniometer and position sensitive detector PSD.

The composites were produced by slip casting supported by ultrasonic vibration (suspension of the Si powder in petroleum) and nitrided under the same conditions as the fibers (see above). The influence of various materials (particle size, oxide additives, Fe content) and processing (atmosphere) parameters on the shear strength of the fiber/matrix interface was examined on model specimens by a push-out method [1]. The model specimens consist of a single SiC-SCS-6 fiber in a RBSN matrix.

RESULTS AND DISCUSSION

<u>Thermal Stability of Fibers Under Nitridation Conditions</u>

During the nitridation process the fibers can be damaged due to the temperature-induced crystallization and/or chemical reaction between fiber and matrix or fiber and nitridation atmosphere. Fig. 1 shows the results of thermogravimetric analyses of SiC fibers under N_2 (Fig. 1a) and (N_2+5 vol% H_2) atmosphere (Fig 1b).

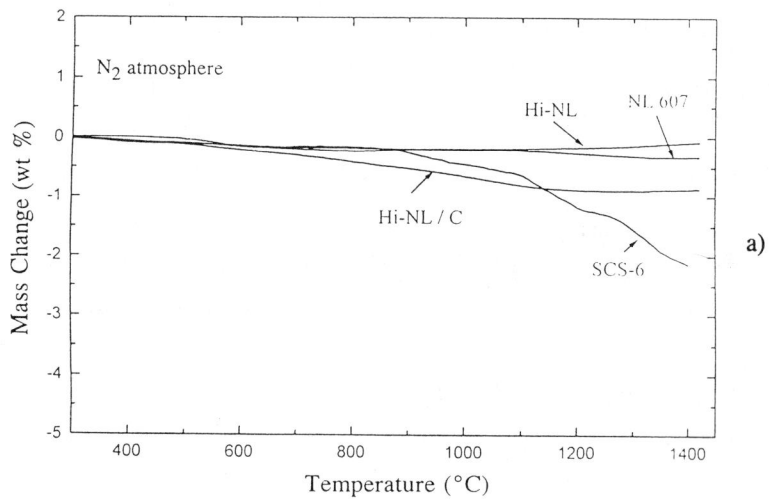

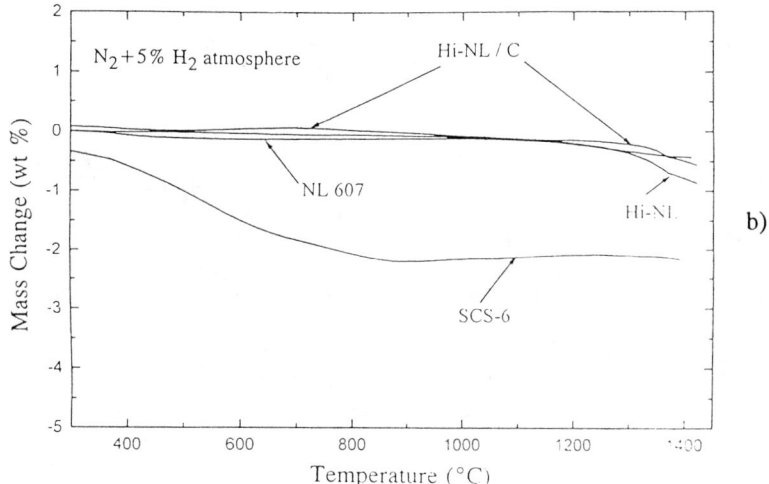

Figure 1. Mass change of various SiC fibers during thermal treatment in a) N_2, b) (N_2+5% H_2) atmosphere.

In N_2 atmosphere the uncoated fibers are more stable than the C-coated fibers under the conditions applied here. The mass loss of the C-coated fibers is obviously caused even by small O_2 impurities of the atmosphere.

In H_2 containing atmosphere the mass changes of the various uncoated and coated Nicalon fibers are rather low up to a temperature of 1420 °C. At high temperatures only small variations were observed between the different fiber types. In contrast, the SCS-6 fibers exhibit already at low temperature a relatively high mass loss (2.3 wt % at 800 °C). This may be caused by the reaction between carbon and oxygen which was present as an impurity in the N_2+H_2 atmosphere.

The X-ray diagram of NL 607 fibers in the as-received state shows peaks of ß-SiC and a high background indicating an amorphous phase (Fig. 2). After 60 h treatment at 1380°C the SiC peaks are getting a bit smaller while the peak intensities increase slightly. This may be explained with the growth of SiC crystallites or/and with an increase of the SiC content.

Different compositions of the NL 607 fibers are given in literature: 55 wt% ß-SiC, 40 wt% $SiO_{1.15}C_{0.85}$, 5 wt% free C and 0.2 wt% H_2 or 57 wt% ß-SiC, 31 wt% $SiO_{1.09}C_{0.45}$ and 12 wt% free C [2]. The O_2 content calculated from these formulae is 13 wt% and 10.6 wt% respectively. During heat treatment at a temperature of about 1400 °C the free carbon can react with the amorphous phase in fibers following the overall scheme [2].

$$0.7\ SiO_{1.15}C_{0.85\ (s)} + 0.42\ C_{(s)} \rightarrow 0.24\ SiO_{(g)} + 0.56\ CO_{(g)} + 0.46\ SiC_{(s)}$$

As a result of this reaction, the mass of fibers decreases and the crystalline SiC content increases, which is in agreement with the results of TG and X-ray examinations. However, the theoretical mass loss of 25 wt% could not be reached because the short exposure time at the temperature of 1400 °C.

The comparison of the NL 607 fibers with the Hi-Nicalon fibers shows that the newly developed fibers are more crystalline in the as-received state, as demonstrated by the relatively small peaks in the X-ray diagram (Fig. 2). After 60 h treatment at 1380 °C the structure of the fibers change only insignificantly. From these results it may be concluded that the crystallization of the Hi-Nicalon fibers is limited during the nitridation process with a maximum temperature of 1380 °C. This may be explained in this fiber type with the higher SiC content and the lower amount of amorphous phase which is related to the low O_2 content (according to the producers information, Hi-Nicalon fibers exhibit only 0.5 wt% O_2).

The SCS-6 fiber consists of a central amorphous carbon core which is coated with turbostratic carbon. The intermediate layer is composed of chemical-vapor-deposited SiC. The SCS-6 monofilament is coated with a 3 μm thick carbon-rich coating [3].

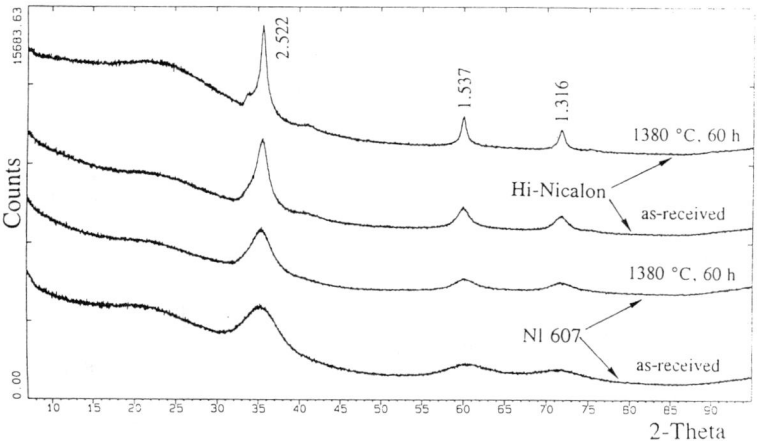

Figure 2. X-ray diagrams of SiC Nicalon fibers in the as-received state and after treatment under nitridation conditions.

Improvement of Powder Processing and Nitridation Procedure

High fiber content and satisfactory homogeneity of the composite material regarding fiber distribution and low-porous microstructure of the RBSN matrix is very important to achieve improved mechanical and thermo-mechanical properties of SiC/RBSN. In order to optimize the processing of composites reinforced with continuous fiber bundles, theoretical considerations were performed under the aspect of powder packing around the fibers, assuming cubic and hexagonal fiber packing. As result, for high fiber volume of about 40 vol %, the ratio fiber-fiber distance to fiber radius of about 1, and a high particle packing between the fibers, the ratio of powder particle to fiber radius should be <0.05. This means, in the case of SiC-Nicalon fibers with a diameter of 14 µm the Si-powder particles should be smaller than 0.7 µm. These fine powders can be obtained by special powder milling [4] or by chemical routes. It was expected that small particle sizes additionally affect the interfacial characteristics fiber/matrix (see next chapter).

A further aim in improving processing of SiC-reinforced RBSN is to reduce the nitridation temperature because of the limited thermal stability of the SiC fibers. This may be achieved by using again small Si particles as starting material.

Another possibility is to optimize the nitridation atmosphere by adding H_2. The (N_2+H_2) atmosphere could be used because during nitridation no chemical degradation of the fibers was observed [4]. The nitridation of Si in H_2-containing atmosphere is very intensive. For example at 1300 °C the mass increase in $N_2+5\%H_2$ is 43 wt% compared to only 12 wt% in N_2 atmosphere. The nitridation process at temperatures lower than 1380 °C is beneficial for the fiber stability, however, the matrix is not fully nitrided and exhibits still some metallic Si which may react with the C-coating of fibers. As a result, this in principle reasonable procedure can only be applied if well-suited coatings are available.

Characteristics of Fiber/Matrix Interface

As well known, the interfacial characteristics control the macroscopic mechanical properties. In this paper various parameters were investigated which may affect interfacial characteristics. In this scope the influence of Si-powder particle size, chemical composition (impurities), additives and nitridation atmosphere on the shear strength of the interface fiber/matrix was examined. The investigations were carried out with SCS-6 microcomposites (single fiber in RBSN matrix) by push-out tests [5]. Table 1 shows the results of push-out tests carried out with SCS-6 microcomposites. The shear strength of the interface fiber/matrix in the system investigated is controlled by the particle size of the Si-starting powder.

Table 1. Si-powders characteristics and shear strength of the interface fiber/matrix of model composites (nitridation cycle see Experimental Procedure)

Symbol	Si-powder	Particle size, µm	Fe content wt %	Additives 9 wt % Y_2O_3 3 wt % Al_2O_3	Atmosphere	τ_d* MPa
MC-1	Sic. 4c	2.0	0.09	-	N_2	15.0±3.4
MC-2	Sic. 4c	1.4	0.09	-	N_2	16.4±5.0
MC-3	Sic. 4c	2.9	0.09	Y/Al	N_2/H_2	6.0±2.1
MC-4	Sic. 4c	1.1	0.09	Y/Al	N_2	5.6±1.7
MC-5	SiMP 1	1.3	0.27	Y/Al	N_2	9.2±3.2
MC-6	SiMP 1	0.4	0.27	Y/Al	N_2	18.0±5.8

* mean value from 15-30 push-out tests

The shear strength increases with decreasing particle size. This dependence was observed for composites without additives (MC-1 and MC-2) and with ($Al_2O_3+Y_2O_3$) additives (MC-5 and MC-6). In general, oxide additives decrease the shear strength of the interface fiber/matrix (MC-2 and MC-4). The interface bonding is stronger in composites with higher Fe content (specimens MC-4 and MC-5 differ only in the Fe content). H_2 in the nitridation atmosphere does not influence the shear strength if the temperature range in which H_2 is used, is well adjusted to the fiber stability in (N_2+H_2) atmosphere (MC-3).

CONCLUSIONS

The Hi-Nicalon fibers are more stable under nitridation conditions than the NL 607 fibers. This is obviously due to the high SiC content and to the only small amount of an amorphous phase in the newly developed fiber.

Fiber content and -distribution as well as the RBSN matrix density can be improved by using very fine Si particles. The H_2-containing nitridation atmosphere and the fine powder particles result in a reduction of the nitridation temperature (the latter point is not shown here, see [4]). This result is important for damage-free processing of the SiC/RBSN composites.

The bonding stress of the interface fiber/matrix can be varied between 5.6 and 18.4 MPa by changing the materials and processing parameters. Based on the knowledge of the relationship between processing parameters, interface shear strength and macroscopic properties, the composites reinforced with the continuous SiC-fiber bundles can be optimized.

ACKNOWLEDGEMENT

The authors would like to thank the Deutsche Forschungsgemeinschaft (DFG) for financial support.

REFERENCES:

1. G. Rausch, B. Meier and G. Grathwohl, A Push-out Technique for the Evaluation of Interfacial Properties of Fiber-Reinforced Materials, J. Eur. Ceram. Soc. 10 (1992) 229-235
2. C. Vahlas, P. Rocabois and C. Bernard, Thermal Degradation Mechanism of Nicalon Fiber: A Thermodynamic Simulation, J. Mater. Sci. 29 (1994) 5839-5846.
3. X. J. Ning, P. Pirouz and S. C. Farmer, Microchemical Analysis of the SCS-6 Silicon Carbide Fiber, J. Am. Ceram. Soc. 76 (1993) 2033-2041
4. G. H. Wroblewska and G. Ziegler, Optimization of RBSN Matrices for the Development of Continuous Fiber-Reinforced Composites, 8th CIMTEC, Florence, July 1994
5. G. H. Wroblewska and G. Ziegler, Investigations of Interfacial Bonding and Friction in SiC/RBSN Model-Composites by Push-out Technique, J. Am. Ceram. Soc., to be submitted

MECHANISM OF INTERFACE FORMATION IN A SILICON CARBIDE FIBER-REINFORCED MAGNESIUM ALUMINOSILICATE

Atul Kumar and Kevin M. Knowles[*]
Department of Mechanical Engineering, Naval Postgraduate School, Monterey, CA 93943, U.S.A., [*]University of Cambridge, Department of Materials Science and Metallurgy, Pembroke Street, Cambridge, CB2 3QZ, U.K.

ABSTRACT

The formation of sliding interfacial layers is a major key to the success of fiber-reinforced glass-ceramics. This paper reports the mechanism of formation of fiber-matrix interfaces during oxidizing heat treatments in a SiC fiber-reinforced magnesium aluminosilicate.

1. INTRODUCTION

A variety of fiber-reinforced glass-ceramics have been developed over last two decades in an attempt to improve the performance of structural ceramics for high temperature applications (1). In the literature it has been shown that a carbon-rich interlayer in SiC (Nicalon type, Nippon Carbon Company of Japan) fiber-reinforced glass-ceramics is generally associated with good mechanical properties, particularly damage tolerance (2,3). It has been suggested that the carbon-rich interfacial layer forms on oxidation of Nicalon SiC fibers during hot-pressing of composites according to the following reaction (4,5):

$$SiC(s) + O_2(g) = SiO_2(s) + C(s) \qquad (1)$$

This reaction is thermodynamically favorable up to 1500°C compared with other oxidation reactions for SiC (4). However, there is no direct evidence in the literature to show that the formation of silica and carbon interlayers occurs on oxidation of these composites.

The aim of the present paper is to describe how the fiber-matrix interfaces are formed during oxidizing heat treatments in a SiC fiber-reinforced magnesium aluminosilicate. As-fabricated samples of a SiC fiber-reinforced magnesium aluminosilicate glass-ceramic were heat treated in air at temperatures between 1000°C and 1200°C. Analytical transmission and scanning electron microscope techniques have been used to study the effects of oxidation on the morphology, structure and chemistry of the fiber-matrix interfaces. The underlying mechanisms of oxidation have been discussed in the light of these experimental observations.

2. EXPERIMENTAL PROCEDURE

The details of material and processing parameters have been reported elsewhere (6). Briefly, the matrix contained 22.3% MgO, 21.3% Al_2O_3, 53.4% SiO_2 by weight with small amounts of P_2O_5 (2.0 wt%) and B_2O_3 (1.0 wt%). SiC (Nicalon, NL 202, Nippon Carbon Company of Japan) fibers were used as reinforcement. The composite plates were produced by hot pressing composite prepregs at 920°C in argon which was followed by ceraming in air at 1150°C for 1 hour. The fiber volume fraction was 0.40 (6). Oxidation experiments were carried out on samples of sizes 50 x 5 x 3 mm. The samples were polished to 1 μm finish prior to heat treatments. Heat treatments were carried out in a tube furnace in flowing (flow rate ≈ 100 cc/minute) dried air at temperatures between 1000°C and 1200°C for a total time of 120 hours. The fiber-matrix interfaces were analyzed using transmission and scanning electron microscope techniques, parallel electron energy-loss spectroscopy (PEELS) and ultra thin window energy dispersive X-ray analysis (EDS). The details of the techniques and equipment used in this work have been reported elsewhere (6).

3. RESULTS AND DISCUSSION

Examination of polished cross-sections of oxidized samples in optical and scanning electron microscope revealed that the matrix and fibers reacted during the oxidation heat treatments. The extent of oxidation reaction increased with time and temperature of oxidation. Fig. 1 shows polished cross-section of a sample oxidized at 1200°C for 120 hours. The dark rings around the fibers were identified as silica by quantitative EDS analysis in a scanning electron microscope (SEM). The thickness of this layer varied from the edge to the center with the fibers having thick layers generally located near the edge of the sample. In the samples oxidized at 1000°C, this layer was associated only with the fibers near the edge. Radial cracks in the silica layer and debonding at the interface between the fiber and the silica layer are also seen in Fig. 1. The radial crack pattern seen in this micrograph is typical of the cracking caused by the high (β) to low (α) cristobalite phase transformation and would suggest that crystalline silica formed at some stage during oxidation heat treatments. SEM examination of polished longitudinal sections of oxidized samples revealed the presence of pores at the fiber-layer and matrix-layer interfaces. It should also be noted that pores were not observed at the center of the samples. The presence of pores along the interfaces suggests the evolution of gases at the fiber-matrix interfaces.

TEM examination of thin-foils prepared from the regions close to center and surface of the oxidized samples revealed several important interfacial features. A bright field image typical of the fiber-interface-matrix region close to the center of samples oxidized at 1200°C is shown in Fig. 2(a). Four distinct interfacial layers (I1 to I4) are seen within an approximately 1.2 μm thick interfacial region between the matrix and the fiber. EDS spectra from the fiber and these layers are shown in Fig. 2(b) and Figs. 2(c)-(e) respectively. Interlayers I2 and I4 were identical. Interlayers I1 and I2 (and I4) showed an increase in the O/Si ratio compared with that in the fibers. The increased O/Si ratio together with the absence of C peak in interlayers I2 and I4 suggest that these layers contain silicon oxide. Further analysis by PEELS confirmed that these layers consisted of silica. The EDS spectrum from layer I3 shows the presence of C, O and Si (Fig. 2(e)). The C/Si ratio is very high compared with that in the fiber and other interlayers. Since no C was observed in interlayers I2 and I4, the carbon peak can only be from interlayer I3. The observed Si and O peaks are likely to be from the adjacent interlayers I2 and I4. This was confirmed by the presence of the C peak alone in the PEELS spectra from interlayer I3. Further examination of interlayer I3 in a high resolution electron microscope showed a highly entangled network of lattice fringes of turbostratic carbon (7). The formation of the network of the lattice fringes of carbon layers can be described by the progressive coalescence of the basic structure units (BSUs) of turbostratic carbon (8,9).

A different interface morphology was observed in the thin foils obtained from regions very close to the surface of samples oxidized at 1200°C (Fig. 3). Two different interlayers, K1 and K2, are seen between the fiber and the matrix. Interlayer K1 is similar to the interlayers seen next to the fibers in the foils from both the center of the oxidized sample (interlayer I1) and the as-received composites (6). EDS and PEELS analyses of interlayer K2 established that this consisted of pure silica. Some transverse cracks (perpendicular to the fiber axis) are seen in the silica layer (see double arrows). These cracks can be identified as the radial cracks that were seen in the silica layer around the fibers in SEM studies. An additional feature of the interface in Fig. 3 is the absence of a carbon interlayer. Instead, elongated pores and voids lying parallel to the interface between the interlayers K1 and K2 are observed. Some of these pores are bridged by silica. The elongated pores at the interface between interlayers K1 and K2 are likely to arise from oxidation of the carbon layer such as interlayer I3 seen in Fig. 2(a). Furthermore, close examination of interlayer K1 reveals a thin gray layer (see double arrows in Fig. 3) near the pores. EDS analysis of this layer showed a decrease in carbon peak compared with that in the lighter regions within K1, which suggests that this layer is likely to be silica.

The structure and chemistry of the interfacial regions seen in the thin foils obtained from the regions both close to the surfaces and the center of samples oxidized at 1000°C and 1100°C were very similar to those seen in the samples oxidized at 1200°C. However, a silica layer similar to interlayer I4 was not observed in the thin foils obtained from the areas close to the center of the samples oxidized at 1000°C. This is consistent with the SEM observation that no silica was observed around the fibers in the interior of samples oxidized at 1000°C.

4. MECHANISM OF INTERFACE FORMATION

A mixed silica and carbon interlayer was present next to the fibers in the as-received composites (6). The presence of a silica layer adjacent to the matrix (e.g., I4 and K2) in the oxidized composites suggests that this layer forms by oxidation of the carbon which diffuses from the mixed interlayer towards the matrix and/or a free surface. Since the equilibrium partial pressure of oxygen for the oxidation of carbon to CO (by the reaction $C + 1/2\ O_2 = CO$) is higher than that for the oxidation of SiC to silica and carbon (by reaction (1), up to $\approx 1500°C$), it is expected that the fibers will also be oxidized to form more silica and carbon. This is supported by the presence of the mixed silica and carbon interlayer next to the fibers (e.g., I1 and K1) in all the oxidized samples, irrespective of the region from where the thin foils for TEM examination were obtained.

Elongated pores and voids at the interface between the silica layer adjacent to the matrix and the mixed silica and carbon interlayer next to the fibers (see Fig. 3) were observed in TEM examination of near-surface regions of oxidized samples. The presence of these pores can be attributed to the oxidation of the carbon interlayer to CO gas. If this is so, a carbon layer must first form at the interface between the silica layer and the mixed silica and carbon interlayer. Carbon enrichment at this interface can occur from the diffusion of carbon from the mixed interlayer towards the silica interlayer. The driving force for this is the concentration gradient of carbon between the two interlayers. The stability of carbon at this interface will now depend on the availability of oxygen at the interface between carbon and silica. If the partial pressure of oxygen at this interface is higher than the equilibrium partial pressure for the oxidation reaction of carbon ($C + 1/2\ O_2 = CO$), carbon will oxidize to CO gas and subsequently pores and voids may form at this interface provided that the rate is limited by the diffusion of CO away from the interface. In addition, the enrichment of carbon at the interface between the mixed interlayer and the pure silica layer is expected to give rise to a silica-rich layer within the mixed interlayer. Under these conditions, an interface morphology such as fiber–silica+carbon–silica–pores–silica–matrix will be expected. Such an interface morphology has been observed during TEM examination of areas close to the surface of the oxidized samples.

On the other hand, if the partial pressure of oxygen is lower than the equilibrium partial pressure for the oxidation reaction of carbon to CO gas, the carbon interlayer will remain stable at the interface between the mixed and the silica interlayers. The continuous oxidation of the fibers will again form a mixed silica and carbon layer and an interface morphology of the type fiber–silica+carbon–silica–carbon–silica–matrix will be expected. This interface morphology was observed during TEM examination of areas close to the center of the samples oxidized at 1100°C and 1200°C. An interface morphology of the type fiber-silica+carbon-silica-carbon-matrix was observed in the areas close to the center of the samples oxidized at 1000°C. The absence of the silica interlayer adjacent to the matrix in these samples can be attributed to the incomplete oxidation of carbon which diffuses towards the matrix from the mixed silica and carbon interlayer.

It is evident from the preceding discussion that the morphological evolution of the interfacial layers is governed by the local oxygen flux. Therefore, the presence of interconnected porosity, cracks and the diffusion of oxygen along the interfaces at the exposed ends of fibers (the so-called pipe-line oxidation, cf. ref. 10) may change the interface morphologies by changing the local oxygen flux.

5. Conclusions

The fibers oxidized to form silica and carbon which were present in an aggregate morphology. Subsequent oxidation of this layer and the fibers gives rise to complex interface morphologies. It has been demonstrated that the interface morphology varies with temperature and from one region to another because of the variation in the local oxygen potential within the composite. This implies that the interface developed during hot-pressing of composites will depend upon the temperature, time and the environment. A study of interfacial regions as a function of time and temperature of oxidation in combination with rigorous thermodynamic modeling is therefore required to be able to understand fully the morphological evolution of interfaces in these composites.

Acknowledgements

We would like to thank Pilkington plc for the provision of samples of fiber-reinforced glass-ceramics. AK would like to thank the Cambridge Commonwealth Trust, Cambridge, U.K. for financial support.

References

(1) K.M. Prewo, J.J. Brennan and G.K. Layden, "Fiber Reinforced Glasses and Glass-Ceramics for High Performance Applications," *Am. Ceram. Soc. Bull.*, **65** [2] 305-13 (1986).
(2) J.J. Brennan, and K.M. Prewo, "Silicon Carbide Fibre Reinforced Glass-Ceramic Matrix Composites Exhibiting High Strength and Toughness," *J. Mat. Sci.*, **17**, 2371-83 (1982).
(3) J. J. Brennan, "Interfacial Chemistry and Bonding in Fiber Reinforced Glass and Glass-Ceramic Matrix Composites, " pp. 387-99 in *Ceramic Microstructure'86* . Edited by J.A. Pask and A.G. Evans. 1986.
(4) R.F. Cooper and K. Chyung, "Structure and Chemistry of Fibre-Matrix Interfaces in Silicon Carbide Fibre-Reinforced Glass-Ceramic Composites: An Electron Microscopy Study," *J. Mat. Sci.* **22**, 3148-60 (1987).
(5) L.A. Bonney and R.F. Cooper, "Reaction-Layer Interfaces in SiC-Fiber-Reinforced Glass-Ceramics: A High-Resolution Scanning Transmission Electron Microscopy Analysis," *J. Am. Ceram. Soc.* **73** [10] 2916-21 (1990).
(6) A. Kumar and K.M. Knowles, "Microstructure-Property Relationships of SiC Fibre–Reinforced Magnesium Aluminosilicate: Part I: Microstructural characterisation," submitted to *Acta. Metall.*

(7) A. Kumar and K.M. Knowles, "High temperature thermo-mechanical behavior of a SiC fibre-reinforced magnesium aluminosilicate: Part-I Oxidation kinetics and mechanism of interface formation" submitted to *J. Am. Ceram. Soc.*
(8) A. Oberlin, "High-Resolution TEM Studies of Carbonisation and Graphitisation"; pp. 1-143, in Chemistry and Physics of Carbon, Vol. 22. Edited by P. A Thrower, Marcel Dekker, New York, 1989.
(9) J.N. Rouzaud and A. Oberlin, "Structure, Microtexture, and Optical Properties of Anthracene and Saccharose-Based Carbons," *Carbon,* **27** [4] 517-29 (1989).
(10) M.W. Pharaoh, A.M. Daniel and M.H. Lewis, "Stability of Interfaces in Calcium Aluminosilicate Matrix/Nicalon SiC Fibre Composites," *J. Mat. Sci. Let.,* **12**, 998-1001 (1993).

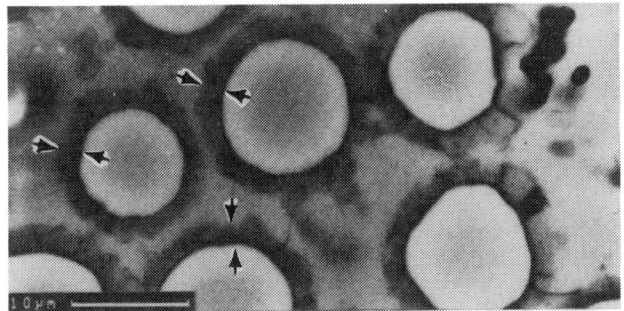

Figure 1: A backscattered electron image showing silica layers (arrowed) around the fibers in the polished cross-section of a sample oxidized at 1200°C. Radial cracks within the silica layer and debonding at the fiber-matrix interface are also seen.

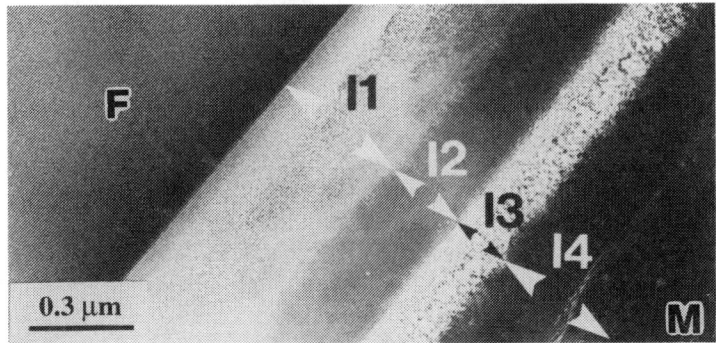

Figure 2: (a) Bright field TEM image of fiber-interface-matrix region close to the (cont.)

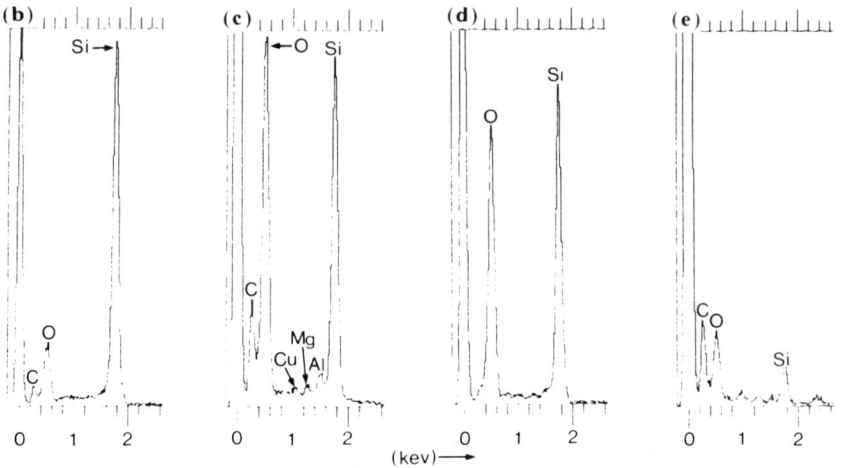

Figure 2: (cont.) center of a composite sample oxidized at 1200°C, (b)-(e) EDS spectra from fiber, interlayers I1, I2/I4 and I3 respectively. Note that EDS spectra are plotted on the same full-scale intensity. F and M refer to fiber and matrix respectively.

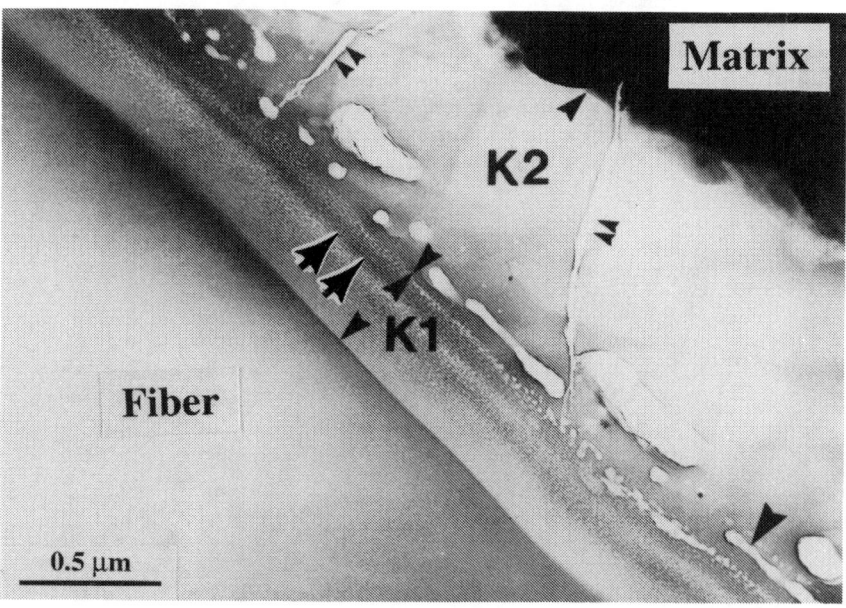

Figure 3: Bright field TEM image of fiber-interface-matrix region close to the surface of a sample oxidized at 1200°C. The arrowed regions are referred to in the text.

FILAMENT WINDING OF SiC$_{fibre}$/Si$_3$N$_4$ COMPOSITES - PROCESS AND CONTROL OF FIBRE DISTRIBUTION

A. Kristoffersson, E. Laarz, R. Carlsson
Swedish Ceramic Institute, Box 5403
S-402 29 Göteborg, Sweden

R. Lundberg
Volvo Aero Corporation, Advanced Materials
S-461 81 Trollhättan, Sweden

ABSTRACT

Filament winding of a model system of SiC$_{fibre}$/Si$_3$N$_4$ has been studied with regard to the microstructure. Processing parameters such as speed, fibre roving tension and spacing, together with slurry properties have been varied. Fibre rovings were completely impregnated in the immersion step but voids formed in the winding step. Fewer voids formed for the water-based slurries compared with slurries with petroleum spirit. Layered structures were formed when a low winding speed was used as the surface dried before the next roving was applied. This type of inhomogeneity can be quantified using image analysis.

INTRODUCTION

Filament winding is a technique that has been used mainly for producing polymer composites but recently also for the manufacture of ceramic matrix composites. The technique is especially suited for rotational symmetric shapes such as rocket nozzles and gas turbine combustor cans. Whilst the knowledge and process control of polymer filament winding are extensive (1), there still remain areas to be investigated and understood for ceramic filament winding.

The technique comprises several steps. The fibre roving is passed through a ceramic slurry and is thereby impregnated and coated with the slurry. The fibre roving is then wound onto a rotating mandrel. After the winding is completed the component is dried and released from the mandrel. Filament winding is a relatively complex technique and several parameters can be varied. The winding speed and the fibre tension have a large influence on the final result. The fibre angle and the lay up of the fibres can vary from simple cylindrical to intricate interlocking helical patterns. The slip properties are also important.

In this work, viscosity, viscoelastic behaviour and solids content of the slip, and parameters such as winding speed, fibre tension and their effect on the microstructure were studied. A model system with a Si$_3$N$_4$ matrix and SiC fibres (Nicalon NL 607) was used. The rheology of slips with different solids content giving different viscoelastic behaviour was studied with a controlled stress rheometer. The filament winding was performed at varied speeds and fibre tensions. The microstructures of the composites were studied with regard to the fibre volume fraction and fibre distribution. SEM/image analysis and optical microscopy were used to characterize the porosity, fibre volume fraction and fibre distribution.

EXPERIMENTAL PROCEDURE

A model system with a Si_3N_4 matrix and carbon-coated SiC fibres was chosen. This system can be sintered by HIPing at reasonably low temperatures (1550°C) without extensive reactions between fibre and matrix (2). Silicon carbide fibres, grade Nicalon NL-607 from Nippon Carbon Company, Ltd., Japan were used as fibre material. A Si_3N_4 powder, UBE E05, with a narrow particle size distribution and a mean particle size of 0.4 µm was chosen to reduce the crack sensitivity of the material in the drying stage (3). As sintering aids an Al_2O_3 powder, Sumitomo AKP 30, and an Y_2O_3 powder, HC Starck fine, were used. The powder composition was 92 wt% Si_3N_4, 6 wt% Y_2O_3 and 2 wt% Al_2O_3 in all experiments. Two different solvents were used, water and petroleum spirit, of which petroleum spirit has a lower surface tension which gives reduced capillary pressure during drying (3). The dispersants were a polyacrylic acid, Dispex A40, for the water-based slips and KD3 from ICI Chemicals for the solvent-based slips. The slips were prepared by ball milling.

The rheological properties of the slips with different solvents and varied solids content were characterized with a controlled stress rheometer (Stresstech Rheometer, Reologica Instruments Sweden). The viscosity was measured over a shear rate interval of 0.04 to 462 s^{-1}. An oscillation stress sweep at 1 Hz from a stress of 0.026 to 0.300 Pa was used to characterize the viscoelastic properties in the linear viscoelastic region and at higher stresses when the structure within the slip is broken down (4). Finally, the thixotropic behaviour or time-dependent structure formation in the slurry was studied in an oscillation experiment. The slurry was sheared at high stress, 10 Pa, for 2 min and then oscillated at 1 Hz and 0.1 Pa while the change in storage modulus with time was recorded. The measurement conditions were similar to those used during the process, with stirring in the impregnation step of the roving followed by low stresses in the winding step.

Impregnation experiments of the fibre rovings at varied winding speeds and fibre tensions for different slurries were performed. Rovings were pulled through the slurry, kept stirring at 5 s^{-1}, but not wound onto the mandrel. The slurry take up and any remaining voids were studied.

In the filament winding experiments processing parameters such as winding speed, fibre tension and spacing were studied together with different types of slurries, i.e. varied solids content and solvent. Cylinders, 50 mm in diameter and 40 mm in length, were wound with a 90-degree angle. A numerically controlled filament winding machine (type 4-axis Josef Baer) at SICOMP (Swedish Institute of Composites) at Piteå, Sweden was used.

The microstructures of the wound composites were studied with optical microscopy and SEM. Prior to the sample preparation the composites were presintered at 1350°C for 1 h in a N_2 atmosphere. The microstructures of the composites were studied with regard to the overall structure, the pore content, the pore size and the homogeneity of fibre distribution and fibre content. The pore content and fibre distribution were analysed with image analysis.

RESULTS AND DISCUSSION

The results from the rheological measurements are presented in Table 1. All of the slurries had a pseudoplastic behaviour. The degree of pseudoplasticity was higher for the slips with petroleum spirit and increased with the solids content. As can be seen in Table 1 the viscosity level was lower for the water-based slip compared with the petroleum-containing slips. As expected the viscosity increased with increasing solids content. The same trend was seen for the complex modulus, G^*, and storage modulus, G' in the linear viscoelastic region. The time needed, after stirring, for structure build up, in this case the time until the storage modulus reached a constant value, decreased with increasing solids content.

Table 1. Summary of the rheological properties of the slurries.

Rheological properties	Slurries, type of solvent and solids content in vol%			
	Water 39.4%	Petroleum 35.4%	Petroleum 40.0%	Petroleum 44.0%
Viscosity at 0.1 s^{-1} (Pa·s)	0.28	2.3	3.0	13.3
Viscosity at 5 s^{-1} (mPa·s)	50	130	196	620
G^* (Pa)	0.5	8.0	10.7	35
G' (Pa)	0.3	7.6	10.4	33
Phase(°)	53	19	19	18
Time (s)	> 300	> 300	130	50

In the impregnation experiments the fibre roving was pulled through a slurry bath at varied speed and fibre tension using different slips. The slurry take-up increased with increasing speed, slip viscosity and decreased fibre tension. At a winding speed exceeding 120 mm/s there were indications that the take-up had reached a plateau level. Cross sections of dried pre-sintered rovings showed almost no large pores irrespective of slurry type and speed (Figure 1). For low fibre tensions < 0.5 N pores were more frequent and the cross section shape was more irregular. The cross sections from the water-based slips were circular in shape probably because of the lower viscosity and the higher surface tension of these slips.

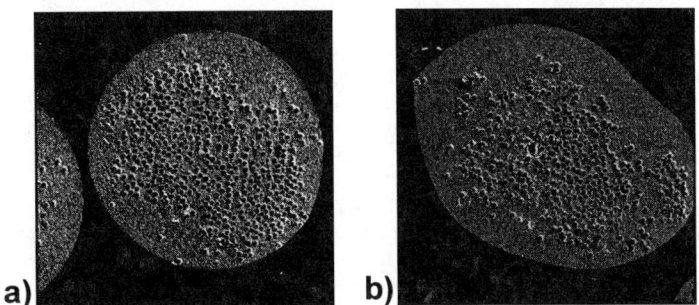

Figure 1. SEM-images of fibre roving cross-sections wound at 24 mm/s
a) water-based slip, and b) petroleum spirit slip.

In the composites from the filament winding experiments the total fibre content in the green state varied between 23 and 49 vol%. The fibre content of the filament-wound composites increased with increasing winding speed (Figure 2). This result is the reverse of the result from the impregnation experiments and can be explained by a dripping of the slurry at winding speeds exceeding 21 mm/s. During winding a portion of the slurry was squeezed out to the surface. The higher the winding speed the thicker the layer, because of higher slurry take-up at higher winding speeds. At a critical thickness rotational and gravitational forces exceeded the cohesive forces and the slurry started to drop from the cylinder. However, consistent with the impregnation results the fibre content decreased with increased viscosity of the slurry.

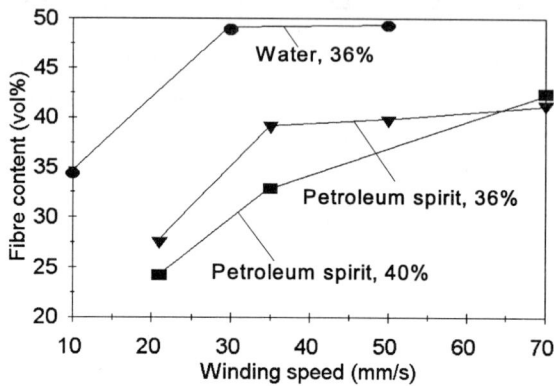

Figure 2. Fibre content of composites wound with 1 mm spacing/revolution at varied speeds for different slurries with solids content in vol%.

In an experiment in which both speed and fibre tension were varied no significant effect on fibre content of fibre tension in the range 0.1-1.0 N could be seen.

The amount of larger pores in the composites was studied by image analysis at 150× magnification of 20 squares 0.36 mm^2 in area (Table 2). Pores smaller than 27 μm^2 were discriminated. This method for measuring the porosity gave a relatively large scatter and therefore the results should be looked upon as trends. However, the technique is simple and can be automated so that large areas can be covered. The amount of porosity was in the 5 to 14% range. The porosity increased with increasing winding speed and water-based slips gave less porosity compared with the ones with petroleum spirit. This can either be an effect of the lower viscosity of the water-based slips and/or the more regular shape and size of the impregnated rovings. The major part of the porosity was to be found between two roving layers and was probably formed when a roving met a wet layer. Air was then entrapped which created voids. The more elastic behaviour of the slip, the harder the voids were to remove. Pores were formed to a lesser degree when the wet roving met a dried surface. This was probably a result of better spreading because of capillary forces.

At low winding speeds layered composites were formed as the top layer had time to dry before the next wet roving was applied. Inhomogeneity was also formed due to pores between fibre layers.

Table 2. Porosity in samples measured by image analysis.

Slurry	Porosity at different winding speeds (%)				
	21 mm/s	30 mm/s	35 mm/s	50 mm/s	70 mm/s
Petroleum, 36%	8±3			14±3	12±4
Petroleum, 40%	9±5		7±3		13±4
Water		5±2			5±3

The homogeneity of fibre distribution was measured by image analysis. A circular section of 512 pixels in diameter with a line mesh with 1 pixel line width was superimposed on binary images of wound cross sections in 150× magnification. The area fraction of fibres in each line was detected. The line mesh was rotated in 15° steps from 0 to 180°. For each angle the mean area fraction and its standard deviation were calculated. A plot of normalized standard deviation versus the angle reflects the degree of homogeneity of the investigated microstructure. The homogeneity within a sample varied and tended to be higher in the middle of the sample compared with the edges, as depicted in an example below (Figure 3). With this technique the homogeneity within a sample and between composites can be compared. It can also give an idea of what type of inhomogeneity is present, and it was especially suited to detect layered structures.

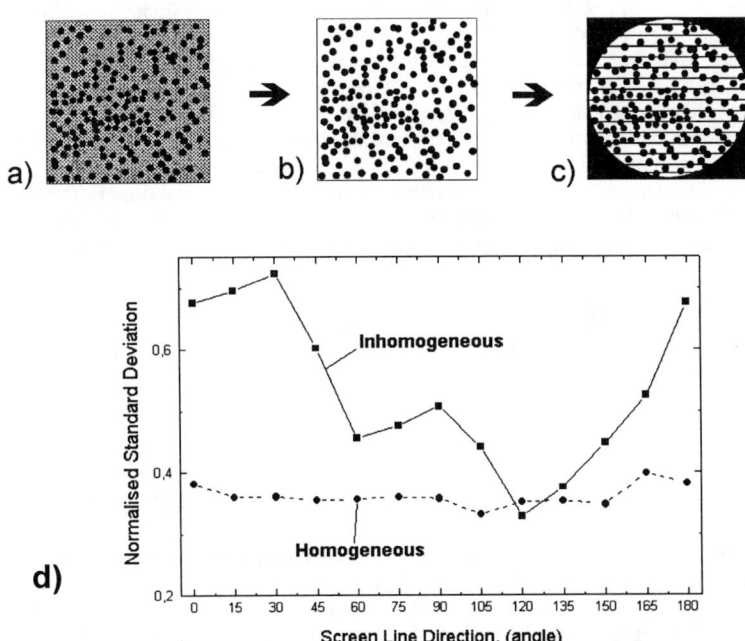

Figure 3. Schematic drawing describing the fibre distribution measurement and the resulting curve of homogeneity; a) SEM-image, b) binary image, c) binary image with screen mesh at 0° angle, and d) fibre distribution measurement of an inhomogeneous area at the edge and a homogeneous area in the middle of the composite (sample PS 36% 21 mm/s).

To obtain a reproducible process the winding speed should be kept low to avoid dripping due to excessive slip and rotational and gravitational forces. The formation of a layered structure is a problem when using a low winding speed but this can be minimized if several rovings are wound at the same time. One way of increasing the winding speed could be the use of a solvent with a higher vapour pressure. A higher winding speed could be balanced by a more rapid drying. If excessive slip is gathered at the surface without dripping off this can cause a cracking problem due to differential shrinkage, especially for the water-based slurries.

In filament winding of polymer composites excess resin is removed from the roving after the impregnation step by a set of squeeze rollers or a small orifice (1). The impregnation equipment used in these experiments was not equipped with this type of wiping device. By wiping off excess slip with this technique several advantages can be foreseen, such as better control of the band width and flatness of the roving, higher winding speed and better control of the fibre content. The fibre tension that could be used with the present impregnation equipment was limited to approximately 2 N. A higher fibre tension can have a positive influence on the pore formation in the winding step.

CONCLUSIONS

Filament winding is a promising technique for manufacturing of rotational symmetric ceramic long fibre composites. Rovings were completely impregnated in the immersion step but voids were formed in the winding step. Fewer pores were observed in the composites from water-based slips which had a lower viscosity compared with petroleum-based slips. The amount of pores increased with increasing winding speed for the petroleum-based slips. Layered structures were formed at a low winding speed as the surface dried before the next roving was applied. When a high winding speed was used dripping of slurry occurred which resulted in a high fibre volume fraction. Future work should be directed towards minimising the pores formed in the winding step and a better control of the fibre volume fraction. Both the porosity and the fibre distribution were quantified by image analysis. This can be a useful tool in the development of the process since process parameters can be correlated to the microstructure.

ACKNOWLEDGEMENTS

We would like to thank Kurt Olofsson and Runar Långström of SICOMP for sharing with us their expertise on filament winding. We are also indebted to Lars Eklund for his image analysis work.

REFERENCES

1. Mallick, P.K., Fiber-Reinforced Composites, 2nd edition, 393-403, Marcel Dekker Inc., New York 1991.
2. Lundberg, R., Pompe, R., Carlsson, R., "HIPed Carbon Fibre Reinforced Silicon Nitride Composites", Ceram. Eng. Sci. Proc., **9**[7-8] 901-906 (1988).
3. Scherer, G.W.,"Theory of Drying", J. Am. Ceram. Soc., **73** [1] 3-14 (1990).
4. Barnes, H.A., Hutton, J.F., Walters, K., An Introduction to Rheology, 1st edition, Elsevier Science Publishers B.V., Amsterdam 1989.

INTERACTION BETWEEN CAPILLARY FLOW AND MACROSCOPIC SILICON CONCENTRATION IN LIQUID SILICONIZED CARBON/CARBON

Frank H. Gern

German Aerospace Research Establishment, DLR
Institute of Structures and Design
Pfaffenwaldring 38–40
70569 Stuttgart, Germany

ABSTRACT

This article describes a model for the numerical simulation of liquid silicon infiltration into porous carbon/carbon preforms. Macroscopic silicon concentration has been calculated from capillary flow equations. As a result, time dependence of silicon concentration during infiltration as well as silicon distribution in the ceramic end product can be calculated. Simulation values of silicon concentration after infiltration are in good accordance with experimental measurements.

1. INTRODUCTION

Ceramic matrix composites are promising materials for the realization of hypersonic aircraft and re-entry vehicle hot structures. Mainly liquid silicon infiltration is a fast and low-cost manufacturing process for structural components based on carbon/silicon carbide composites. Showing great potential with regard to short manufacturing cycles now cost effective production of other high performance ceramic composites like brake discs seems to be possible.

As a key point of the entire manufacturing route, process optimization necessitates profound understanding of infiltration dynamics of liquid silicon into the porous carbon/carbon preform. Due to high furnace temperatures and vacuum conditions direct observation of silicon infiltration seems to be impossible. Therefore, numerical process simulation is the only way to find closer access to infiltration dynamics and, in addition, allows variation of process parameters for optimization.

2. MICROSTRUCTURE OF THE POROUS PREFORM

The infiltration of liquid silicon into porous carbon/carbon preforms is a fast CMC manufacturing process. Being well-described in literature the in-house developed fabrication route can be devided into three process steps [1, 2]:
- CFRP manufacturing of the preform by RTM or autoclave technology,
- conversion of the polymeric matrix into carbon by pyrolysis thus forming a porous carbon/carbon preform,
- ceramization of the matrix by melt infiltration of liquid silicon and formation of the ceramic end product called C/C-SiC.

The utilization of polymeric precursors with high carbon yields leads to the formation of a characteristic microstructure of the porous carbon/carbon preform after pyrolysis showing discrete translaminar capillary channels (Fig. 1). In these channels capillary force driven fluid transport allows good penetration of the preform by liquid silicon and a fast infiltration may be realized.

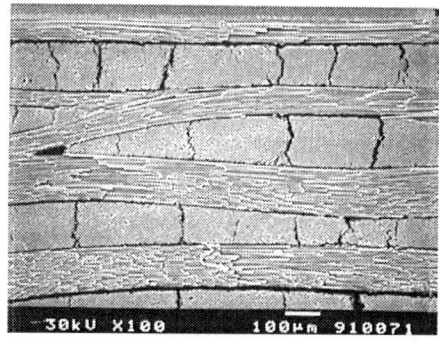

Figure 1: Porous microstructure of the carbon/carbon preform

The high geometrical orientation of the capillary channels leads to a significant orthotropic infiltration behaviour of the preform. Therefore, common models for the description of infiltration of porous media like the diffusion model [3] based on stochastic filling mechanisms of pore capillaries are not suitable in this case.

3. SIMULATION OF CAPILLARY FLOW

For all fluid systems the description of flow necessitates a solution of the – in this case onedimensional – Navier Stokes equation:

$$\rho \frac{du}{dt} = \rho \left(\frac{\partial u}{\partial t} + u \cdot \nabla u \right) = -\nabla p + \eta \nabla^2 u + \rho F. \tag{1}$$

Taking into account capillary, mass and friction forces the basic equation of motion in terms of infiltration length h is defined by the nonlinear differential equation:

$$\ddot{h} = \frac{4\sigma \cos \Theta}{\rho \, v_w \, d_K} \cdot \frac{1}{h} - \left(\frac{64\eta}{\rho \, d_K^2} \cdot \frac{h}{\dot{h}} + 2.32 \right) \frac{v_w}{2} \cdot \frac{\dot{h}^2}{h} - \frac{g}{v_w} \, . \tag{2}$$

For simulation of infiltration dynamics of a single capillary system a solution of equation (2) has to be found. The first contact of molten silicon leads to the immediate formation of silicon carbide layers at the capillary walls [4]. Due to this fact chemical reaction of silicon carbide formation is mainly controlled by diffusion of silicon atoms through the SiC layer into the reaction zone. Therefore, after initial silicon carbide formation, chemical reaction is negligable and capillary diameters d_K may be considered constant for short infiltration times [5].

Inserting the physical properties of molten silicon numerical solution of equation (2) is possible for different capillary diameters d_K (Fig. 2). For this integration a constant diameter along the entire length of the capillary channel has been assumed. As shown in fig. 1 this assumption is only true for an idealized single capillary system. However, in case of nonuniform capillary diameters, as they are found in reality, energy dissipation effects at variable cross sections can be neglected for low microscopic infiltration velocities i.e. viscous flow. For instance, Reynolds number Re_d has to be less than 2 for each capillary diameter [6]:

$$Re_d = \frac{v_w \, d_K \, \rho_{Si}}{\eta_{Si}} \cdot \dot{h} < 2 \, . \tag{3}$$

As shown in fig. 3, this requirement is fulfilled for each capillary diameter at any time during silicon infiltration.

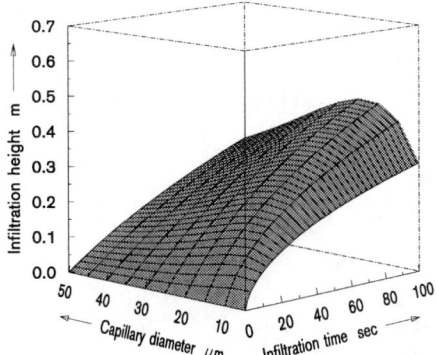

Figure 2: Infiltration dynamics of the single capillary system

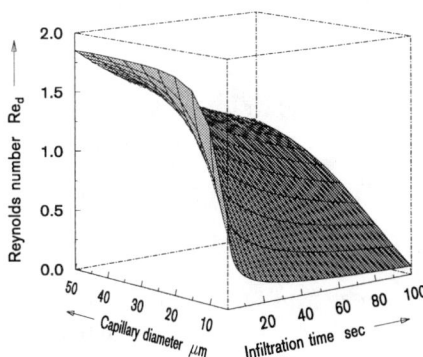

Figure 3: Reynolds number distribution of the single capillary system

4. MACROSCOPIC SILICON CONCENTRATION

Infiltration dynamics of entire parts like plates, tubes or other structural components are not only influenced by the infiltration behaviour of the single capillary system but also by interaction of all the capillaries forming the porosity of a component. One main characteristic of this totality of capillary channels is given by the statistical frequency p_i of each diameter $d_{K\,i}$ in the system.

Since it is of great evidence to have a statistical distribution function of high reliability, detailed microstructural investigations of the porous carbon/carbon preforms have been carried out [7]. In addition, mathematical error correction of experimental measurements using least square approximation was conducted (Table 1).

With energy dissipation at variable cross sections being negligable, capillaries may be divided into discrete segments of constant diameters. It is now possible to establish three capillary systems of different microstructure but all showing an identical macroscopic infiltration behaviour (Fig. 4). In case of three dimensional interconnection local blockage of small capillaries due to silicon carbide formation is bridged.

Table 1: Statistical frequencies of capillary diameters

$d_{K\,i}$ [μm]	p_i [%]	$d_{K\,i}$ [μm]	p_i [%]
5.0	8.5	30.0	5.3
10.0	27.4	35.0	2.6
15.0	27.2	40.0	2.0
20.0	17.0	45.0	1.0
25.0	8.0	50.0	1.0

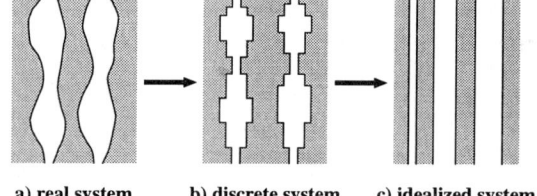

a) real system   b) discrete system   c) idealized system

Figure 4: System idealization for calculation of silicon concentration

Macroscopic silicon concentration in the idealized system can now be calculated using the infiltration lengths of the single capillary system (Fig. 2) and the statistical frequency p_i of each capillary diameter $d_{K\,i}$ (Table 1):

$$\frac{V_{Si\,(t,h)}}{V_{Porosity\,C/C}} = \frac{\sum\limits_{i=1} p_i \cdot d_{K\,i}}{d_K}. \qquad (4)$$

Time dependence of silicon concentration in the porous carbon/carbon preform has now to be calculated by numerical integration of equation (2) for each capillary diameter $d_{K\,i}$ and summation in correspondence to the appropriate statistical frequency p_i. As a result, silicon concentration in the preform during infiltration is shown in fig. 5 for integration steps of 1 sec.

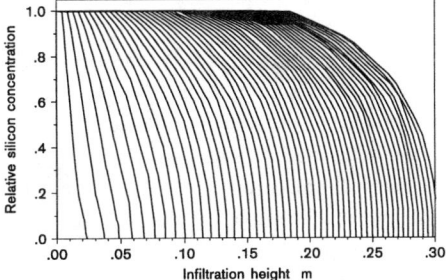 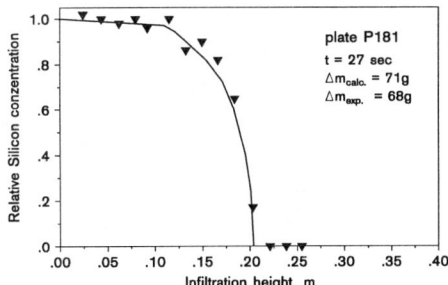

Figure 5: Time dependence of macroscopic silicon concentration ($\Delta t = 1$ sec)

Figure 6: Correlation of calculated and experimental silicon concentration

5. EXPERIMENTAL RESULTS

Comparison between simulation and experimental investigation is based on measurement of silicon concentration in the siliconized part. For this purpose, siliconized plates are cut into small samples for determination of density. Using the densities of the carbon/carbon preform and of the C/C-SiC sample it is possible to calculate experimental silicon concentration:

$$\frac{V_{Si}}{V_{Porosity\,C/C}} = \frac{\rho_{C/C-SiC} - \rho_{C/C}}{\rho_{Si} \cdot e'_{C/C}}. \tag{5}$$

Fig. 6 illustrates the good correlation between simulation (continuous line) and measured silicon concentration (triangles) for the plate P181. SEM micrographs show the lower region of the plate which is completely infiltrated (Fig. 7), whereas the upper region shows decreasing silicon concentration (Fig. 8).

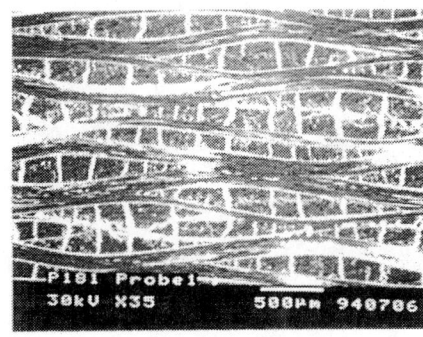 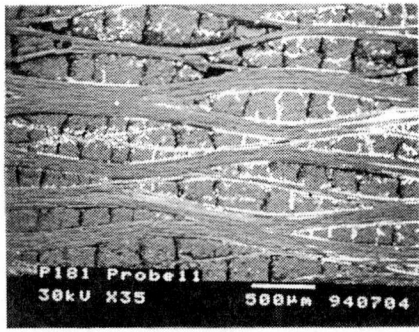

Figure 7: Plate P181, lower region (completely infiltrated)

Figure 8: Plate P181, upper region (decreasing silicon concentration)

6. CONCLUSIONS

Due to the high geometrical orientation of the translaminar capillary channels description of infiltration dynamics necessitates the development of a special simulation model. The model is verified by correlation of experimental and calculated values of silicon concentration. It is shown that the gradient in silicon concentration depends on the statistical distribution of capillary diameters. The region of decreasing silicon concentration in the C/C-SiC material spreads with the range of capillary diameters. As a result, future research work should aim on the development of carbon/carbon preforms with capillary channels of a small variation range in diameters to ensure good infiltration and fair gradients in silicon concentration.

NOTATION

d_K capillary diameter $[\mu m]$ η dynamic viscosity $[Pa \cdot s]$
g gravity acceleration $[m/s^2]$ ρ density $[kg/m^3]$
h infiltration height $[m]$ σ surface tension $[N/m]$
p statistical frequency $[\%]$ Θ wetting angle $[°]$
v_w tortuosity of capillary $[-]$ e' open porosity $[\%]$
V volume $[m^3]$

References

[1] Kochendörfer, R.: Liquid Silicon Infiltration – A Fast and Low Cost CMC Manufacturing Process, *Proceedings of the 8th ICCM*, Honolulu, 1991

[2] Schanz, P., Krenkel, W.: Description of the Mechanical and Thermal Behaviour of Liquid Siliconized C/C, *Proceedings of the Sixth European Conference on Composite Materials, ECCM-6*, Bordeaux, France, 20-24 Sept. 1993

[3] Hillig, W.B.: Melt Infiltration Approach to Ceramic Matrix Composites, *Journal of the American Ceramic Society*, Communications, Vol. 71, No. 2, 1988

[4] Whalen, T.H., Anderson, A.T.: Wetting of SiC, Si_3N_4 and Carbon by Si and Binary Si Alloys, *Journal of the American Ceramic Society*, Vol. 58, No. 9-10, Sept.-Oct. 1975

[5] Krenkel, W., Gern, F.H.: Microstructure and Characteristics of CMC Manufactured via the Liquid Phase Route, *Proceedings of the Ninth International Conference on Composite Materials, ICCM-9*, Madrid, Spain, 12-16 July, 1993

[6] Azzam, M.I.S.: Numerical Simulation of the Problem of Flow in Tubes with Periodic Step Changes in Diameter: Application to Consolidated Porous Media, Ph.D. Dissertation, University of Waterloo, Canada, 1975

[7] Raffke, T.F.: Untersuchung des thermischen Längenänderungsverhaltens von Faserkeramiken, *Institut für Bauweisen- und Konstruktionsforschung der DLR*, Institutsbericht IB 435-92/22, Stuttgart, 1992

FIBRE REINFORCED CERAMIC MATRIX COMPOSITE FABRICATION BY ELECTROPHORETIC INFILTRATION.

S. Kooner, J.J. Campaniello, S. Pickering and E. Bullock.
Institute for Advanced Materials, Joint Research Centre, Commission of the European Communities, P.O. Box 2, 1755 ZG Petten, The Netherlands.

ABSTRACT

Electrophoretic infiltration is a novel technique for the fabrication of fibre reinforced composites. The fibres are arranged as one of the electrodes such that deposition of the colloidal ceramic occurs in the fibre preform. This method has been investigated for the composite system of carbon fibre reinforced Si_3N_4 and has produced green composite microstructures with good infiltration uniformity and fibre distribution and few macro defects.

INTRODUCTION

Ceramic matrix composites (CMCs) are candidate materials for high temperature structural applications, their increased fracture toughness compared with monolithic ceramics being greatest when using long fibre reinforcement [1]. There are a number of techniques commonly employed in the fabrication of long fibre reinforced CMCs, each with associated disadvantages. The traditional densification route involves hot pressing limiting components to small, simple geometries. Densification by low pressure sintering provides a fabrication route for more complex component shapes at lower cost, but essentially requires dense, uniformly infiltrated green bodies with low macrodefect population. A possible means to achieve this is by the regular deposition of ceramic powders on to the fibres under electrophoretic potential.

Electrophoresis is the process by which charged particles in a liquid medium move under an applied potential. Deposition of the particle occurs at the oppositely charged electrode. If the deposition electrode is a conductive fibre preform the phenomena may be used as a novel technique for CMC fabrication.

The composite system studied here is carbon fibre reinforced silicon nitride (C/Si_3N_4), i.e. the fibre preform is conductive. Al_2O_3 and Y_2O_3 are used as sintering aids [2]. However, simultaneous deposition of a mixed powder suspension by electrophoresis would result in segregation of the different powders because of their different electrophoretic mobilities. This difficulty may be overcome by using ceramic powders coated with sinter aid material - in this case Si_3N_4 coated with Al_2O_3/Y_2O_3 [3]. In anticipation of the development of Al_2O_3/Y_2O_3

To the extent authorized under the laws of the United States of America, all copyright interests in this publication are the property of The American Ceramic Society. Any duplication, reproduction, or republication of this publication or any part thereof, without the express written consent of The American Ceramic Society or fee paid to the Copyright Clearance Center, is prohibited.

coated Si_3N_4 powders, the process development work to date has concentrated on using pure Al_2O_3 powders, infiltrated into 1D and 2D fibre tape lay-ups.

In dispersion in polar liquids the charge on a particle arises from either dissociation or adsorption processes which occur when a particle is in contact with a liquid [4]. The sign is determined by the surface chemistry of the particle and adsorbed species which modify it. This surface charge, for reasons of neutrality, must then be balanced by oppositely charged ions or polar molecules and results in the formation of a double layer around the particle. The electrophoretic mobility of the particle is defined by Smoluchowski's equation:

$$\text{Electrophoretic mobility} = \frac{U}{X} = \frac{\varepsilon \zeta}{4\pi\eta} \qquad (1)$$

where U = velocity, X = field strength, ε = dielectric constant, ζ = zeta potential; and η = viscosity.
A suitable suspension for electrophoretic infiltration should have:
- high particle surface charge to increase the mobility of the ceramic
- high dielectric constant of the solvent
- low conductivity of the suspending medium to minimise solvent transport and
- low viscosity to increase particle mobility.

Depending on the sign of the charge on a particle, deposition occurs at the anode or cathode. However, the cell also operates as an electrolysis cell for the solvent, for which the electrolysis product at the electrodes may be disruptive to the powder deposit, especially when gases are evolved. The selection of solvent and particle adsorbent can be tailored to determine at which electrode the powder will deposit and to optimise dense and uniform deposition.

Electrophoretic infiltration as a means of CMC fabrication is performed by using a conductive fibre preform as the deposition electrode in an electrophoretic bath. Infiltration into a porous structure has been investigated by Gal-Or et al and has shown to be enhanced by high field strength, high dispersion concentration and high particle mobility [4], however, with time there is a build-up of surface deposit preventing further penetration. Most of the work to date that has been carried out using electrophoresis has involved some form of coating process [5,6] or the production of monoliths [6]. Nicholson et al have fabricated laminar microcomposites by electrophoretically depositing layer by layer [8]. In all cases this involves a simple deposition process and not infiltration. Recently however, electrophoresis has been used to fabricate SiO_2 matrix composites [9] and it is here that the problem of infiltration versus deposition is highlighted, i.e. there is residual porosity within the weave because the surface deposit inhibits full infiltration.

EXPERIMENTAL

A novel electrophoretic route for fabrication of green forms has been developed as an alternative to an equivalent slurry infiltration/hot processing process using the same ceramic system [10].
The project splits into two distinct areas:
a) suspension optimisation, and
b) electrophoretic deposition/infiltration.

The surface charge of the particles has been engineered by the addition of surfactants, to give a high charge and the required sign, and since they will be retained in the green body, the chosen surfactants are clean burning. Suspensions have been characterised by use of a Matec Applied Sciences' Electrokinetic Sonic Analysis System™ [11]. A suspension of sufficiently high electrophoretic mobility has then been used for infiltration studies. Green body cross-sections were prepared by vacuum epoxy impregnation and examined to see the extent of infiltration.

Figure 1 shows the basic set up for the electrophoresis cell. For convenience, all electrophoretic infiltration has been done in constant voltage mode. Carbon fibres are mounted in an insulated clamp so as to give electrical contact to each filament, but to exclude electrical contact between clamp and suspension. This is to avoid preferential deposition on any exposed metallic surfaces. The opposite electrode is a flat stainless steel plate. The distance between the electrodes, important as it determines the field strength, was kept constant at 1.5cm. 1D and 2D fibre preforms have been used. Early experiments showed the importance of eliminating stray fibres in lay-ups since these are attracted to the opposite electrode and experience a higher field strength, attracting preferential deposition. 2D weaves, while presenting more difficult problems of access for the infiltrating ceramic, are easier to handle and are technologically more relevant and have therefore been used for the majority of the work. The 2D weaves are made from Granoc™ XN-50 fibre, with 1000 filaments per tow and a tensile modulus of 490 GPa. These have to be desized prior to use by washing in hot acetone. 1D lay-ups consisted of single fibre tows, with typical volume fraction 35%, whereas the 2D preforms result in a fibre volume fraction of ~50%.

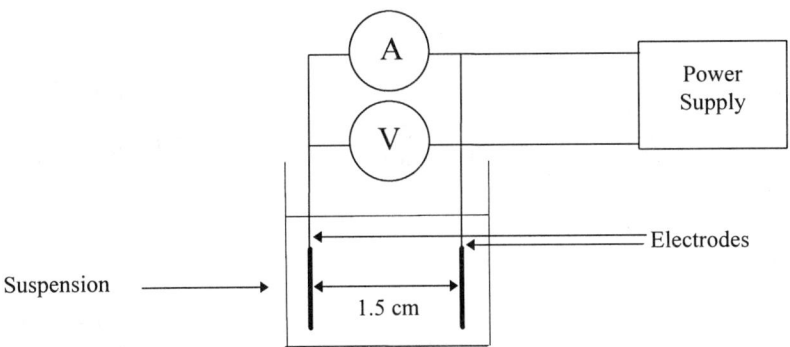

Figure 1. Schematic of the electrophoretic cell set-up.

RESULTS AND DISCUSSION

Process Validity Trial

A first trial experiment was made with a Si_3N_4 (Ube SN-E-10) slurry mix to assess the viability of using a standard suspension [10] of varying concentration with well-known commercial dispersion and wetting agents Darvan C™, an ammonium polymethacrylate, and Aerosol OT™, an acid sulphonate, as additives and water as a suspending medium with 1D fibre preforms. Deposition occurred at the cathode implying that the particles carry a positive charge. A constant potential of 3.5V was used, to avoid gas evolution by water electrolysis (4V), with time of deposition being varied. Cross-sectional examination showed that there was full penetration, i.e. no fibre bunching and no macro flaws, for the electrophoretically infiltrated specimens. Fibre bunching is the main problem for slurry infiltration resulting in macroporosity at these sites. The applied voltage prevents this happening in the case of electrophoresis due to the electrical repulsion between fibres.

These experiments confirm the validity of using electrophoretic deposition for infiltration but revealed many difficulties with this particular mix. Water electrolysis at voltages › 4V is a serious limitation, despite the advantage of a high dielectric constant (78). The surfactant mix is complex and deposition is slow.

Suspension Optimisation

The success of the validity trial, and recognition of its limitations led to a series of experiments with suspensions using ethanol (dielectric constant 24) as the suspending medium. The predicted difficulty of concomitant deposition of several powder species stimulated an investigation of the effect of different surfactants using both Si_3N_4/ethanol and Al_2O_3/ethanol (Alcan RA207) with initial volume fraction of 5%, 1% and 0.5%. In anticipation of the development of coated Si_3N_4 powders, the rest of the work has concentrated on using pure Al_2O_3 powders. Table 1 shows the maximum dynamic mobility possible using each surfactant for Al_2O_3/ethanol 5 vol %. The sign of the charge cannot be determined by the Matec system and must be determined independently e.g. by simple electrophoretic deposition.

Table 1. Effect of surfactants with Al_2O_3/ethanol 5 vol %.

Surfactant	Dynamic Mobility m^2/V.s x 10^8	pH	Conductivity µS/cm
Aerosol OT	+ 0.1	10	110
H_2SO_4	- 0.2	5.5	4
HNO_3	+ 0.15	5	100
NaOH	+ 0.2	13	150
Citric acid and triethylamine	- 0.45	9	6

Figure 2 shows the different effects of progressive H_2SO_4 and HNO_3 addition to the dispersion. HNO_3 addition produces a typical titration curve giving an isoelectric point of pH 9.5. In contrast the addition of H_2SO_4 past the maximum mobility decreases the mobility value, Figure 2 and also increases the conductivity of the solvent. This is a saturation effect in that maximum SO_4^{2-} adsorption has occurred. Further addition of H_2SO_4 tends to neutralise the adsorbed negative charge by H^+ and leads to an increase in the conductivity due to excess ions in solution. This effect is not observed for any of the other surfactants investigated.

Thus, H_2SO_4 and citric acid/triethylamine confer negative charge to the particles while Aerosol OT, HNO_3 and NaOH convey a positive charge. The surface charge of the Al_2O_3, without surfactants, in ethanol is negative. This slight negative charge, with addition of surfactants then may attract positively charged species e.g. Na^+ in the case of NaOH and Aerosol OT and H^+ in the case of HNO_3, to form a positively charged surface layer. However, in the case of H_2SO_4 and citric acid/triethylamine, where there are also positively charged ions present, this is not the case.

It is suggested that a complexing mechanism occurs between the Al^{3+} sites on the Al_2O_3 surface and the negative ions, SO_4^{2-} and carboxylic acid groups for H_2SO_4 and citric acid/triethylamine respectively, giving an overall negative charge. This is confirmed by the initial decrease in conductivity observed on addition of these two surfactants.

Electrolysis of ethanol yields ethane at the cathode and acetaldehyde, which oxidizes to acetic acid, at the anode. Deposition of ceramic particles at the anode avoids the inevitable disruption by gas evolution. Hence, suspensions using H_2SO_4 and citric acid/triethylamine as surfactants have been used for electrophoretic infiltration giving deposition at the positive electrode. It must be noted that these additives also give the lowest solvent ion conductivity readings at maximum particle mobility.

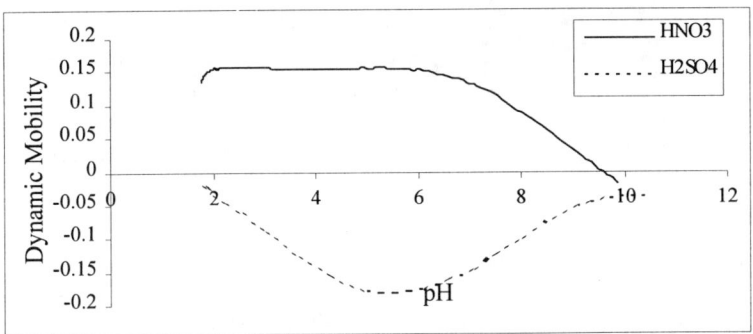

Figure 2. The effect of H_2SO_4 and HNO_3 addition to Al_2O_3/ethanol, 5 vol %

Electrophoretic infiltration using Al_2O_3/ethanol.

Following the optimisation of Al_2O_3/ethanol suspensions an electrophoretic cell was set up. The suspensions used were 5, 1 and 0.5 vol % Al_2O_3 in ethanol using H_2SO_4 and citric acid/triethylamine as surfactants. The fibre preforms used consisted of single, double and triple plies of 2D weaves. Time and voltage effects were studied for the different concentration suspensions. At 50V there was no infiltration at tight points in the weave at any of the concentrations investigated, although the outer layer increased in thickness with time. Higher voltages led to improved infiltration but not to full penetration.

Full infiltration was achieved by the superposition of ultrasonic vibration which may aid infiltration by disrupting the dense surface deposit, encouraging particle penetration and the opening up of tight points of the weave. Figure 3 shows such an infiltration for a triple 2D weave lay-up, conditions being 200V for 5 mins with high power ultrasonic vibration provided by a standard commercial ultrasonic bath (35 kHz). Similar conditions, but with 150 V also gave good infiltration but with a higher macrovoid population.

The best infiltration into a 2D weave was achieved using a combination of high particle electrophoretic mobility, high field strength and ultrasonic vibration.

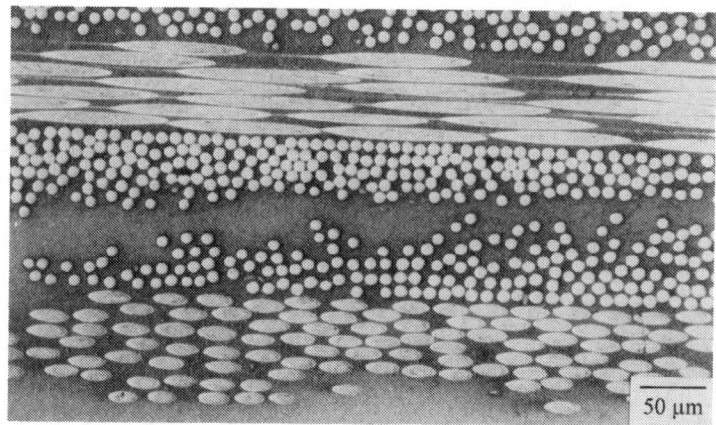

Figure 3. Optical micrograph showing cross-section of infiltrated triple 2D weave lay-up made using 5vol% Al_2O_3/ethanol using citric acid/triethylamine as a surfactant.

CONCLUSIONS

It has been shown that electrophoretic infiltration is a feasible technique for fabrication of fibre reinforced CMCs. It is possible to achieve uniform infiltration of 2D fibre weaves using high voltage and ultrasonics coupled with tailored surface charged particles. Further work will involve the use of coated powders and a sintering study.

REFERENCES
[1] K.M.Prewo, J.J.Brennan, 'High strength SiC fibre-reinforced glass-ceramic matrix composites exhibiting high strength and toughness', J. Mat. Sci. Vol. 17, pp 2371-83 (1982).
[2] K.Negita, 'Effective sintering aids for Si_3N_4 ceramics', J. Mat. Sci. Let. Vol. 4 pp755-58 (1985).
[3] B.Djuricic, I.J.Davies, S.Pickering, D.McGarry, E.Bullock, M.Verwerft, P.M.Bronsveld and J.Th.M.De Hosson, 'Study of particle coatings for the design of intergranular phases in engineering ceramics', 3rd Int. Conf. of the Ceramic-Ceramic Composites Soc, Mons, Belgium, October (1994).
[4] L.Gal-Or, S.Liubovich, S.Haber, 'Deep electrophoretic penetration and deposition of ceramic particles inside porous substrates', J. Electrochem. Soc. Vol. 139 No. 4, pp 1078-81 (1992).
[5] A.Das Sharma, A.Sen, H.S.Maiti, 'Effectiveness of various suspension media for electrophoretic deposition of YBCO superconductor powder', Ceramics International, Vol. 19 pp 65-70 (1993).
[6] W.J.Dalzell, D.E.Clark, 'Thermophoretic and electrophoretic deposition of sol-gel composite coatings', Ceram. Eng. Sci. Proc. Vol 7, 7-8, pp 1014-26 (1986).
[7] K.Moritz, T.Reetz, 'Electrophoretic deposition of aluminium nitride', Proc. 3rd Euro Ceram Soc Conf, Madrid, Eds. P.Duran and J.F.Fernandez, Vol 1 pp 425-30 (1993).
[8] P.S.Nicholson, P.Sarkar, X.Haung, 'Electrophoretic deposition and its use to synthesize ZrO_2/Al_2O_3 micro-laminate ceramic/ceramic composites', J.Mat.Sci. Vol 28 pp6274-78 (1993).
[9] J.Illston, C.B.Ponton, P.M.Marquis, E.G.Butler, 'The manufacture of woven fibre ceramic matrix composites using electrophoretic deposition', Proc. 3rd Euro Ceram Soc Conf, Madrid, Eds. P.Duran and J.F.Fernandez, Vol 1 pp 419-424 (1993).
[10] C.Grenet, L.Plunkett, J.B.Veyret and E.Bullock, 'Carbon fibre reinforced silicon nitride composite fabricated by slurry infiltration', to be published in HT-CMC-2, (1995).
[11] T.Oja, G.L.Petersen, D.W.Cannon, 'A method for measuring the electrokinetic properties of a solution.' United States Patent #4,497,207, (1985).

PROCESSING AND FLEXURAL BEHAVIOR OF ALUMINA MULTILAYER STRUCTURES

P. Letullier[°], K. Debray[*], E. Martin[*], J.M. Quenisset[°*] and J.M. Heintz[°]

[°] : ICMCB, Université Bordeaux I, Av Dr A. Schweitzer, 33600 Pessac (FRANCE)
[*] : LGM, IUT A, Université Bordeaux I, 33405 Talence Cedex (FRANCE)

Abstract : The processing and the sintering of an alumina multilayer structure constituted of a stacking of dense and porous layers is presented. Influence of the processing parameters on the cracking of the porous layers are analyzed and discussed. The flexural behavior of this material is described. Numerical simulations show that the Young's modulus of the porous layer and therefore its density control the fissuration path.

I. INTRODUCTION

Ceramics are known to behave as brittle materials under stress. One of the main reason put forward is the presence of many defects associated to a large notch sensitivity. These flaws induce catastrophic failure for a critical stress related to the size of the defects. In order to limit this brittle behavior, a useful concept deals with the control of the crack path using a discontinuity of properties between the different parts of a material [1,2]. More precisely, this paper deals with the development of a ceramic with an homogeneous chemical composition where interfaces were introduced to generate a discontinuity of elastic properties due to a change of microstructure. Such a material is in fact constituted of a stacking of dense and porous alumina layers. A sintering treatment allows the density of the porous layers to be controlled so that their elastic properties and finally the intensity of the discontinuity at the interface can be adjusted [3, 4].

In a first part, the preparation process is presented. Especially, the role of the processing parameters is studied to avoid cracking of the porous powder layer embedded between dense substrates. The sintering kinetics of the layer is also discussed and explained. In a second part the flexural behaviour of the whole laminated material is described and analysed. The influence of the thickness and of the stiffness of the porous layer is also investigated using a finite element analysis.

II. PROCESSING

Techniques such as casting of a slurry or a sol-gel precursor on a dense substrate seem adequate to prepare a material constituted of a stacking of porous layers and dense substrates. However, without hot pressing, these processes lead to systematic cracking of the powder layer during drying or sintering [5]. In order to solve this problem, a simple and new method was implemented. A stable slurry was first obtained by ultrasonication of a submicronic pure alumina powder (Baikowski 99.99 %) in acid medium. This desagglomerated slurry was then sprayed on a dense alumina substrate (Coors 96 %) maintained at a temperature where water can be immediately evaporated ($120 \leq T \leq 180°C$). Constrained drying is therefore avoided. Uniform powder films were obtained and their thicknesses were adjusted depending on the experimental parameters. The final multilayer structure was obtained by cold pressing two or more substrates on which a powder film was previously deposited and then sintered at 1400°C [6].

Control of the preparation of these materials was achieved using a statistical design method. First, the influence of five variables (powder concentration, pH of the slurry, gas flow, substrate temperature and pressing pressure) on the powder deposition rate was investigated. The statistical analysis of the experiences showed that the two most important parameters were the powder concentration of the suspension and the gas flow. Other parameters and second order interactions were less significant. In a second step, we looked at the way to promote homogeneity within the powder layer and therefore to limit occurence of cracking during sintering. Recent papers have shown that conditions of film cracking depend not only on a constraint (sintering pressure e.g.) and on the properties of the film but also on its thickness [7]. This study was then performed using the critical cracking thickness (CCT) concept. A powder film deposited on a substrate with a varying thickness was sintered and the thickness corresponding to the beginning of cracking was determined. The studied parameters were gas flow, substrate temperature and applied pressure. Gas flow and second order interactions showed little influence on CCT. On the contrary, temperature and pressure were important : CCT increased when the temperature was lowered or when pressure was increased. It has been shown using thermodynamical and mechanical considerations that a crack in a powder film will not propagate spontaneously when its thickness is lower than a critical value [5]. Now, the presence of flaws within the powder film greatly depends on the processing parameters. When a low pressure was used, many defects were present in the film whereas none of them were visible after pressing at higher pressure. We also observed that the lower the temperature, the more homogeneous the powder layer. In both cases, dimensions of the remaining flaws had to be smaller than the critical value which prevented occurrence of further cracking when the film thickness was lower than the CCT.

The sintering behavior of powder layers constrained by dense substrates (here sapphire to limt chemical contamination) was investigated. It was compared to the sintering of non constrained samples called free samples. Green relative density of both types of specimens were 44 % and powder film thickness was lower than 200 µm, i.e. lower than the CCT (here 245 µm). Grain sizes were measured from SEM micrographs of fractured samples. Equivalent pore sizes were deduced by image analysis of the same SEM micrographs.

Examples of evolution of the relative density of free and constrained samples are presented in Fig. 1. Final densities of the constrained films are lower than that of the free pellets. The difference in densification kinetics occurs mainly during the first hour of the sintering treatment. Afterwards, densification rates of both samples become low and the difference in density remains constant. Grain growth was also lower for constrained samples than for free samples.

It has been shown, based on Coble's model, and taking into account the evolution of pore size, that the main mechanism involved in the densification of this alumina powder was grain boundary diffusion [6]. The difference in the sintering behavior appears related to porosity evolution. The constraint due to the dense substrates leads to an increase of the average pore size in the powder layer, as it has been observed. Mobility of grains is highly reduced due to the adhesion of grains close to the substrates. Therefore, pore opening or pore coalescence is susceptible to occur around inhomogeneities initially present in the powder film. Since the diffusion flux is inversely proportional to the size of the pores, the densification rate is consequently reduced.

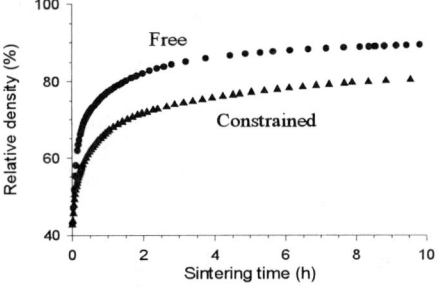

Fig. 1 : Evolution of relative densities of free and constrained samples versus sintering time.

III. FLEXURAL BEHAVIOR

The samples were constituted of two dense substrates bonded with a porous ceramic layer sintered at 1400°C for periods between 5 min. to 20 h. Thicknesses of the green powder layer were lower than the CCT and were adjusted to ensure that the final thicknesses were about the same after sintering treatment (porous thickness layer ≈ 0.16 mm). The typical size of samples was 45x7x1.43 mm^3. The flexural behavior of Single Edge Notched Bending (SENB) specimens was investigated using four-point bending tests.

Fig. 2 shows representative results of SENB tests carried out on samples densified 5 min. (a), 3 h. (b), 10 h. (c) and 20 h. (d). Each sample presents at first a linear elastic be-

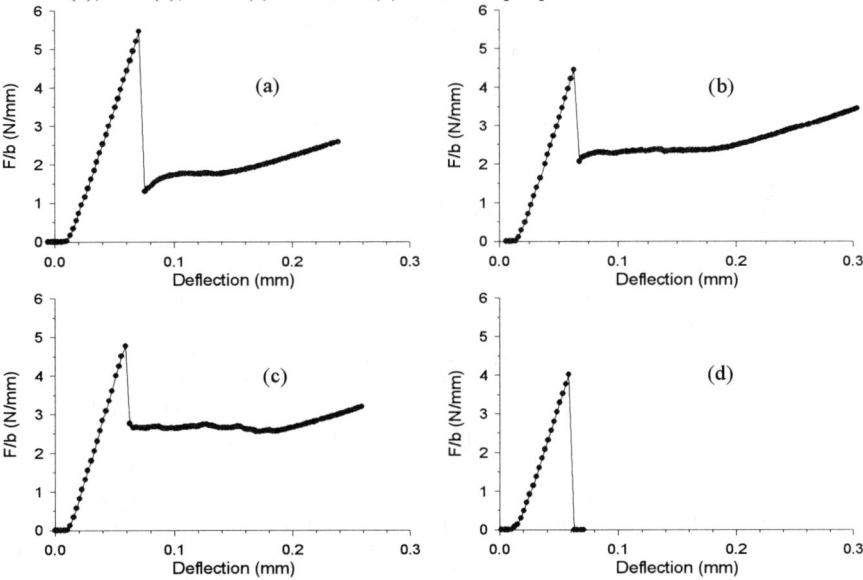

Fig. 2 : SENB tests of dense/porous/dense structures sintered (a) 5 min., (b) 3 h., (c) 10 h., (d) 20 h.

havior. Crack growth begins from the notch as the maximum load is reached. Afterwards the behavior depends on the density of the porous layer. The most densified sample (d) presents a catastrophic failure whereas the other specimens show a drop of load followed by a plateau and a final load rise. Plateaux correspond to stable crack propagation by delamination along the porous layer/dense substrate interface, i.e. the second interface with respect to the propagation direction, as it was experimentally observed by optical microscopy. This delamination ends when the interfacial crack tips reach the inner loading points. The catastrophic failure of sample (d) corresponds to a straight mode I propagation of the crack through all the interfaces. The value of the load corresponding to the plateau depends on the toughness of the interface (G_{ic}). Since crack growth along the interface is stable, an estimation of the value of this parameter can be derived from the area under the load-deflection curve and from compliance determinations following a procedure detailed elsewhere [6]. This area is equal to the energy of new surfaces and to an increase of stored elastic energy. The related values of G_{ic} are given in Table I. The main incertitude comes from the estimation of the surface really cracked. G_{ic} is strongly correlated with the densification of the ceramic powder layer. This is in agreement with the fact that G_c depends on the quantity of cracked material which is directly related to the density of a porous ceramic [8].

Table I : G_{ic} values calculated from Fig. 3.

Sintering time	G_{ic} (J.m^{-2})
5 min.	4.5 ± 0.6
3 h.	6.3 ± 0.8
10 h.	10.3 ± 1.0

IV. NUMERICAL SIMULATIONS AND DISCUSSION

Numerical simulations concerned the system previously described. Thickness of dense substrates had a fixed value (t_d) while thickness of porous layers (t_p) was a variable expressed as a fraction of t_d. In a same manner, elastic modulus of porous layers (E_p) was a fraction of that of dense substrates (E_d). The Poisson's ratio of porous medium was assumed to be the same as that of dense alumina, i.e. 0.21 [9]. When the material is under stress, a crack is allowed to propagate perpendicularly to the substrate plane from the tensile to the compressive face. Two modes of cracking were considered at the interfaces : normal propagation or double deflection along the interface.

Geometric considerations allowed the analysis to be performed on half of the beam. Triangular elements made of 6 nodes constituted the mesh. The number of elements depended on the thickness ratio t_p/t_d and on the crack length a/t_d. The strain energy release rate, G, was computed with the help of the crack closure method assuming a plane strain state [10]. The accuracy of the numerical simulations was checked on samples made of two identical materials separated by an ideal interface. The simulation gave results with less than 2 % of error compared to the analytical ones [11]. When a crack tip is approaching an interface between two materials, 1 and 2, it has been shown that G tends to 0 if the elastic modulus of material 2 is higher than that of 1, and tends to infinite in the reverse case. However, these singularities are only encountered in the case of an ideal discontinuity of elastic properties at the interface. A more realistic model has been proposed based on a continuity of the elastic properties through a thin layer [12]. On the other hand, crack extension cannot be considered smaller than the characteristic length of the microstructure (flaws, grain size, ...). Here, G was calculated with the help of a crack extension assumed equal to a characteristic length. Moreover, the same crack extension was used in the two crack path directions in order to perform a meaningful comparison of G at interfaces.

Computed results are presented as a normalized ratio $[G/F^2]$ where F is the applied load. G_o/F_o^2 is the normalized strain energy release rate when the porous layer has the same elastic modulus than that of the dense substrates. It can be seen in Fig. 3 that the introduction of a low elastic modulus ceramic layer between two high modulus layers leads to a large increase of the normalized energy release rate near the first interface

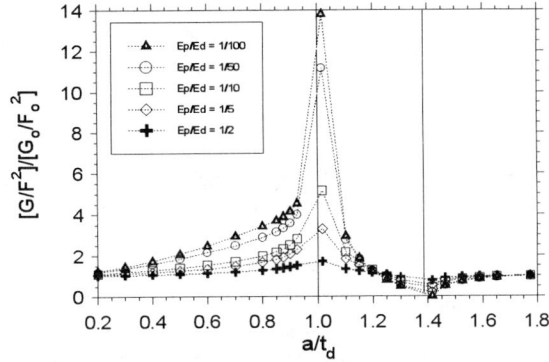

Fig. 3 : Evolution of normalized G versus crack length and E_p in the normal direction $(t_p/t_d = 0.4)$.

followed by a rapid decrease within the porous layer down to a minimum value at the second interface. This evolution is even more significant when E_p is smaller. The study is then focused on the second interface where $[G/F^2]/[G_o/F_o^2]$ is minimum which may lead to the wished behavior. Fig. 4 shows values of normalized G versus E_p/E_d for the two modes of cracking at this second interface. It is worthy to note that the value of G_o/F_o^2 for crack propagation in the normal diraction is higher than that for interfacial proapagation. $[G/F^2]/[G_o/F_o^2]$ is always lower than 1 and decreases whith the in-between layer compliance. In the case of a normal crack propagation, $[G/F^2]/[G_o/F_o^2]$ tends to zero for the lowest values of E_p/E_d although in the case of a crack propagation along the interface the normalized energy release rate never tends to zero.

Once the crack reaches an interface, the propagation path is determined by the minimization of the free energy of the system. The difference between G and its critical value must be positive and maximum. From these thermodynamical considerations, an interfacial deflection criterion has been proposed and justified elsewhere [13] : $G_n-G_i < G_{nc}-G_{ic}$ where G_i and G_n are the strain energy release rate respectively along the interface and normal to the interface. G_{nc} and G_{ic} refer to critical values of the material and of the interface respectively. This criterion can be verified for low values of E_p. In fact, numerical simulations have shown that very low values of E_p generate negative values of G_n-G_i because G_n tends to zero while G_i remained strictly positive. Flexure tests (a), (b) and (c) actually correspond to this configuration since the densification of the porous layer remained low and induces a low elastic modulus. Simultaneously, the lower the densification of the porous layer, the lower the value of G_{ic} as

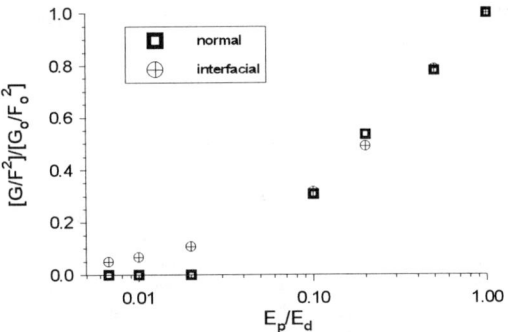

Fig. 4 : Evolution of normalized G versus E_p/E_d through the second interface for the two directions of crack propagation.

shown in Table I, while G_{nc} is not changed by the sintering process. It results that for very low densification of the porous layer, the deviation at the second interface is quasi obvious. On the other hand, for higher densification, G_n-G_i increases with E_p as well as G_{ic} so that a catastrophic failure can be expected when the criterion is no longer verified. These evolutions of G_n, G_i and G_{ic} as a function of densification explain the low toughness of sample (d). Besides, increasing energy dissipation during laminate failures requires high values of G_{ic} although favoring crack deviation. This contradiction points out a need of optmization which can be based on numerical simulations.

V. CONCLUSION

Alumina structures constituted of a stacking of dense and porous layers were processed by spraying and sintering a submicronic Al_2O_3 powder between dense substrates. The study of the processing parameters allowed to obtain homogeneous powder layers, free of cracks with a controlled thickness. Single edge notched bending tests showed the efficiency of this multilayer design to deviate crack provided the delamination critical energy release rate was maintained lower than a threshold. Numerical simulations of crack propagation tests justifies the experimental results and constitutes a useful tool for further improvements of laminated ceramics.

REFERENCES

[1] J. R. Rice, "Elastic Fracture Mechanics Concepts for Interfacial Cracks", J. Appl. Mech., **55** [1], 98-103 (1988)
[2] M. Y. He and J. W. Hutchinson, "Crack Deflection at an Interface Between Dissimilar Elastic Materials", Int. J. Solids Structures, **25** [9] 1053-67 (1989).
[3] N. Ramankrishnan and V. S. Arunachalan, "Effective Elastic Moduli of Porous Ceramic Materials", J. Am. Ceram. Soc., **76** [11], 2745-52 (1993).
[4] A. R. Boccaccini and G. Ondracek, "On the Effective Young's Modulus for Porous Materials, J. Mech. Behavior Mater., **4** [2], 119-28 (1993).
[5] J. M. Heintz, O. Sudre and F. F. Lange, "Instability of Polycrystalline Bridges than Span Craks in Powder Films Densified on a Substrate", J. Am. Ceram. Soc., **77** [3], 787-91 (1994).
[6] P. Letullier, "Design and processing of a damageable ceramic material", PhD thesis, University Bordeaux I, 1994
[7] A. G. Evans, M. D. Drory and M. S. Hu, "The Cracking and Decohesion of Thin Films", J. Mater. Res., **3** [5], 1043-49 (1988).
[8] D. C. Lam, "Processing Control and Mechanical Properties of Porous Ceramics", PhD. Thesis, University of California, Santa Barbara (1991).
[9] L. J. Gibson and M. F. Ashby, "The Mechanics of Three-Dimensional Cellular Materials", Proc. R. Soc. London, **A 382**, 43-59 (1982).
[10] E. F. Rybicki and M. F. Hanninem, "A Finite Element Calculation of Stress Intensity Factors by a Modified Crack Closure Integral", Eng. Frac. Mech., **9** [9], 931-38 (1977).
[11] G. Charalambides, "Development of a Test Method for Measuring the Mixed Mode Fracture Resistance of Bimaterial Interfaces", Mech. Mater., **8**, 269-83 (1990).
[12] K. Kaw and G. H. Besterfield, "Comparison of Interphase Models for a Crack in Fiber Reinforced Composite", Theo. Appl. Fract. Mech., **17**, 133-47 (1992).
[13] E. Martin, N. Piquenot, K. Debray and J. M. Quenisset, "Numerical Simulation of the Fracture Behavior of a Fragile Microcomposite", submitted to Comp. Interface (1994).

Fiber Reinforcement of Reaction Bonded Oxide Ceramics

R. Janssen, J. Wendorff, N. Claussen
Technische Universität Hamburg-Harburg
Advanced Ceramics Group
D-21071 Hamburg, Germany

ABSTRACT

Pure oxide ceramic matrix composites (CMC) offer an encouraging route for materials capable of maintaining excellent stability in oxidizing atmospheres at high temperatures. The present paper deals with fiber-reinforced oxide matrix ceramics fabricated by reaction bonding.
In order to produce a weak interface between fibers and matrix, two different approaches are discussed: a) fiber/matrix debonding by adjusting the porosity of not fully densified matrices and b) through formation of a weak fiber coating as a third component of the composite system.

INTRODUCTION

Current research still addresses the need for ceramic matrix composites (CMC) suitable for high temperature applications in oxidizing environments. Materials based on C, SiC, or Si_3N_4 have to be protected against their oxidation through suitable coating of the fibers or the whole composite structure. Many problems such as long-time stability or crack sensibility are not yet solved sufficiently for these systems [1, 2].
Plain oxide systems represent an interesting alternative due to their inherent oxidation resistance. Such CMC's are thermodynamically stable in oxygen at high temperatures even in the case of interconnected porosity or crack networks.

The matrices used are Reaction Bonded Aluminum Oxide (RBAO) and Reaction Bonded Mullite (RBM) which allow easy manufacturing of composites by simple powder technologies [3-7]. In this process, metal-ceramic precursors containing reactive fine-grained Al particles are oxidized and subsequently sintered. The plasticity of metallic Al allows an excellent green state compactability and the superior superplastic behavior of sintered RBAO makes this technology very attractive for composite fabrication [8, 9]. Due to their good high-temperature behavior, we focus on reinforcement of these matrix materials with continuous sapphire fibers in order to produce composites.

EXPERIMENTAL

To produce the RBAO matrix material, a mixture of Al, Al_2O_3, and ZrO_2 was attrition milled in a liquid medium, dried and sieved. In the case of mullite, fine grained SiC powder was added instead of ZrO_2. Only some amount of ZrO_2 was introduced into the powder through wear debris of the milling media.

The sapphire fibers (Saphicon Inc.) with diameters of ~ 125 μm were incorporated into the powder which was pressed uniaxially and cold isostatically to form one dimensionally reinforced green bodies.

Oxidation and sintering was carried out in a standard box furnace in air with a two step heating cycle. First, up to 1100 °C at slow heating rates, the metallic Al was oxidized forming fine-grained alumina particles. In the second step, sintering at different temperatures (1250 - 1550 °C, 30 - 60 min) lead to samples with controlled porosity (1 - 15%). In the case of mullite, a dwell time at 1150 °C was added between reaction bonding and sintering to ensure almost complete oxidation of the SiC particles.

For porous matrices, uncoated sapphire fiber were used. In almost dense composites, fiber coatings of metallic Ti with a thickness of 1 - 3 μm were used to form a weak Al_2TiO_5 interface during heat treatment by reaction with oxygen and the matrix alumina. To accelerate the reaction to Al_2TiO_5 and therefore prevent possible fiber degradation, first experiments were carried out with coatings consisting of a mixture of Ti and Al. The fiber/matrix interfaces were investigated using SEM and TEM. The interface properties were characterized by single fiber push out tests. In order to investigate the temperature dependence of the properties and the effect of residual stresses, tests will be performed at room temperature as well as at elevated temperatures using a new developed experimental setup [10].

RESULTS AND DISCUSSION

Sintering at low temperatures and short sintering times (1300 °C / 30 min) lead to densities of about 90 %TD. These porous RBAO composites exhibit the desired fracture behavior with fiber/matrix debonding, as shown in **Fig. 1**. Debonding lengths of about 500 µm should lead to a significant increase in fracture toughness.
No detrimental reactions between fibers and matrix have been observed.

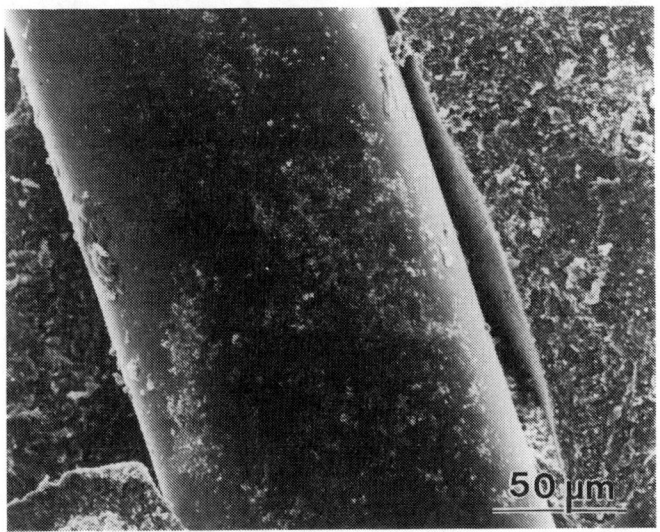

Fig. 1: Fracture surface (detail) of a porous sapphire/RBAO composite, showing fiber/matrix debonding

First push-out experiments using porous composites at room temperature (see **Fig. 2**) resulted in debonding stresses in the range of 40 - 60 MPa.

Fig. 2: Sapphire fiber, pushed out of a porous RBAO matrix composite (RT)

Considering the fracture behavior, the interfacial strength data seem to be suited for this material. Tests at elevated temperatures are in progress to investigate a possible change of the debonding stress.

Further sintering of the material (1550 °C / 60 min) results in almost dense composites (~ 98 %TD) without cracks or voids even in between fiber bundles. These results confirm that the RBAO material is well suited as a matrix for oxide composites.
The plastic deformation of Al particles allows the production of defect free green bodies. Due to the small grain size (0.2 - 1 µm) of oxidation formed alumina particles [9], RBAO ceramics behave superplastically. Therefore, homogeneous composites with large inclusions - such as continuous sapphire fibers - can be produced. Additionally, the volume expansion of aluminum during oxidation reduces the shrinkage on sintering and therefore prevents cracking of the composite material [8].

In contrast to porous matrices, dense RBAO composites show no crack deflection and fiber push-out is no longer possible. This behavior seems to be due to the nearly equivalent values of Young's modulus of fibers and matrix, although microscopic investigations again show no reaction between fibers and matrix. Coating of the fibers with metallic Ti should result in a weak interface layer through formation of Al_2TiO_5 [11, 12]. As shown in **Fig. 3**, long time annealing of Ti coated fibers in an RBAO matrix leads to the desired porous - and therefore weak - interface. Fiber degradation, however, may be possible.

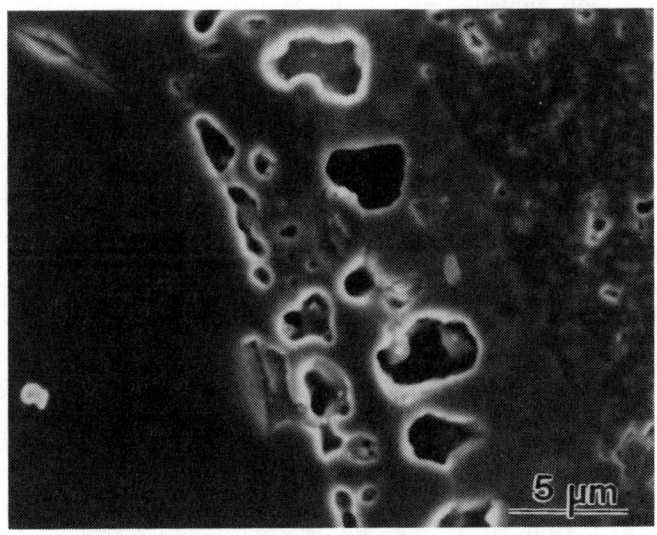

Fig. 3: Porous interface between sapphire fiber (left), originally Ti coated, and RBAO matrix (right) after annealing for 50 h at 1300 °C

In order to prevent a detrimental strong bonding, first experiments are in progress with a two component coating of metallic Ti and Al. This mixture should prevent fiber degradation due to the presence of aluminum in the original coating layer

With mullite matrix composites, the interfacial strength seem to be adjustable through the shrinkage on sintering of the matrix material. Experiments to vary the heating cycle and therefore the SiC oxidation will be done in the next future to produce a sapphire fiber / RBM composite where an additional fiber coating is no longer necessary.

CONCLUSIONS

- Fiber reinforced oxide ceramics can easily be manufactured through the RBAO process by simple powder processing.
- Alumina (RBAO) and mullite (RBM) matrix systems are most suitable for sapphire fiber reinforcement.
- Porous RBAO as well as RBM composites with uncoated fibers show fiber / matrix debonding.
- Dense RBAO materials show the need for fiber coating to form a weak interface. First results indicate that the formation of Al_2TiO_5 through a Ti- or a Ti/Al coating of the fibers is a promising way.
- Further investigation of these systems is in progress to specify the high temperature properties of these materials.

ACKNOWLEDGMENT

We acknowledge the financial support by the German Research Foundation (DFG) under contract no. Ja 655/2-3.
We like to thank L. Lew and A. Argon, M.I.T., for Ti-, as well as R. Bormann and C. Michaelsen, GKSS-Germany, for Ti/Al coating of sapphire fibers.

REFERENCES

[1] Prewo, K.M., "Fiber Reinforced Ceramics: New Opportunities for Composite Materials", Am. Ceram. Soc. Bull. **68** (1989) 395
[2] Hüttinger, K.J., and Greil, P., "Ceramic Composites for High Temperature Applications", (in German: "Keramische Verbundwerkstoffe für Höchsttemperaturanwendungen"), cfi/Ber., DKG **69** [11-12] (1992) 445
[3] Claussen, N., Travitzky, N.A., and Wu, S., "Tailoring of Reaction-Bonded Al_2O_3 (RBAO) Ceramics", Ceram. Eng. Sci. Proc. **11** (1990) 806
[4] Wu, S., and Claussen, N., "Reaction-Bonding of ZrO_2-containing Al_2O_3 ", pp. 293-300 in *Solid State Phenomena*, Vol. **25 & 26**, edited by A.C.D. Chaklader and J.A. Lund (1992)
[5] Wu, S., Holz, D., and Claussen, N., "Mechanisms and Kinetics of Reaction-Bonded Aluminium Oxide Ceramics", J. Am. Ceram. Soc. **76** (1993) 970
[6] Holz, D., Wu, S., Scheppokat, S., and Claussen, N., "Effect of Processing Parameters on Phase and Microstructure Evolution in RBAO Ceramics", J. Am. Ceram. Soc. **77** [10] (1994) 2509

[7] Holz, D., Pagel, S., Bowen, C., Wu, S., and Claussen, N., "Fabrication of Low-to-Zero Shrinkage Reaction-Bonded Mullite Composites", to be published in J. Europ. Ceram. Soc., (1995)

[8] Wendorff, J., Garcia, D.E., Janßen, R., Claussen, N., "Sapphire-Fiber Reinforced RBAO", Ceram. Eng. Sci. Proc. **15** (1994) 364

[9] v. Minden, C., Boutz, M.M.R., Scheppokat, S., Janssen, J., and Claussen, N., "Superplastic Deformation of Zirconia/Alumina Composites Produced by Reaction Bonding", to be published in: Proc. of "Plastic Deformation of Ceramics", eds. Bradt, R.C., Routbort, J., and Brookes, C., Aug. 7-12, (1994)

[10] Wendorff, J., Janßen, R., Claussen, N., "The Fiber-Push-Out-Test at High Temperatures for Interface Characterization of Ceramic/Ceramic Composites", Poc. ECCM-CTS2, Sept. 13-15, Hamburg, (1994) 69

[11] Staudt, T.,"Properties, Corrosion and Possible Applications of Aluminum Titanate", (in German: "Eigenschaften, Korrosionsverhalten und mögliche Anwendungen von Aluminiumtitanat"), Ph. D. thesis, RWTH Aachen, Germany, (1988)

[12] Hori, S., Kurita, R., "Sintering of CVD Aluminum Oxide - Titanium Dioxide Powders", Int. J. High Technology Ceramics **1** (1985)59

FABRICATION OF CONTINUOUS FIBER-REINFORCED CERAMICS WITH A NANOSIZED MULLITE PRECURSOR

O. Reese, B. Saruhan, B. Kanka and H. Schneider
German Aerospace Research Establishment (DLR), Institute of Materials Research,
D-51140 Cologne / Germany

ABSTRACT

Chemically synthesized mullite precursor powders which are suitable materials for the production of continuous fiber-reinforced mullite composites, owing to their high sintering activity at relatively low processing temperatures were used as a matrix material. Since commercially available polycrystalline mullite fibers become instable at high temperatures, optimized slip-casting and sintering conditions were used which allowed hot-pressing of the composites at temperatures lower than 1250°C. A strong interfacial bonding between fiber and matrix has been observed due to the preferential grain growth which starts on the fiber surfaces and extends into the matrix.

1. INTRODUCTION

Mullite is a very attractive phase, due to its low thermal conductivity, low thermal expansion coefficient and high creep resistance (1). Mullite/mullite-ceramic composites recently gained great importance for high-temperature applications where oxidation and creep resistance under thermo-cycling and long-therm use are required.
Commercially available polymer-derived mullite fibers are stable nowadays up to 1250°C, although new fibers with better properties are under investigation (2). These fibers have very small diameters (~10 µm) and are delivered in bundles consisting of approximately one thousand single fibers. This makes the fabrication of continuous fiber-reinforced mullite composites with the traditional fabrication routes very difficult. In order to achieve a homogeneous distribution ot the fibers, each fiber should be individually surrounded by the matrix. Slip casting is a suitable fabrication route for such requirements, however an optimum slurry viscosity and wettability should be adjusted.

To the extent authorized under the laws of the United States of America, all copyright interests in this publication are the property of The American Ceramic Society. Any duplication, reproduction, or republication of this publication or any part thereof, without the express written consent of The American Ceramic Society or fee paid to the Copyright Clearance Center, is prohibited.

Chemically synthesized, nanosized mullite precursor powders (3,4,5) are attractive materials as matrix, however, fabrication of continuous fiber-reinforced composites with these powders is difficult due to their high surface areas and sintering activities. Chawla and coworkers have developed a technique to fabricate such composites by stacking and hot-pressing (1300°C, 35 MPa for 1h in vacuum) the impregnated fiber tapes (6). It turned out that the uncoated mullite fibers (Nextel 480) were totally degradated in the mullite matrix under these conditions. Fibers were able to survive if a double-coating (BN/SiC) was applied.

This paper presents the results on the fabrication of a continuous fiber (Nextel 440) reinforced mullite matrix composite where the as-received fibers remain intact after processing.

2. MATERIALS AND METHODS

2.1. Fabrication of Composites

The mullite precursor used in this study was provided by Condea Chemie (Brunsbüttel/Germany) and labelled Siral. The precursor powder consists of ultra-fine, nanometer-sized particles. Slurry preparation of such powders by the conventional slip-casting techniques displays difficulties, due to the lack of particle mobility (high viscosity) and wettability in the slurry, which is caused by the extremely high specific surface areas. Calcination of the precursor powders is a suitable way to decrease their surface areas and thereby to improve the properties of the slurry. Slurries for matrix were prepared as water suspensions and contain 1 wt.% of dispersion aids and binders, based on the solid content of the slurry. The solid content of the slurries was increased successively up to the saturation point where the slurry is no longer castable.

Fiber bundles provided by 3M corporation (Nextel 440) were pulled through the optimized slurry and were wound on a plaster of Paris mandrel and left to be dried, before cutting and forming. Finally the green bodies were hot-pressed at 35 MPa and 1225°C for 1h in air.

2.2. Characterization of Composites

X-ray powder diffraction (XRD) patterns were measured at room temperature by means of a computer-controlled Siemens D5000 powder diffractometer using Ni-filtered $CuK_{\alpha 1}$ radiation. Diffraction patterns were recorded in the 10°-80°, 2θ range, in the step scan mode (3s 0.02°, 2θ).

Microstructural observations were carried out both on the polished sections with an optical microscope and on the fracture surfaces with a Philips 525M scanning electron microscope.

3. RESULTS AND DISCUSSION

3.1. Optimization of Sintering Conditions and Crystallization Behaviour During Hot-Pressing in Air

BET-measurements of the powders calcined at 860, 1050, 1150 and 1200°C show that the surface area decreases sharply above 1050°C (Table I). X-ray diffraction data of the powders given in Table I indicate that the as-received mullite precursor consists of pseudo-boehmite of poor crystallinity and a coexisting SiO_2-rich amorphous phase. Pseudo-boehmite converts to transition alumina (γ-Al_2O_3) above about 400°C and is present up to 1150°C. Above 1250°C, mullite is the only crystalline phase. As the calcination temperature increases from 860 to 1200°C, the maximum solid content in the slurry where it is still castable increases too, and as a result, the green density of the samples produced from these slurries gradually becomes higher (Table I).

Table I: Properties of slurries prepared from mullite precursor powders after different calcination temperatures.

	Calcination Temperature (°C) for 5h				
	as-rec.	860	1050	1150	1200
Phase Assemblage	PB, A	γ, A	γ, A	γ, A	M, δ, A
Weight Loss (%)	-	18.4	18.6	19.1	19.5
Surface Area (m²/g)	332	274	236	146	50
Slurry Solid Content (%)	-	30	30	45	57
Green Density (% A.S.D.[+])	-	18.8	17.2	32.0	35.9

A.S.D.[+]: Apparent solid density is experimentally determined and includes the combined volume of solid material and sealed pores (7). It is assumed that the samples in the green form contain only open porosity. **PB** : Pseudo-Boehmite, γ : γ-Al_2O_3, δ: δ-Al_2O_3, **A** : Amorphous SiO_2-rich phase, **M** : Mullite

Figure 1 exhibits that the sintering activity of the green bodies decreases slightly up to 1150°C. Above this temperature a drastic loss was observed, indicating that the densification is hindered by the reduction of the specific surface area, mainly due to the formation of hard aggregates, as well as the crystallization of mullite above 1150°C.

In order to achieve high densifications at lower sintering temperatures, uniaxial hot-pressing assisted sintering techniques were applied. Thereby special care was taken to avoid fiber damage during processing. Densities of the samples, hot-pressed at various temperatures (1200 to 1250°C) and pressures (25 to 35

MPa) are presented in Table II. At a given sintering temperature, the degree of the densification is strongly affected by the uniaxial pressure: For instance, at a sintering temperature of 1250°C, the density of a sample which had a green density of 32 % A.S.D. increases from 87.5 % under 25 MPa uniaxial pressure to 92.1 % under 35 MPa. In turn, the variation of the sintering temperature from 1200°C to 1250°C under the same uniaxial pressure has no visible effect on the densification.

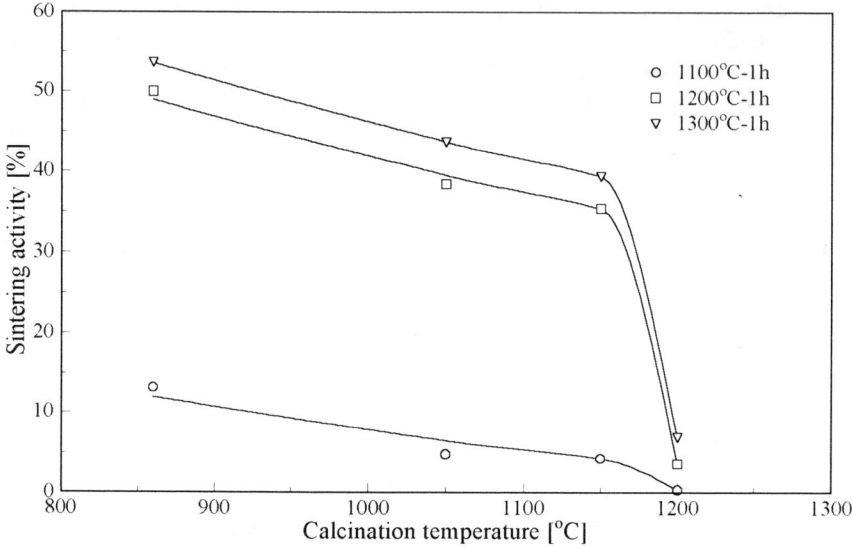

Figure 1: Variation of sintering activity with calcination temperature. Sintering activity is the difference between the relative sintered and green densities.

Table II: Relative densities (% A.S.D.) of the matrix samples for various hot-pressing temperatures and pressures

	Temperature (°C) for 1h		
Uniaxial Pressure (MPa)	1200	1225	1250
25	88.2	87.8	87.5
35	92.8	92.8	92.1

Green bodies were produced from mullite precursors calcined at 1150°C (5h)

3.2. Phase Development and Microstructural Observation

X-ray diffraction (XRD) studies on the pressureless sintered and hot-pressed samples indicate that uniaxial hot-pressing accelerates mullite formation as well as yielding higher densities (Figure 2).

Microstructural observations on the polished sections of the hot-pressed bodies show that the individual wetting and homogeneous distribution of the fibers have been achieved by the described fabrication route (Figure 3).

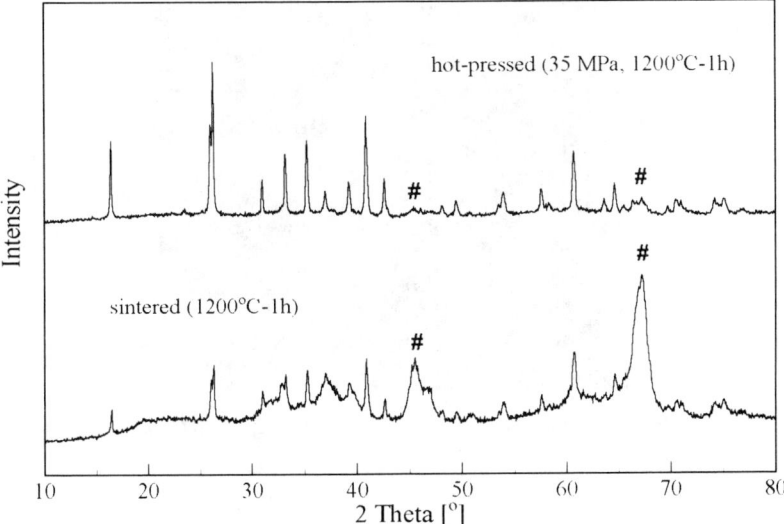

Figure 2: X-ray diffraction patterns of the samples pressureless sintered (lower) and hot-pressed (upper) under 35 MPa, sintering conditions in both cases: 1200°C, 1h. [# δ-Al_2O_3; all other peaks belong to mullite]

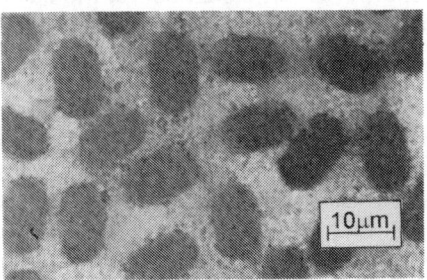

Figure 3: Optical micrograph of the composite, showing homogeneous fiber distribution (chemically etched polished section).

Scanning electron microscopic (SEM) studies demonstrate that a reasonable densification at the relative low hot-pressing temperatures has actually been achieved. Despite of the presence of interfacial reactions between the fiber and matrix a complete degradation of the uncoated fibers could be avoided (Figure 4). However, in order to achieve the production of damage-tolerant mullite composites with a relatively weak fiber-matrix bonding, the fibers must be coated prior to composite processing.

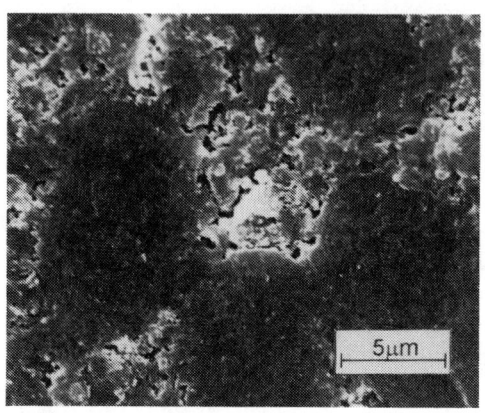

Figure 4: Scanning electron micrograph showing interfacial and interspacial relations in the composite (fracture surface).

References

1. Schneider, H.; Okada, K. and Pask, J.A., Mullite and Mullite Ceramics, Chapter 2, Crystal Chemistry of Mullite, pp. 4-77, John Wiley and Sons, Chichester, (1994).
2. Technical Data Sheet of 3M Corporation, St. Paul, Minnesota, USA.
3. Okada, K. and Otsuka, N., Characterization of the spinel phase from SiO_2 - Al_2O_3 xerogels and the formation process of mullite, J. Am. Ceram. Soc., **69** (9) 652-56 (1986).
4. Schneider, H.; Saruhan, B.; Voll, D.; Merwin, L. and Sebald, A., Mullite Precursor Phases, J. of European Cer. Soc., **11** 87-94 (1993).
5. Saruhan, B.; Albers, W. and Schneider, H., Improved densification and mullitization by seeding of reaction-sintered mullite with a sol-gel-derived precursor, J. of Material Science Letters, **12** 1812-1814 (1993).
6. Ha, J.S.; Chawla, K.K. and Engdahl, R.E., Effect of Processing and Fiber Coating on Fiber-Matrix Interaction, Mullit Fiber - Mullite Matrix Composites, Mat. Sci. and Eng. A 161 303-308 (1993).
7. British Standards Institute, No. 1902, Part IA, (1966 & 1982).

PROCESSING AND MICROSTRUCTURE OF CONTINUOUS FIBRE REINFORCED $MoSi_2$-BASED COMPOSITES

A.R.Bhatti, A.J.Pritchard and B.Mortimer
Structural Materials Centre, DRA Farnborough, Hampshire, GU14 6TD, UK.

ABSTRACT

High density continuous reinforced $MoSi_2$-based composites have been produced by hot-pressing technique. Both SiC and single crystal Al_2O_3 fibre reinforcements were investigated. Modification of $MoSi_2$ matrix with sialon powder helps to minimises the thermal expansion mismatch with the SiC fibre, thereby suppressing the matrix cracking. No matrix cracking was observed in Al_2O_3 reinforced $MoSi_2$ composites due to good fibre/matrix thermal expansion match.

1 INTRODUCTION

Molybdenum disilicide ($MoSi_2$) is an attractive matrix material for various high temperature aerospace and industrial applications due to a combination of high temperature oxidation resistance, high melting point, good thermal conductivity and ease of fabrication [1,2]. Mechanically, it undergoes a brittle-to-ductile transition at 1000°C. Creep effects could reduce its high temperature strength.

Recent work on the addition of SiC whiskers to $MoSi_2$ [3], has shown small improvements in both fracture toughness and high temperature strength. Substantial improvements are necessary to achieve the required levels of damage tolerance and strength for high temperature engineering applications. Reinforcing $MoSi_2$ with continuous fibres could achieve substantial improvements in mechanical properties. The use of Nb fibres in a $MoSi_2$ matrix has demonstrated significant improvements in toughness [4]. However, this type of fibre tends to react with the matrix at high temperatures, thus necessitating additional coatings in order to minimise reaction effects [5].

To the extent authorized under the laws of the United States of America, all copyright interests in this publication are the property of The American Ceramic Society. Any duplication, reproduction, or republication of this publication or any part thereof, without the express written consent of The American Ceramic Society or fee paid to the Copyright Clearance Center, is prohibited.

In past work, the use of SiC based, Textron SCS-6 SiC reinforcement to $MoSi_2$ matrix was investigated [6]. Matrix cracking occurred during the high temperature processing of this material which was attributed to the large thermal expansion mismatch between the SCS-6 SiC fibre (4.5×10^{-6} $°C^{-1}$) and the $MoSi_2$ matrix ($7-10 \times 10^{-6}$ $°C^{-1}$). This paper presents the results of modifying the composition of the matrix by the addition of sialon powder to reduce the thermal expansion mismatch between the SiC fibre and the matrix. The work was extended to incorporate Saphikon single crystal Al_2O_3 fibre. The thermal expansion of Al_2O_3 is ~ 8×10^{-6} $°C^{-1}$, a reasonably close match to that of $MoSi_2$ matrix.

2 EXPERIMENTAL

2.1 Fabrication of SiC fibre-reinforced $MoSi_2$/Sialon

A composite with 10 vol% fibre (Textron, SCS-6), reinforcing a matrix, comprising of 60 vol% $MoSi_2$ (Starck, Grade C) and 40 vol% Sialon (Vesuvius Zyalons, 101), was prepared via a three stage route. A SiC filament was wound onto a drum and coated with a slurry, consisting of the blended ceramic powders and dispersants with a polymer binder in an alcohol solvent. The sheet produced was dried, cut and stacked to produce the green composite which was then heated to remove the binder. The compact was then hot pressed in atmosphere in a graphite die at 1650°C under 18MPa for 1 hour. The composite had density >99% of theoretical.

2.2 Fabrication of Al_2O_3 fibre-reinforced $MoSi_2$

Single crystal Al_2O_3 fibre (Saphikon), reinforced $MoSi_2$ matrix composites were prepared using 100% $MoSi_2$ powder (Starck, Grade C). A slurry was prepared and doctor blade cast onto the wound up fibres. After drying, sheets were punched out, laid up, pressed and heated to remove the binder. The final consolidation was carried out by hot pressing as before. Two blocks of composites, with 16 and 20 vol% fibre, were produced. Both the composite blocks had density close to theoretical density.

2.3 Microstructural examination

The composites were examined by optical metallography, X-ray diffractometry (XRD) and scanning electron microscopy (SEM) with Energy Dispersive X-ray Analysis (EDAX).

3 RESULTS AND DISCUSSION

3.1 Microstructure of SiC fibre reinforced $MoSi_2$/Sialon

Fig 1. shows a typical optical micrograph of a transverse section of a hot pressed, SiC fibre (10 vol%) reinforced $MoSi_2$-40 vol% Sialon composite. The results of EDAX and XRD showed that the composite contained mainly $MoSi_2$ and β-Si_3N_4 with a small amount of SiC. Since the sample was not etched, the glassy phase [7] which presumably existed between the β-Si_3N_4 grains could not be seen. The composite matrix did not show any SiO_2 even though the starting $MoSi_2$ contained a significant amount of this phase. Some may have reacted with carbon from the binder material to give the small amount of SiC picked up in the matrix analysis. The remainder may have been consumed in the formation of the glassy phase.

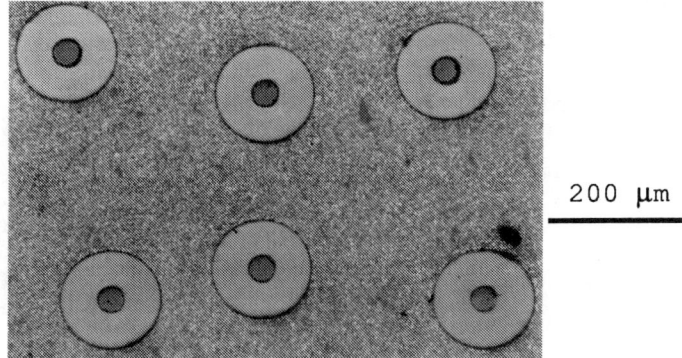

Fig 1. Optical micrograph of a transverse section of the hot pressed SiC/$MoSi_2$-40 vol% sialon composite.

There is no evidence of extensive cracking in the matrix as observed in the SiC-$MoSi_2$ composites investigated previously [6]. This is attributed to the reduction in the coefficient of thermal expansion mismatch between the $MoSi_2$ matrix and the fibre, resulting from the addition of sialon powder to $MoSi_2$. The CTE of a 40 vol% Sialon-$MoSi_2$ block was measured and found to be 7×10^{-6} °C^{-1} as expected for such a mixture of these components. The C-rich surface coating is observed as a ring of black contrast. This implies that no severe reaction has occurred between the C-rich surface of the fibre and the matrix during hot pressing cycle. The preservation of C layer is important as it is known to provide crack deflection, fibre pull-out and hence toughening.

3.2 Microstructure of Al$_2$O$_3$ fibre-reinforced MoSi$_2$

Fig 2. is a typical SEM micrograph showing a transverse section of a hot pressed Al$_2$O$_3$/MoSi$_2$ composite containing 16 vol% fibres. There is no evidence of any matrix cracking during processing of the composite. The matrix consisted of three different contrasting phases, light grey, grey, and dark grey. This change in contrast is due to the different Mo/Si ratios and was determined from EDAX analysis to be 0.8, 0.5 and 0.3 respectively. Using XRD, MoSi$_2$ was positively identified and is probably the grey phase. It was not possible to identify the other two phases with XRD and EDAX analysis. The phases do not match with any known phases from the phase diagram. Further work will involve the use of Electron Backscattered Diffraction (EBSD), TEM and associated analytical spectroscopies to identify these phases. The SiO$_2$ phase normally found in this type of matrix was not detected in either XRD or EDAX analysis. This can be explained by the volatilisation and escape of SiO$_{(g)}$ through the die and initially porous compact during the formation of the Mo rich phase.

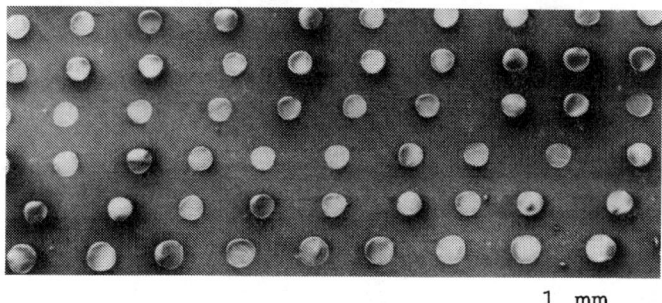

Fig 2. SEM micrograph of a transverse section of the hot pressed Al$_2$O$_3$/MoSi$_2$ composite.

The microstructure of the fibre/matrix interfacial region along with the compositional profile across the interface, are shown in figure 3. There is no evidence of any severe reaction at the interface. The 20 vol% fibre composite has essentially the same interfacial and microstructure features as the composite containing 16 vol% fibre.

Study of creep in monolithic and SiC-whisker reinforced MoSi$_2$ [8] has demonstrated that SiO$_2$ degrades high temperature strength due to grain boundary sliding of MoSi$_2$. The formation of SiO$_2$-free composites in the present work is therefore likely to have beneficial effect on high temperature mechanical properties. Work is underway on the mechanical evaluation of these materials.

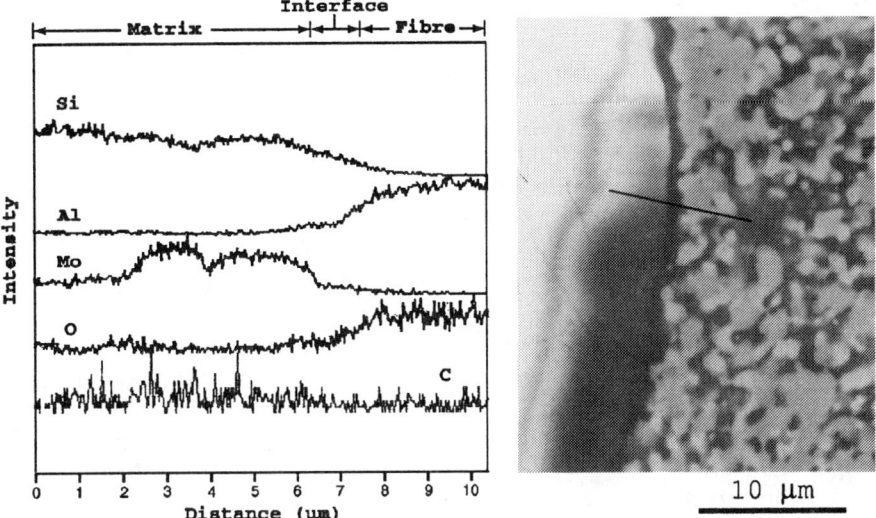

Fig 3. SEM micrograph showing the interfacial region and the compositional profile in the hot pressed $Al_2O_3/MoSi_2$ composite.

Finally it should be noted that thermomechanical stability of a fibre is a major requirement for high temperature structural composites. SCS-6 SiC and single crystal Al_2O_3 fibres were selected due to their potential for use at high temperatures. SCS-6 SiC fibres are structurally more stable than Nicalon and Tyranno fibres. However, they suffer from grain growth above 1400° C which reduces their strength. Single crystal Al_2O_3 fibre overcomes this problem of grain growth and retains strength to above 1400° C. The mechanism for the slow strength loss may be related to thermal effects on dislocation densities.

4 CONCLUSIONS

The main conclusions of the present work are :

1. The problem of matrix cracking during the processing of SCS-6 $SiC/MoSi_2$ may be overcome by the addition of 40 vol% sialon to $MoSi_2$, reducing the fibre/matrix overall thermal expansion mismatch.

2. Single crystal Al_2O_3 fibre-reinforced $MoSi_2$ composites with 16 and 20 vol% fibres can be successfully produced without any matrix cracking. This is attributed to good thermal expansion match between the Al_2O_3 fibre and the $MoSi_2$ matrix.

REFERENCES

1. P.J.Meschter and D.S.Schwartz, "Silicide-matrix materials for high temperature applications" JOM 52-55(1989)
2. J.Schlichting, "Molybdenum disilicide as a component of modern high temperature composites" (in German), High Temp-High Press., 10[3]3 241-269 (1978)
3. W.S.Gibbs, J.J.Petrovic, and R.E.Honnell, "SiC whisker-$MoSi_2$ matrix composites", Ceram Eng Sci Proc., 8[7-8] 645-648 (1987)
4. L.Xiao and R.Abbaschian, "Microstructure and properties of $MoSi_2$/Nb interfaces with and without alumina coating", Mater Res Soc Symp Proc 239 567-573 (1992)
5. H.E.Deve, C.H.Weber and M.Maloney, "On the toughness and creep behaviour of fibre reinforced $MoSi_2$ intermetallic", Mater Sci Eng A153, 668-675 (1992)
6. A.R.Bhatti, "Processing and characterisation of monolithic molybdenum disilicide and silicon carbide fibre-reinforced $MoSi_2$ matrix composites" Ceram Eng Sci Proc., Sep-October, 1068-1075 (1994)
7. A.R.Bhatti, M.H.Lewis, R.J.Lumby and B.North, "The microstructure of sintered Si-Al-O-N ceramics", J Mat Sci 15, 103-113 (1980)
8. K.Sadananda, H.Jones, J.Feng, J.J.Petrovic and A.K.Vasudevan, "Creep of monolithic and SiC-whisker reinforced $MoSi_2$", Ceram Eng Sci Proc. 12 [9,10], 1671-1678 (1991)

DEVELOPMENT OF Si-Ti-C-O FIBER REINFORCED SiC COMPOSITES BY CHEMICAL VAPOR INFILTRATION AND POLYMER IMPREGNATION & PYROLYSIS

S.Masaki, K.Moriya, T.Yamamura[1], M.Shibuya[2], H.Ohnabe[3]
[1] Research Institute of Advanced Material Gas-Generator (AMG), 4-2-6 Kohinata, Bunkyo-ku, Tokyo 112, Japan
[2] Ube Industries, Ltd., 1978-10 Kogushi, Ube-shi, Yamaguchi 755, Japan
[3] Ishikawajima-Harima Heavy Industries Co., Ltd., 3-5-1, Mukoudai-cho, Tanashi-shi, Tokyo 188, Japan

ABSTRACT

Continuous Si-Ti-C-O fiber reinforced SiC matrix composites were developed and evaluated. Three dimensional woven fabrics with carbon interface prepeared by chemincal vapor deposition (CVD) were densified by the combined process of chemical vapor infiltration (CVI) and polymer impregnation and pyrolysis (PIP). The composite exhibited anisotropy depending on the amount of fibers. The maximum tensile strength was more than 500 MPa at room temperature. Two kinds of *in situ* carbon rich surface layers were formed by fiber-processing. One is ultra-thin 10 nm graphite layer on graded carbon and the other one is graded carbon layer covered with thin silicon oxide. Both of them worked as effective interfaces.

INTRODUCTION

Future high performance advanced gas-generator requires an increase in temperature and pressure. The ceramic matrix composite (CMC) is an atractive material to apply to the componets of that. Major subjects to develop such CMC are (1)development of heat resistant fibers (2)fabrication process to achieve the highest yield of fiber performance (3)stable interface against the oxidation (4) finally, reduction of fabrication costs of CMC components.

The SiC/SiC composite by CVI (SiC/SiC$_{CVI}$) /1/ is well known, and this may be one of the most reliable CMC for high temperature use today. However, the properties of SiC/SiC$_{CVI}$ are strongly restricted by the CVI matrix which is stiffer and more brittle than the fibers. It results in quasi-isotropic of the composite /2/ and limits the fiber potential. Ishikawa et al ./3/, developed orthogonal three dimensional (3-D) woven fabric and its SiC/SiC composite by polymer impregnation & pyrolysis (SiC/SiC$_{PIP}$). SiC/SiC$_{PIP}$ showed remarkable tensile strength which was almost twice of ordinary plain woven or satin woven two dimensional (2-D) SiC/SiC$_{CVI}$. Meanwhile the matrix formed by CVI is stiffer than the fiber and far more than the matrix by PIP. The combination of SiC$_{CVI}$ and SiC$_{PIP}$ could provide another tailorable feature in the elastic properties.

The interface limits the oxidation resistance of CMC. Naslain et al ./4-6/reported the self-healing of SiC/SiC$_{CVI}$ to form SiO$_2$ generated by fiber and matrix. Thinner interface is more resistant against oxidation. The problem is to produce and to control extremely thin interfaces.

The present study aims to develop the tailorable SiC/SiC by introducing the controlable matrix. The composite strength is to be determined by the fiber ratio in the desired direction, and the modulus is also due to the combined matrix formed by CVI and PIP .

To the extent authorized under the laws of the United States of America, all copyright interests in this publication are the property of The American Ceramic Society. Any duplication, reproduction, or republication of this publication or any part thereof, without the express written consent of The American Ceramic Society or fee paid to the Copyright Clearance Center, is prohibited.

This is also an initial study to develop extremely thin interfaces formed by heat treatments of fiber and to assess the mechanical properties and oxidation resistance of these *in situ* interfaces.

1. MATERIALS

Orthogonal 3-D (X,Y,Z) fabrics were woven by Shikibo Co. with the continuous small diameter yarn fiber, Tyranno™ LOX-M (Si-31.6C-2.0Ti-12.4O ; % by weight) from Ube Co.. Woven fabrics ($100^W \times 200^L \times 4^T$ mm) nominally contained about 40 vol % fibers.

CVI process was carried out to coat 3 to 4 μ m SiC on fibers. Fibers in each strand were consolidated and bonded by this process. CVI was followed by PIP with "poly-titano-carbosilane" to fill the cavities among the strands. Polymer was cured at 1200° C in nitrogen atmosphere after each impregnation step. 4 PIP cycles were selected to shorten the process time in the present study. The volume fraction of the porosity after CVI and 4 PIP cycles was 15 to 20%. The secondary SiC phase was formed in the matrix by PIP and this phase is less stiffer and much weaker than the primary phase which was formed by CVI.

Composites were cut into test specimens and the surfaces were grinded. The dimensions of the gage area of test specimens were $8^W \times 3^T \times 40^L$ mm. Tensile properties at room temperature were mainly examined.

To evaluate oxidative stability, no seal coatings were carried out on the grinded surfaces. Some test specimens were thermally aged in air at 1200° C for 100 hours and tensile tested at room temperature..

Details of composites developed and evaluated in the present study are listed in Table I .

2. RESULTS AND DISCUSSION

2.1 Tailorable 3-D Composite with CVD Carbon Interface

The maximum tensile strength of developed composite was 540 MPa and almost all rupture strains were 0.8% or more. Typical stress-strain curves of all types are illustrated in Fig. 1.

Orthogonal 3-D CMC by the combined process are characterized by following three stages.
Stage I : the early linear slopes
Stage II : the successive non-linear slopes
Stage III: the relative linear slopes

Acoustic emissions (AE) were monitored during tensile tests. The first and largest event occurred at 0.05 % strain, which corresponds to the transition from stage I to stage II. According to the SEM observation on the fracture surfaces, the first AE event and transition in tensile moduli are considered to be mainly caused by the cracking or fracture between Y-Z planes which are perpendicular to the tensile direction X. Other breakage events can happen, such as cracking in strands X, debonding strands X from Y and Z strands in transverse directions. The transition stresses and strains from stage I to stage II are strongly related to the fiber ratio. The following non-linear behavior in stage II must result from the cracking of the matrix, especially occurring in the CVI matrix, and the debonding at the interface.

Tensile strength of composites are from 48 to 54% of fiber strength measured by organic-bonded fibers. Only fibers are effective on the properties in the final stage III.

Figure 2 shows the relation of experimental tensile moduli in stage I (E1) to the theoretical values calculated by simple "parallel-serial model". They show good agreement. Here, moduli assumed are 190 GPa for the fiber, 300 GPa for the CVI matrix and 60 GPa for the PIP matrix. The tensile moduli in stage III (E3) are similar to those of fibers orientated in the tensile direction. According to the model, theoretical E1 values for various fiber ratios are calculated as illustrated in Fig. 3 versus CVI matrix volume ratio in matrices. High modulus CVI matrix and low modulus PIP matrix provide CMC which has the widely valiable modulus.

The oxidative test at 1200° C for 100 hours showed a 31% reduction in the tensile strength, a 12% reduction in the tensile modulus, and the rupture strain decreased to 0.5%. This retention after oxidation is superior than the ordinary SiC/SiC$_{CVI}$ without a seal coating by CVD. Not only thinner carbon interface but also the matrix formed by poly-titano-carbosilane are considered to be effective on the improvement of oxidation resistance.

2.2 *In situ* Thin Carbon Interfaces

a) Thin carbon interface

Figure 4 gives the Auger depth profile of the surface of Lox-M fiber as fabricated and shows the 50 nm thin graded carbon rich surface layer. TEM thin foil analysis were also conducted on Lox-M fiber as fabricated and on the fully processed composite. Figure 5 shows the microstructure of the interface layer in the composite. It indicates thin lamellar carbon structure and no detectable SiO_2. A Lox-M fiber as fabricated does not show the lamellar carbon structure by TEM analysis, therefore this lamellar carbon interface was formed during the composite fabrication process.

The fabric was directly filled by CVI without CVD carbon interface to it. The density of the composite was 1.76 g/cm^3 nominally and contained 40 vol % porosity. The typical stress-strain curve is shown in Fig. 6.
SiC/SiC$_{CVI}$ with *in situ* interface showed slightly lower tensile strength and lower rupture strain. The behavior of both composites are quite different in stage II. It is obvious that this extremely thin carbon layer acts as an effective interface at room temperature.

Tensile strength, tensile moduli and rupture strain after oxidation were similar to that of as fablicated composite.

b) SiOx/carbon interface

Mild oxidation process was developed for Tyranno fiber. Depth profile by Auger analysis, as shown in Fig. 7A, indicates that a thin SiOx surface layer (<10 nm) is formed over the 40 nm thin graded carbon inner layer /7/. By this heat treatment, the tensile strength of Lox-M fibers were improved from 3.3 GPa to 4.2 GPa. The fabric with improved Lox-M was densified by the combined process and by the PIP process. Experimental data of the combined process were rather different, and further improvement is necessary for CVI.

All specimens evaluated were densified by PIP (8 cycles). TEM thin foil analysis indicated 10-20 mm produced thin lamellar carbon layer between fibers and the matrix formed by PIP.

Figure 6 shows a typical stress-strain curve of this composite by PIP. Tensile modulus (E1) is 60 GPa which is a half of the composite by the combined process. The stress-strain curve of the composite by the combined process looks like the combined curve of those by CVI and PIP. It starts as SiC/SiC_{CVI} and shifts to SiC/SiC_{PIP} in stage II. Tensile strength of SiC/SiC_{PIP} with SiO_X/C interface was 330 MPa.. This value is lower than expected from the strength of the improved fibers. The fracture mechanism of CMC is strongly related to the debonding at interfaces and the strength of composite is not determined only by the fiber strength.

Auger depth analysis on pulled out fibers at the fracture surface shows similar profile in Fig. 7B to that of improved Lox-M, except the SiOx outer layer. It can be explained by that the PIP matrix strongly bonds to the SiOx outer layer, and the quite limited part of carbon rich layer beneath acts as an interface.

Significant degradation in tensile properties was not observed after oxidation test at 1200° C for 100 hours. Both, the thin graded carbon layer covered by SiOx and the matrix generated by poly-titano-carbosilane, cause oxidative stability.

CONCLUSIONS

Orthogonal 3-D woven fabrics were densified by the combined CVI and PIP process and showed anisotropy due to the soft matrix formed by PIP.

The tensile strength of the composites depend on the amount of fibers. The maximum strength was 540 MPa when the fiber volume fraction in tensile direction was 30%. Tensile moduli of these composites are also controllable over wide range due to the cimbination of high modulus CVI matrix and the low modulus PIP matrix. This CMC is tailorable for the components according to the design requirements.

To improve the oxidation resistance of carbon interface by self-healing of SiC, two types of thin *in situ* carbon interfaces were developed and evaluated. It appears that they are more protective against oxidation than SiC/SiC with ordinary thick carbon interface. They also showed that the lamellar 10 nm thin carbon layers are effective as interfaces with respect to the mechanical properties.

REFERENCES

1) R.Warren, Ceramic Matrix Composites, Chapman and Hall, New York (1992)
2) W.S.Steffier, Ceram.Eng.Sci.Proc., 14[9-10](1993)1045-1057
3) T.Ishikawa, M.Shibuya, T.Hirokawa, SAMPE Proc. (1993)185-190
4) P.Lamicq, Japan-Europe Symp.on Comp.Mat.Proc.(1993)4-9
5) L.Filipuzzi, G.Camus, R.Naslain, J.Thebault, J.Am.Ceram.Soc.,77[2](1994) 459-480
6) J.Jouin, F.Lamouroux, R.Naslain, HT-CMC1.Proc.(1993)707-714
7) B.A.Bender, T.L.Jessen, Ceram.Eng.Sci.,Proc.,14[9-10](1993)931-938

Table I. 3D composites used in this study

NO	X:Y:Z	Vf	RVf [type]	Vcvi	Vpip	Interface
#6-1	A (1: 1:0.21)	0.39	0.18 [A(X)]	0.23	0.16	
#6-2	A (1: 1:0.21)	0.39	0.18 [A(X)]	0.23	0.16	
#6-3	B (1:0.6:0.14)	0.44	0.25 [B(X)] 0.15 [B(Y)]	0.21	0.11	CVD-C (30 nm)
#6-5	C (1:0.3:0.13)	0.43	0.31 [C(X)] 0.09 [C(Y)]	0.17	0.08	
#6-8	A (1: 1:0.21)	0.39	0.18 [A(X)]	0.26	-	In situ thin C (10nm)
#8-1	A (1: 1:0.21)	0.39	0.18 [A(X)]	-	0.60	SiOx/C (10nm/40nm)*

RVf ; Fiber volume fraction ratio in tensile direction
[type] ; [Fabric type (tensile direction)]   * ; Obtained by mild oxidation

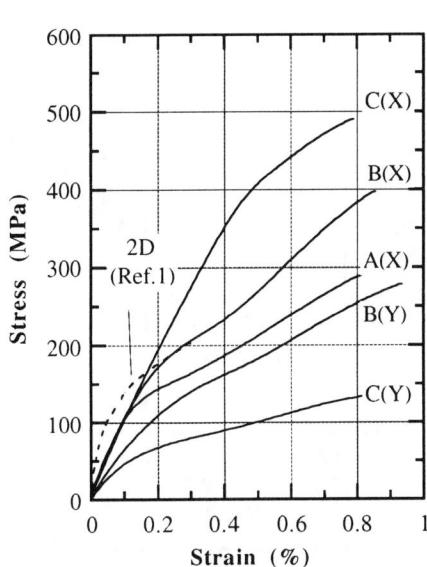

Fig.1 Stress-strain curves of 3D composites

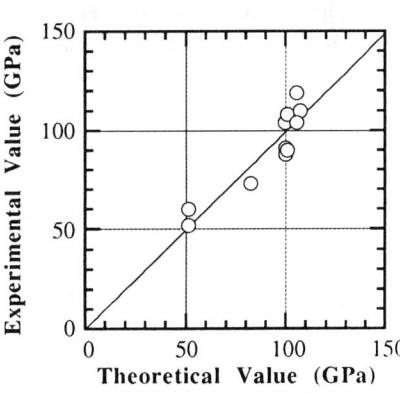

Fig.2 Relationship between theoretical and experimental values of tensile moduli at Stage I (E1)

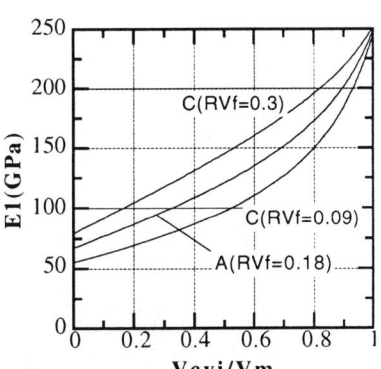

Fig.3 Theoretical E1 values for various fiber ratio versus CVI matrix volume ratio in matrices

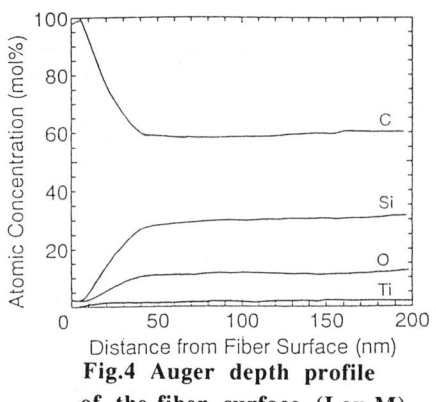

Fig.4 Auger depth profile of the fiber surface (Lox-M)

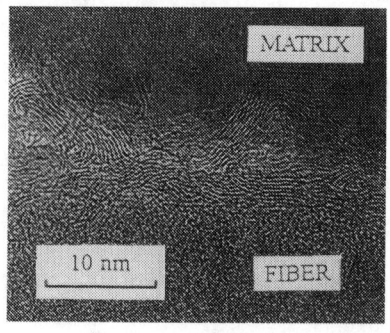

Fig.5 Transmission electron micrograph of the thin in site interface layer

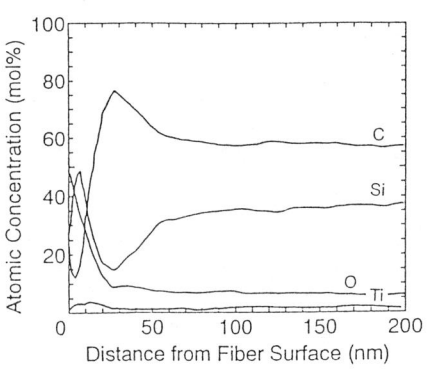

A. Heat-Treated

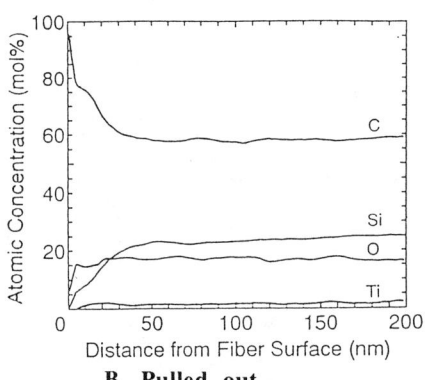

B. Pulled out

Fig.7 Auger depth profile of the fiber with SiOx / carbon interface

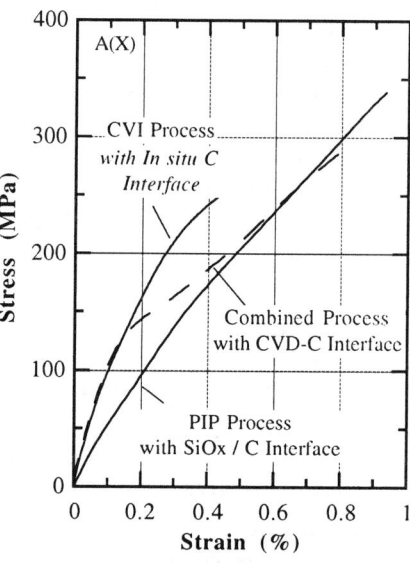

Fig.6 Stress-strain curve of 3D composite with in site interface

LIQUID INFILTRATION AND PYROLYSIS OF SiC MATRIX COMPOSITE MATERIALS

S.Casadio, A.Donato*, C.A.Nannetti, A.Ortona(°), M.Rescio(°°)
ENEA, INN-NUMA, C.R. Casaccia - S.Prov. Anguillarese 301, 00060 Roma;
*ENEA, ERG-FUS C.R. Frascati Roma; (°)Internova S.r.l. Milano;
(°°)CNRSM Brindisi (Italy)

ABSTRACT

SiC matrix composites were prepared by prepregging carbon and Nicalon fibre cloths with polycarbosilane (PCS) solution or nanosized SiC powder dispersion in PCS solution. After consolidation of the stacked cloths and pyrolysis, densification of the matrix was accomplished by multiple infiltration/pyrolysis steps with PCS solution. The pyrolysis behaviour of the SiC nanopowder/PCS matrix material was investigated in comparison to PCS.

INTRODUCTION

Ceramic Matrix Composites (CMC) are used in the aerospace industry for applications under extreme temperature conditions and mechanical loads. New components with improved performances will be available as new manufacturing processes are developed (1,2).
SiC-SiC and C-SiC CMC are interesting for future thermonuclear fusion reactors applications too. CMC's advantages over metal alloys in fusion reactors are: good high temperature properties, low radioactivity by neutron transmutation, low plasma contamination, low specific weight and large availability of raw materials (3).
Building large components needed for fusion reactors (4), by the industrial Chemical Vapor Infiltration (CVI) process, requires very long manufacturing times and high production costs. New processes, including subcomponents joining (5) are needed to produce materials capable to withstand the fusion reactors envisaged operating conditions.
In this paper some preliminary results on the development of woven C and SiC fibres/ SiC matrix 2D composites are reported.
The aim of this work was to evaluate the feasibility of CMC preparation through a process based on a limited number of infiltration/pyrolysis steps using a preceramic polymer and a

---

To the extent authorized under the laws of the United States of America, all copyright interests in this publication are the property of The American Ceramic Society. Any duplication, reproduction, or republication of this publication or any part thereof, without the express written consent of The American Ceramic Society or fee paid to the Copyright Clearance Center, is prohibited.

filler (1) which, in our case, is based on SiC nanosized powders. A relatively short CVI step is foreseen to fill and close the residual open porosity (6).

MATERIALS

Polycarbosilane from Nippon Carbon was employed as matrix precursor material. The polymer (density 1.1 g/cm^3) was dissolved in toluene to perform the infiltration steps.
The almost fully crystalline beta SiC nanopowder was prepared by Laser assisted synthesis from gaseous precursors (7) and presents the following characteristics: Surface Area= 60 m^2/g; Oxygen = 1.0%; Total carbon= 30.6% (by weight), density 3.2 g/cm^3.
Plain weave high modulus carbon fibres (C 003530, Goodfellow, England) and uncoated Nicalon NL 217 fibres with a mass of 200 g/m^2 and nominal thickness 0.22 mm were used.

EXPERIMENTAL

<u>Matrix Characterization</u> - The pyrolysis behaviour of PCS was investigated by I.R. spectroscopy and simultaneous TG/DTA analysis in argon coupled with mass spectrometry of the evolved gases. Significant differences were not found between this PCS and the one (PC 470) reported by Hasegawa (8). To characterize the matrix material, PCS and PCS/SiC dried and powdered blends, obtained by mixing SiC dispersions and PCS solutions in toluene, were compacted at high density (95% of Theoretical Density (T.D.) as evaluated by the rule of mixtures) in form of small pellets. The pellets were then slowly pyrolysed in a graphite resistance furnace in flowing argon (oxygen content 2-3 v.p.m.) up to 500 °C; the flow was then stopped and the temperature raised up to 800 °C. The pellets were then characterized in terms of linear shrinkage, weight loss, apparent density and porosity, total carbon and oxygen content. Only weight loss and chemical analysis could be performed for pyrolysed PCS due to melting with gas evolution and consequently bloating of the pellets.
Weight loss and chemical modifications were attributed only to PCS (effectively a blank SiC powder pellet was found practically unchanged by the thermal treatment); the carbon and oxygen contents, related to pyrolysed PCS in the blends, were thus calculated by the results of chemical analysis of the pyrolysed blends, taking into account the weigth loss too. The results (Tab. 1) show that SiC powder has a marked effect in reducing the linear shrinkage upon pyrolysis, while the relative density of the mixtures shows a less marked increase. Assuming that no macroscopic shrinkage may occur in the matrix embedded in a rigid fibre skeleton and considering a highly densified SiC/PCS green matrix, it was evaluated that the relative density of the pyrolysed matrix may be increased from about 32% to about 36 and 45%, by using, respectively, 25 and 50% by weight of SiC powder in mixture with PCS.

Tab. 1 - Characteristics of pyrolysed PCS and PCS/SiC blends

PCS/SiC blends (% by w.)	ΔL/Lo (%)	Rel. Dens. (% T.D.)#	Rel. Dens. ** (% T.D.)#	Weight loss (%)	Carbon (w. %)	Oxygen (w. %)
100/0	-	-	32.3	32.8	34.7	1.4
75/25	19.9	51.8	36.3	- 39.1*	31.5 32.9*	6.8 9.9*
50/50 (800 °C)	7.4	55.6	44.7	- 41.9*	31.3 33.1*	3.6 8.3*
50/50 (540 °C)	-	-	-	- 36.9*	34.0 36.5*	1.2 2.0*

* *related to pyrolysed PCS* ** *assuming zero macroscopic shrinkage*
\# *Theroetical Density (T.D.) by the rule of mixtures, assuming 2.2 g/cm^3 (9) as the density of pyrolysed PCS.*

In addition the results show that the high surface area SiC powder induces remarkable modifications either in the PCS weight loss upon pyrolysis and in its final carbon and, mainly, oxygen content. Oxygen is picked up from the environment at temperatures higher than 540 °C in the late stage of PCS conversion to SiC, when large amounts of hydrogen and methyl groups in side chains of PCS are evolved and extensive structural rearrangement occurs.

Indeed oxygen was not significantly increased in the absence of SiC powder and in mixtures pyrolysed at 540 °C, as shown in Tab.1. Moreover the oxygen pick-up was found to depend on the SiC powder surface area since by using a lower surface area (~15 m^2/g) commercial SiC powder, the final oxygen content increased less with respect to blends containing the same amount of high surface area SiC Laser powder.

<u>Composites processing</u> - Nicalon NL 217 or carbon fibres were prepregged by dipping in highly concentrated PCS solutions or PCS/SiC solutions/dispersions in toluene. SiC powder was previoulsy dispersed in toluene, then mixed with PCS solution.

After solvent evaporation under controlled conditions, prepreg layers (14-16 layers) were stacked into a mould and consolidated under vacuum at temperatures above the melting point of PCS (250-300 °C). A low mechanical pressure (1 MPa) was imposed to control the final thickness of the consolidated stack. The laminate was finally pyrolysed at 800 °C. Further infiltration steps, each one followed by pyrolysis, were performed under vacuum using highly loaded (60% by weight) PCS solutions. Intrusion of the PCS solution into the open porosity of the material was aided by subsequently imposing a small positive gaseous overpressure. Up to four further infiltration and pyrolysis steps were performed, evaluating weight and density increase of the composite after each step. The typical evolution of the characteristics of the composites prepared by using carbon fibres and PCS solution for both the prepregging and further infiltrations steps is reported in Tab. 2.

Tab. 2 - Evolution of the characteristics of a typical composite versus processing steps

Processing Steps	Infiltr. Efficiency (vol. %)	Apparent Density (g/cm^3)	Theor. Density (g/cm^3)	Porosity (%)	Matrix Porosity (%)
Consolid. Pyrolysis	79.1	1.1	1.91	42	72
I Infiltr. Pyrolysis	73.5	1.25	1.94	36	60
II Infiltr. Pyrolysis	58.7	1.4	1.96	29	51
III Infiltr. Pyrolysis	47.4	1.48	1.98	25.3	43
IV Infiltr. Pyrolysis	32.8	1.52	1.99	23.6	40

The apparent density values have been evaluated by mass and dimensions and are likely to be somewhat underestimated mainly due to weaviness and roughness of the specimen surface. The theoretical density and consequently the residual porosity have been evaluated by the rule of mixtures taking into account the fibres mass and density (1.79 g/cm^3 as measured by helium picnometry) and assuming, for the density of the matrix material 2.2 g/cm^3 (9). The matrix porosity has been computed by relating the whole porosity to the apparent volume available for the matrix. It is to be noticed that the density increase is reduced quite significantly in the fourth re-infiltration and pyrolysis step, not only due to the decreasing amount of available open porosity, but to a marked decrease of infiltration efficiency too. Practically, at this stage of processing, the total residual porosity is about 20%. It is to be noticed that if the whole porosity is related to the matrix apparent volume, a significantly higher value is to be taken into account (Tab. 2). The pore size distribution of the composite has been measured by Hg porosimetry. By the cumulative volume intruded by the mercury, the open porosity (16%) results to be somewhat lower than the total porosity shown in Tab. 2 and evaluated by dimensions and mass.

On the other hand, the bulk density corrected for the intruded open porosity (2.01 g/cm^3) is in fair agreement with the theoretical density evaluated by the fibres and matrix mass and density, indicating that closed porosity should be quite low and confirming that the total porosity may be somewhat lower than 20%. The main fraction of pores presents access size (to mercury) between 10 and 70 µm; a limited fraction ranges between 0.2 and 10 µm.

Transverse sections of the composite of Tab. 2 are shown in Figs.1 and 2; the matrix material results quite homogenously distributed; the interspaces among single fibres appear well filled too. Microcracks are quite uniform in size and distribution and are visible mainly in the matrix material separating the single layers. On the other hand the presence of a large amount of SiC nanopowder in the PCS solution used for the pre-pregging step (40% by weight with respect to PCS), prevents the infiltration inside the carbon fibre bundles, even if a further infiltration step with PCS solution is performed, as shown in the composite of Fig. 3. By using a lower amount of SiC nanopowder (less than 20% by weight with respect to PCS) and Nicalon fibres having a larger size (~15 μm) in comparison to carbon fibres (~7μm), infiltration inside the fibres bundles is not strongly prevented as shown in Fig. 4.

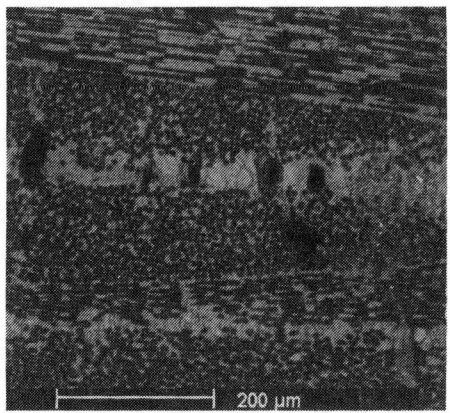

Fig. 1 - Tab. 2 composite. Transverse section (SEM)

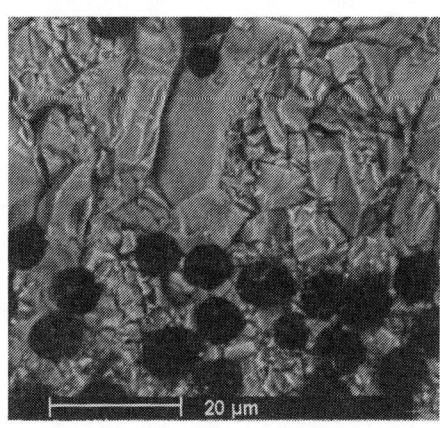

Fig. 2 - Tab. 2 composite. Transverse section (SEM)

Uncoated Nicalon fibres were used for the preparation of the composite shown in Fig. 4. The strong bond between fibres and matrix is clearly evident by the paths of the cracks developed in the matrix during its conversion to SiC. The stress imposed on the strongly bonded fibres by the shrinking matrix was locally so high to break many fibres both radially and axially. This strong bond may have influenced the cracks size and their uneven distribution too.

Fig. 3 - Comp. showing no infiltration inside fibres bundles (SEM)

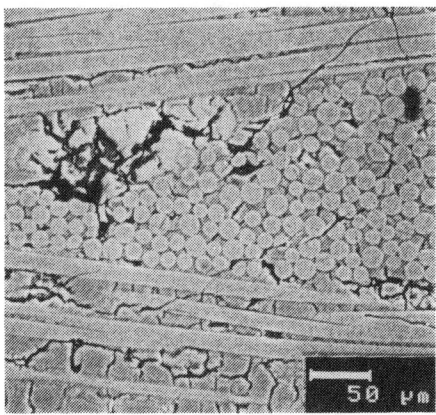

Fig. 4 - Comp. showing infiltration but fibres/matrix bond

CONCLUDING REMARKS

The pyrolysis behaviour of PCS is affected by the presence of high surface area SiC powder; remarkable amounts of oxygen are picked up from the environment, possibly due to SiC surface interaction with the pyrolysing PCS, "catalyzing" oxygen absorption from the gas phase.

High concentrations of SiC nanopowder in the PCS solution employed in the pre-pregging step, prevent the intrusion of the matrix precursor material inside the fibres bundles, at least when using carbon fibres. Four re-infiltration/pyrolysis steps, following the pre-pregging/consolidation step, seem to be well enough to obtain composites with a residual porosity around 20%; these composites seem suitable to be subjected to a final relatively short CVI treatment to partially fill and close the residual open porosity.

Work is in progress to explain and control the oxygen uptake in presence of SiC nanopowder, to optimize the prepregging and consolidation steps with reference to the optimum SiC nanopowder content and to its dispersion in the PCS solution in toluene and alternative solvents. Carbon and carbon coated Nicalon fibres will be used to get materials suitable to be evaluated in terms of mechanical properties both after processing by infiltration and pyrolysis and after the final foreseen CVI treatment.

REFERENCES

1) T. Haug, R. Ostertag, W. Schafer, R. Knabe, U. Ehrmann, J. Woltersdorf, "Processing and mechanical properties of CMCs by the infiltration and pyrolysis of Si-polymers" 6th EACM HT-CMC1, eds. R. Naslain, J. Lamon, D. Doumeingts, Woodhead Publ. Ltd, Abington Cambridge - England, 1993, 767.
2) A. Ortona, F. Cipri, "Structural Ceramic Composite Materials for High Temperature Applications", ATA-MAT 94, Oct. 1994, Turin (Italy).
3) A. Donato, R. Andreani, "Material Requirements and Perspectives for Future Thermonuclear Fusion Reactors", Fusion Technology, in press.
4) The ARIES-I Tokamak Reactor Study, Final Report 1991 UCLA-PPG-1323
5) A. Donato, P .Colombo, M.O. Abdirashid, "Joining of SiC to SiC using a preceramic polymer", present Conference.
6) Y. Wook Kim, J. Soo Song, S. Whan Park, J. Gunn Lee, "Nicalon-fibre-reinforced composites via polymer solution infiltration and chemical vapour infiltration", J. Mat. Sci., **28**, (1993), 3866.
7) E. Borsella, S. Botti, M.C. Cesile, A. Nesterenko, R. Giorgi, S. Martelli, G. Zappa, "Laser synthesis and microstructural characterization of ceramic nanosized powders", in "Laser Materials Processing: Industrial and Microelectronic Applications", E.Beyer, M.Cantello, A.V.La Rocca, L.Laude, F.O.Olsen, G.Sepold Eds., Proc. SPIE-2207, Apr. 5-8, 1994, Wien (Austria).
8) Y. Hasegawa, M. Iimura, S. Yajima, "Synthesis of continuous silicon carbide fibre: Part 2" J. Mat. Sci., **15**, (1980), 720
9) G.D. Soraru, F. Babonneau, J.D. Mackenzie, "Structural evolution from polycarbosilane to SiC ceramics", J. Mat. Sci., **25**, (1990), 3886.

INTERFACE AND MECHANICAL PROPERTIES OF CERAMIC FIBER REINFORCED SILICON NITRIDE COMPOSITES PREPARED BY A PRECERAMIC POLYMER IMPREGNATION METHOD

K. Sato, H. Morozumi, A. Tezuka, O. Funayama and T. Isoda
Corporate Research and Development Laboratory, TONEN Corporation,
1-3-1 Nishi-tsurugaoka, Ohi-machi, Saitama 356, Japan

ABSTRACT

The influence of fiber coatings on the mechanical properties were examined for unidirectional reinforced silicon nitride composites. To obtain excellent strength by a preceramic polymer impregnation (PCPI) method, the bonding between fiber and coating should be stronger than between coating and matrix. If this condition is not fulfilled, the coating layer peeled from the fiber during fabrication process. It caused direct bonding between the fiber and matrix.

INTRODUCTION

Ceramic fiber reinforced ceramics have been developed as durable materials for mechanical uses, which have heat resistance and oxidation resistance. Preceramic polymer impregnation (PCPI) method is one of the most promising processes to obtain near-net-shaping parts by two dimensionally reinforced composite with excellent strength. Authors have developed a PCPI method by using silazanes as a matrix precursor. The low viscosity, high wettability and high conversion yield to ceramic of perhydrosilazane or methylhydrosilazane showed advantages in obtaining strong composites, because dense materials were easily fabricated. [1,2,3]

The control of the interface between a fiber and a matrix is important for PCPI composites. Many studies on the mechanical properties of the interface have been performed for other composites. [4,5,6,7] This study shows the importance of controlling the interfacial strength both between coating and matrix and between coating and fiber for the composites which made by perhydrosilazane and methylhydrosilazane.

EXPERIMENTAL PROCEDURE

Matrix Precursor

Two types of silazanes, perhydrosilazane (PHPS) and methylhydrosilazane (SNC), were used as matrix precursors (Table 1).

The basic structural unit of PHPS is -(SiH$_2$NH)$_n$-. PHPS converts to silicon nitride by pyrolyzing in an inert or nitrogen atmosphere. SNC is a random copolymer, which is formed by -(SiH$_2$NH)- and -(SiMeHNH)-. This polymer converts to a mixture of silicon nitride and silicon carbide.

These polymers are low viscous liquids (10-80 mPa•s at 30°C) and cured by 100-300°C heating. By pyrolyzing up to 1000°C, the polymers convert to amorphous ceramics with char yield of 70-90 wt%, and density increases from 1.2 x 10^3 kg/m^3 to 2.5 x 10^3 kg/m^3. The amorphous phase is maintained to 1250 and 1450°C respectively for PHPS and SNC. Low molecular weight type of PHPS contains excess Si than stoichiometric silicon nitride, so it crystallizes easily.[3]

Table 1 Matrix Precursors

Polymer	Molecular weight (Mn)	Density (x10^3 kg/m^3)	Viscosity (mPa•s, at 30°C)	Composition (wt%)
PHPS	800-1,000	1.1-1.2	10-50	Si: 62.4, N: 28.2 C: 0.3, O: 1.9
SNC	800-1,000	1.1-1.2	40-80	Si: 56.4, N: 21.3 C: 14.5, O: 1.3

Reinforcement

Si-N fiber (SNF, TONEN Corporation, strength: 2.1 GPa, elastic modulus: 165 GPa) was used as reinforcement.

SNF was coated with carbon by CVD. SiCl$_4$-CH$_4$-N$_2$ and CH$_4$-N$_2$ gases were used as precursor for type A and type B coating respectively. Both of coatings were consisted with only carbon by the analysis of AES. The crystallinity and orientation of microcrystals were slightly different. By observation of the selected-area electron diffraction (SAED) patterns of coatings, type A coating consists of microcrystals of graphite which oriented with the c axis perpendicular to the fiber surface (Fig. 1(a)). In contrast, type B coating has no orientation (Fig.(b)). The thickness of both coatings is about 0.1 µm.

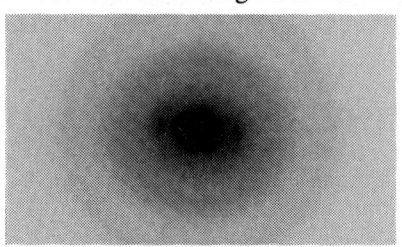

(a) Type A Coating (b) Type B Coating
Fig. 1 SAED Patterns of Carbon Coatings

Fabrication procedure

Unidirectional fiber prepreg was fabricated by winding a strand, which was impregnated with a polymer, on a drum. The prepreg was cut and stacked to aligned to the fibers in the same orientation. The laminate was pressed and cured by a vacuum bagging method as like the forming of fiber reinforced plastics. The curing condition was 0.05-0.1 MPa of pressurized nitrogen gas at 100-300°C. The cured material was pyrolyzed under a nitrogen atmosphere at 1250 and 1350°C respectively for PHPS and SNC. The pyrolysis temperatures were determined to maintain amorphous phase for each silazanes. Because of the shrinkage of the polymer, the density of pyrolyzed body was low. To obtain dense matrix, the impregnation of polymer followed by pyrolyzing was repeated for 5-7 times until saturating the increase of density (Fig. 2).

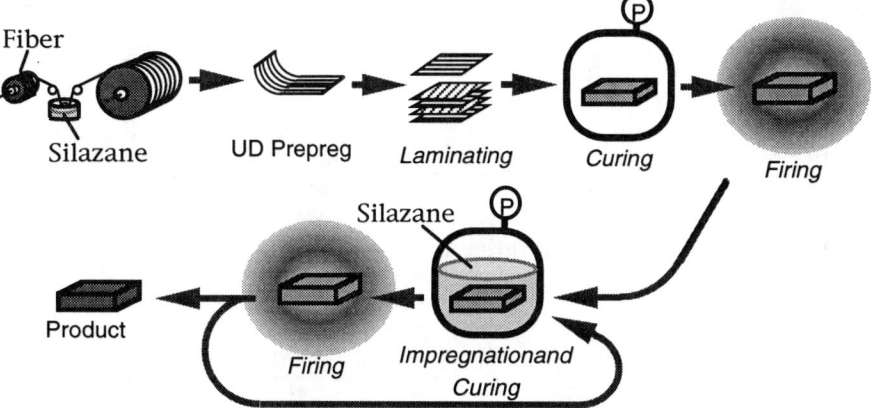

Fig. 2 Preceramic Polymer Impregnation (PCPI) Method

Characterization

The mechanical properties of composites were measured by three points flexural tests of thin beam (σ_{fl}) and inter laminar shear strength (ILSS) tests. The objective of the flexural test on thin beam is to evaluate fracture strength under the tensile stress dominant condition in place of tensile tests. A span was 30 mm, and specimen dimensions were 4 x 40 x 1 mm^3. Other conditions were according to JIS R1601 standard. ILSS was according to JIS K7057 standard (span: 8 mm, specimen dimensions: 8 x 12 x 2 mm^3).

The interfacial shear sliding strength (τ_s) and the interfacial shear debonding strength (τ_d) between fiber and matrix were measured by micro-indentation method. [5,6] SHIMAZU's microhardness tester (DUH-50, Japan) was used. The indentation load and loading rate were respectively 0.2N and 1.3 mN/sec. The displacement of a indenter was corrected by subtracting the displacement of the sample that was prepared with fibers without coating.

The τ_s and τ_d and were calculated using the following formulas: (5,6)

$$\tau_s = F^2 / 4\pi^2 R_f^3 E_f \delta$$

$$\tau_d = F_d / 2\pi R_f^2 \sqrt{2 E_m / E_f (1+v_m)^2 \ln(1/\sqrt{V_f})}$$

(R_f: fiber diameter, E_f: elastic modulus of fiber, E_m: elastic modulus of matrix, v_m: Poisson ratio of matrix, V_f: volume fraction of fiber, δ: displacement of indenter, F: load, F_d: debonding load).

The τ_d was calculated by the assumption that E_m and elastic modulus of SNF were the nearly same, because the precursor of SNF was similar to material with the matrix processors and the both of heat treatment temperatures of SNF and matrix were close. The Poisson ratio of matrix was assumed to 0.2.

The microstructure of composites and fibers were observed by SEM (JXA-8600MX, JEOL, Japan), TEM (JEM-3010, JEOL) and AES (PHI-650, ULVAC, Japan).

RESULTS AND DISCUSSION

Table 2 shows the mechanical properties and the interfacial sliding shear strength of SNF reinforced composites which were made with two types of carbon coatings. The mechanical properties of the composites mainly depend on τ_s and τ_d. The τ_s and τ_d were nearly the same value for the both composites prepared by using type A coating in spite of the difference of matrices. The τ_s of type B coating was two times of type A coating. No debonding load was observed in type B coating.

Table 2 Mechanical and Interfacial Properties of SNF Reinforced Composites

Matrix	Coating	Pyrolyzing temp. (°C)	Bulk density (x 10^3 kg/m^3)	V_f (vol%)	σ_{fl} (GPa)	ILSS (MPa)	τ_s (MPa)	τ_d (MPa)
PHPS	A	1250	2.33	46	0.80	44	59 [12]	~0
PHPS	B	1250	2.46	46	0.22	(17)*1	123 [34]	40-65
SNC	A	1350	2.35	54	1.05	45	66 [8]	~0

σ_{fl}: flexural strength, ILSS: inter laminar shear strength, τ_s: interfacial shear sliding strength, τ_d: interfacial shear debonding strength, []: standard deviation
*1: The sample showed tensile fracture.

The fracture surfaces of composites were quite different depending on their coatings. The fracture surface of type A, associated with low τ_s and high strength of composites, showed remarkable fiber pullout. For type B coating, only a few fibers pulled out and most fibers braked in the matrix crack plane (Fig. 3). The composite used with type B coating shows brittle tensile fracture not only on flexural tests but on shear strength tests.

In order to examine the reasons why two coatings caused quite different results, the pulled-out fibers were observed by AES. The typical AES profile of a pulled-out fiber with type A coating showed the remaining of a carbon coating on the surface (Fig. 4(a)). In contrast, for type B, the carbon coating was absent in most cases (Fig. 4(b)), in some cases the carbon coating was found on the surface of the matrix pit that formed after fiber pull-out. These observations indicate the weak adhesion of type B coating on the fiber and the strong adhesion of type A coating. However, the high τ_d and τ_s value of type B appeared to be contrary to this speculation. This contradiction will be explained by the damage of coatings during densification process of the PCPI method.

(a) Type A Coating (b) Type B Coating

Fig. 3 Fracture Surface of Composites

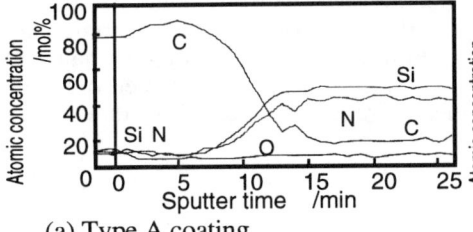 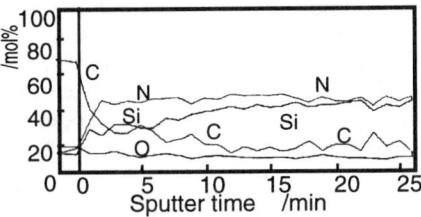

(a) Type A coating (b) Type B coating

Fig. 4 AES Depth Profile of Pulled-out Fiber

PCPI method implies reimpregnation and repyrolysis of the polymer, and the matrix repeats shrinking and cracking due to the conversion from polymer to ceramics. This causes tensile stress between matrix and fiber. If week interface is exist in between the matrix and fiber, the matrix debonds from fiber during pyrolysis. The space generated by the debonding is filled by following reimpregnation. The trace of the debonding and refilling is remained in the microstructure of composite (Fig. 5).

If coating is weakly bond to fiber, the coating is peeled off and naked fiber bonds hardly to matrix by reimpregnation.

If the adhesion between fiber and coating is stronger than the adhesion between coating and matrix, the debonding occurs between the coating and matrix, and the coating remains on the fiber.

In the case of type B coating, the bonding strength to fiber was week enough to be peeled off in some spots. The spotted bonding between the fiber and matrix caused high value of τ_d and τ_s. Other parts of coating remained after fabrication, and observed on fiber or matrix pit on fracture surface.

In order to obtain excellent strength of PCPI composites by designing coating, the bonding between fiber and coating must be stronger than the adhesion between coating and matrix. If multiple layer coating will apply to fiber, the bonding control between multiple layers will be important also.

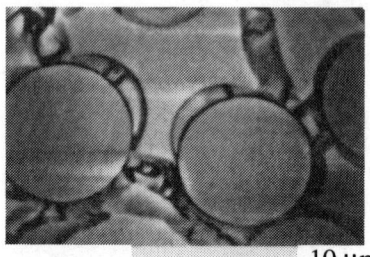

Fig. 5 Microstructure of the Composite

CONCLUSIONS

The importance of controlling the bonding between fiber and coating was revealed for PCPI composites. Because of the shrinkage of the matrix during fabrication process, the coating/fiber bonding must be stronger than the coating/matrix bonding. If this condition is not filled, coating is peeled off from fiber and the controlling of proper interfacial strength between fiber and matrix became difficult.

ACKNOWLEDGMENTS

This work was conducted by the Petroleum Energy Center with financial support from the Ministry of International Trade and Industry.

The authors acknowledge Y. Kagawa and K. Honda at the Institute of Industry Science of University of Tokyo for measuring and discussing the interfacial properties.

REFERENCES

1) K. Sato, T. Suzuki, O. Funayama, and T. Isoda, Ceram. Eng. Sci. Proc., 13, 614-621 (1992)
2) T. Isoda, Proceedings of Euro-Japanese Colloquium on Ceramic, 161-167 (1993)
3) H. Morozumi, K. Sato, A. Tezuka, H. Kaya, T. Isoda, Proceedings of the 8th CIMTEC, Florence, Itary, Jun 28 (1994), in press
4) E. Y. Sun, S. R. Nutt and J. J. Brennan, J. Am. Ceram. Soc., 77 [5] 1329-39 (1994)
5) C. G. Cofer, J. Economy, M. K. Feber and E. Lara-Curzio, Ceram. Eng. Sci. Proc.,Sep.-Oct., 447-455 (1994)
6) D. B. Marshall and A. G. Evans, J. Am. Ceram. Soc., 68- [5] 225-31 (1985)
7) K. Honda and Y. Kagawa, Proc. 2nd Japan International SAMPE Symposium, Dec. 11-14,858-63 (1991)

ADVANCED REFRACTORY CERAMIC-MATRIX COMPOSITES FROM POLYMER PRECURSOR FOR WORKING TEMPERATURES UP TO 1700^0 C

R.A.Rabinovitch, S.L.Gershkohen, L.G.Polyak, N.M.Balagurova, V.I.Bezruchenko, N.A. Vygovskii, E.A. Chernyshev

State Scientific-Research Institute of Chem.& Technol. of Organoelement Compounds, 38, Sh.Entusiastov, 111123, Moscow, Russia

ABSTRACT

A new refractory ceramic-matrix composite of a "ceramic fiber-ceramic matrix" type based on preceramic polyalumoxane siloxane precursor was developed. Identical chemical origin of components of the ceramic-matrix composite resulted in optimal compatibility of the fiber and matrix and in increasing the temperature of operation of such materials up to 1700^0C. By using cheap Al_2O_3-SiO_2 fibers containing up to 60% of aluminium oxide and preceramic polyalumoxane binders containing 100% of aluminum oxide refractory ceramic-matrix composites with enhanced thermal stability as compared to that of constituent fibers, were prepared.

INTRODUCTION

Preceramic polyelementoxanealumoxane binders with a high aluminium oxide content have a high chemical and adhesion activity. Hence, using these binders to produce ceramic-matrix composites based on Al_2O_3 and Al_2O_3-SiO_2

allows one to tailor the chemical and phase compositions of ceramic-matrix composites so as to obtain very stable corundum-mullite or mullite-corundum structures. This would make it possible to fabricate ceramic-matrix composites with thermal stability exceeding that of the constituent fibers. In this paper we present experimental evidence supporting that hypothesis.

EXPERIMENTAL

The technology of refractory ceramic-matrix composites included fiber dispersion in the binder and subsequent molding and annealing at temperatures up to 1500^0C. The first series of experiments was carried out on discrete alumoxane fibers with an aluminium oxide content of 90wt%. These fibers were prepared from fusible preceramic polyalumoxanesiloxane at 150^0C. Molded fibers were dried in wet air at 200^0C and then fired at 1200^0C. The fibers had a diameter up to 5 μm and their phase composition was α-Al_2O_3/mullite. Using the above described technology utilizing water soluble polyalumoxanesiloxane binders these fibers were processed into 400×300×40 mm plates which were subsequently tested at temperatures up to 1700^0C. The second series of experiments was carried out on cheap discrete Al_2O_3-SiO_2 fibers with aluminium oxide contents of up to 60wt%. These fibers were prepared from melt containing a mixture of aluminium oxide and silicon dioxide (which is a conventional way to fabricate kaoline fibers). In this case, we used 100%-aluminium-oxide-rich polyalumoxane binders to produce ceramic-matrix composites. Samples of these composites were further tested at temperatures up to 1500^0C.

DISCUSSION

Ceramic-matrix composites based on alumina fibers

High compatibility of the fiber and matrix having identical chemical origin makes possible the fabrication of products displaying virtually no shrinkage after the heat treatment stages. The linear shrinkage at 1700^0C was less than 1.5%. The phase composition of the material (α-Al_2O_3/mullite) provided high thermal stability and high thermal shock resistance. This lack of shrinkage permits to manufacture products of very complicated configurations allowing for very small tolerances. It also makes possible the fabrication of modular constructions "heater-thermal insulator" for high efficiency energy saving equipment.

Ceramic-matrix composites based on alumosilicates fibers

These fibers, when subjected to heat treatments, displayed linear shrinkage of 6-8%, while under operational conditions the products manufactured from such fibers showed a shrinkage of less than 2% for temperatures of $1500^0 C$. Usually, materials based on Al_2O_3-SiO_2 fibers the aluminium oxide contents of 40-60wt% operate at temperatures of up to $1300^0 C$ [1]. At higher temperatures, the strength is lost due to deformations caused by crystobalite formation. 100%-aluminum-oxide containing binders bind most of the excess silicon dioxide into mullite thus practically neutralizing the crystobalite formation. This results in substantially increased operation temperatures. The phase composition of the material corresponds to the system "mullite-α-Al_2O_3" (without crystobalite). At the same time, X-ray diffraction analysis of constituent fibers heat treated at temperatures higher than $1300^0 C$ always revealed the presence of α-crystobalite lines.

CONCLUSIONS

The achieved results can be considered as a starting point in developing a wide range of relatively cheap refractory ceramic-matrix composites based on various combinations of alumina-containing fibers and certain binders. An important issue to be addressed in further research is to understand the nature of interactions between fibers of various compositions and prerceramic polymer binders and to specify exact conditions under which stable corundum-mullite or mullite-corundum structures are formed.

REFERENCES

1. W.Eitel, "Special silicates sistems", pp 457-463, "The physical chemistry of the silicates", The university of Chicago Press, (1954)

FIBERS AND CERAMIC-MATRIX COMPOSITES FROM POLYMER PRECURSORS: METHODOLOGY OF PRODUCTION AND PROPERTIES

R.A.Rabinovitch, N.M.Balagurova, S.L.Gershkohen, E.A.Chernyshev

State Scientific-Research Institute of Chem.& Technol. of Organoelement Compounds, 38, Sh.Entusiastov, 111123, Moscow, Russia

ABSTRACT

Synthesis of continuous alumina fibers from polyalumoxanes is used as an example to demonstrate an approach allowing to optimize the initial stage of fibers technology development in order to save time and labour.

INTRODUCTION

Polymer ceramic precursors (polyalumoxanes, polycarbosilanes, polysiloxanes) are being more and more widely used to obtain fibers, ceramic-matrix composite, protective coatings and glues. The advantages of such precursors are related to comparatively low temperatures of formation of thermally stable ceramic structures ($1000-1400^0 C$) and to unique possibility to model the ceramic structure by means of chemical synthesis thus allowing to tailor the materials properties to exact needs. The use of ceramic fibers and ceramic matrices of identical polymer origin allows to achieve an optimal fitting at their interface and hence to obtain ceramic-matrix composites with excellent properties[1]. In the present paper we describe the way to organize the initial stage of developing a recipe for a polymer precursor used to manufacture continuous alumina fibers-forceramic polyalumoxane:

$$-[\underset{|}{\overset{OR_1}{Al}}-O-\underset{|}{\overset{OR_2}{Al}}]-$$

Alkoxy radicals in the lateral chains provide for the spinning properties of the polymer and for the solidification of the formed fibers. These radicals are eliminated during the high temperature firing.

EXPERIMENTAL

The aim of this work was to fabricate continuous alumina fibers with tensile strength of 2-2.5 GPa at room temperature and elastic modulus of 200-250 GPa. These fibers should preserve their strength at a level not worse than 80% of the room temperature value when maintained at temperatures up to 1200^0C for at least 10 hours. At the first stage of research a starting variant of precursor was synthesized so as to permit the spinning of a fiber with the fiber diameter of up to 20 micron. This stage ended up by developing the regimes of fiber production and by mechanical testing of the fibers at room temperature. These tests reveal the main macrodefects of the fibers structure and suggest improvements in technology necessary to eliminate gaseous bubbles, inclusions, irregularities of diameter, etc. (Fig. 1)

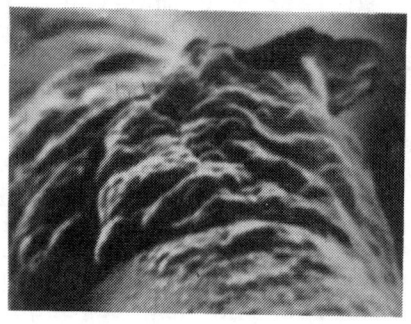 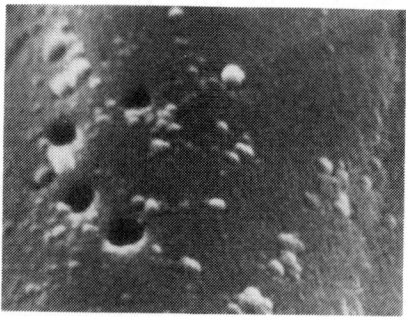

a   1 μm   b

Figure 1. Defects of technological origin:
a) solid phase inclusions, b) pores

The next stage involves high temperature testing of the fibers in order to understand mechanisms of the strength loss and to optimize the structure of our material. For the case in hand the loss of strength observed at high temperatures is caused by chemical interactions, crystallyzation and recrystallyzation of various phases and by phase transitions between several crystalline modifications, e.g. $Al_2O_3 - \gamma Al_2O_3 - \alpha Al_2O_3$ [2,3].

The problem is to define the critically important processes and to introduce corresponding corrections in precursor recipe. It was found that, in our experiments, the critical reason for the loss of strength at temperatures exceeding $1000^0 C$ was related to a phase transition from cubic to hexagonal modifications with subsequent recrystallization of aluminium oxide.

Therefore one needs to shift these adverse processes to the higher temperatures region. For polycrystalline alumoxide systems this can be done by introducing modifying dopants, e.g. silicon dioxide. Taking this into account it was the aim of the second synthesis to introduce such a dopant and to optimize its concentration in respect to the host phase. The additional phase should not alter the fiber spinning properties of the polymer precursor substantially. This consideration allows to sufficiently narrow down further search by making it possible to study the reasons for strength loss using the samples of precursor without processing them into fibers, but by simply imitating the thermochemical treatments employed in fibers fabrication on the precursor itself. With this aim in mind we synthesized the samples of polyalumoxanesiloxanes containing up to 20% of SiO_2 (after firing).

$$\begin{array}{ccc} OR_1 & OR_2 & OR_3 \\ | & | & | \\ -[Al-O-Al]-_n & & -[Si-O]-_m \\ & & | \\ & & OR_3 \end{array}$$

These samples were heat treated in the temperature range of $900-1300^0 C$ for time durations of 1 to 20 hours and then their phase composition and dimensions of aluminium oxide crystallites were studied by X-ray analysis. It was found that the optimal ratio is 90% of aluminium oxide and 10% of silicon dioxide. For such a ratio the hexagonal modification of aluminium oxide appeared at 1200°C only after 10 hours exposure. This composition of the precursor was used to fabricate the samples of continuous alumina fibers fulfilling the above specified requirements. The fine structure of these fibers (Fig.2) is comprised of γ-Al_2O_3 crystallites of about 7 nm in dimensions enveloped by amorphous silocon dioxide.

Testing monofilaments in order to obtain tensile strength remains the most time consuming operation. To save labour at this stage we developed a method of morphology analysis (express prognosis of quality: EPQ test) that allows to estimate the mean strength from information on surface defects. More than 100.000 fracture samples prepared on Instron-1122 breaking machine were analyzed when developing the EPQ test. Simultaneously with the stength measurements we determined for each sample the type and dimensions of critical defects by observing the fractures in Philips SEM 505 scanning electron microscope. Generalization of these results led us to conclude that more than 80% of critical defects are situated at the fiber surface and can be revealed by observation under an optical microscope (Fig.3).

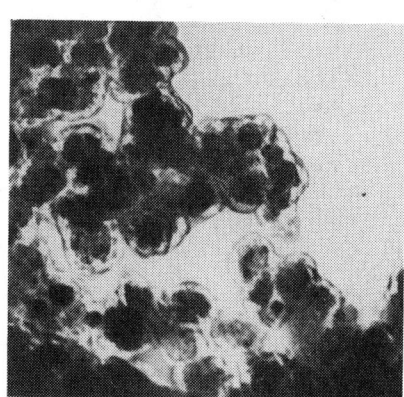
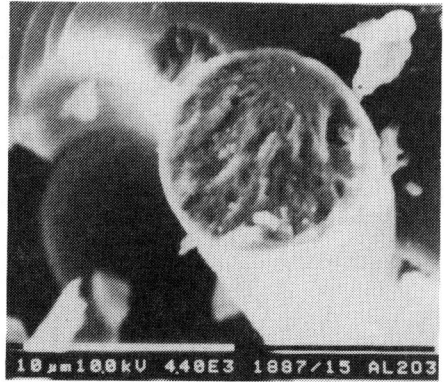

0.1 μm

Fig.2. Fine structure of Al_2O_3(90%)-SiO_2(10%) fibers after firing at 1200°C

Fig.3. Typical fracture of a monofiber after tensile stress measurement

The absense of visible defects at 300✕ magnification points to a high quality of the fiber while the average strength of the samples correlates fairly well with the ratio of "defectless" specimens in the sample. Hence the EPQ test involves the following sequence of operations. A sample of fibers (a yarn) is placed under an optical microscope which is focused at random on any place of the sample. The operator monitors the surface of the specimen and registers the occurrence of certain types of defects using a special classification. He counts the number of detected defects of a certain type and writes it down together with the number of cases when he encountered "defectless" regions. This procedure is repeated 30 times for randomly chosen parts of the sample. The results are summarized and the average strength is calculated using special tables. To estimate average strength from the EPQ test one needs about an hour. The discrepancy between the strength values obtained in a straightforward manner from tensile stress measurements and from EPQ test does not exceed 30%. EPQ test can serve as rapid assessment of the fibres quality including quality control at intermediate stage. However, this test, naturally, cannot substitute conventional methods when the product certification is needed.

CONCLUSIONS

Thus by using the approach outlined above one can optimize the initial stage of precursor recipe development in order to be able to fabricate fibers with specified parameters. These paramaters are achieved by correlating the precursor characteristics and the structure of ceramic which, in the final run, determines the properties.

ACKNOWLEDGEMENTS

The authors with to thank doctor Tatjana Chaplanova for providing the photographs of fine structure fibers.

REFERNCES

1. N.Korneev,R.Rabinovitch,S.Gershkohen,N.Balagurova, N.Vygovsky,Y.Pronin,E.Chernyshev, "Light weight Al_2O_3-Al_2O_3 ceramic composites from polymer precursors: properties and application", pp 181-184, in

"High temperature ceram. matrix composites HT-CMC-1" (R.Naslain,V.Lamon,D.Doumeingts, eds.), Bordeaux, (1993).
2. W.Eitel, "Special silicates sistems", pp 457-463, "The physical chemistry of the silicates", The university of Chicago Press, (1954)
3. C.Lysak, L.Dergaputskay, I.Kalynovskay, A.Gaody, E.Karjakina, "Phase transitions of fiber Al_2O_3", Neorganitheskie materialy, v.21, No 5, pp 802-807, (1985).

SIC - AND SI$_3$N$_4$-MATRIX COMPOSITES ACCORDING TO THE HOT-PRESSING ROUTE

K.Nakano,(NIRI-Nagoya), K.Sasaki, H.Saka(Univ.Nagoya), M.Fujikura(JUTEM-Gifu) and H.Ichikawa(Nippon Carbon Co.) Japan

ABSTRACT

1D Carbon- or BN-Coated Hi-Nicalon fiber reinforced SiC and Si$_3$N$_4$ matrix composites were fabricated by the slurry route followed by hot pressing. The mechanical properties such as flexural strength and fracture toughness as well as structures in the composites were characterized. In the case of carbon-coated fiber reinforced composites the strength and fracture toughness were nearly constant up to a critical temperature, then they abruptly decreased above the critical temperature.
Room temperature strength and fracture toughness of the composites reinforced with BN-coated fiber were considerably higher than those of high temperature.

INTRODUCTION

Silicon carbide (SiC) and silicon nitride (Si$_3$N$_4$) have excellent characteristics for use in high temperature structural materials and mechanical parts because of their heat-stability, superior strength at high temperatures and low density.
However, both materials are brittle in the monolithic state. The major concern in utilizing silicon carbide and silicon nitride in structural materials is improving toughness.
Fiber reinforcement has recently been employed as an effective method of toughening ceramics. Several processes have been employed for fabricating fiber reinforced ceramics. From the cost-effectiveness considerations, two techniques making use of polymer pyrolysis have high potential[1-7]. Whereas one technique utilize an organometallic compound[1,7], the others[2-6] additionally employs a mixture of powders including filler and sintering additive and it is termed a slurry route. In these techniques, either liquid state of the organometallic compound or the slurry are infiltrated into fiber preform. However, the both techniques need repeated infiltrations of the organometallic compound or the slurry for increase of density of a composite body unless pressure

sintering such as hot pressing or hot isostatic pressing are applied[8]).

Carbon or SiC fibers have been used as reinforcement for SiC and Si_3N_4 matrix composites. The former fiber is chemical active in oxygen-containing atmosphere above 500°C, whereas the latter one, which derived from organometallic compound such as Nicalon has high content of oxygen, and is limited for application at high temperature. Recently low-oxygen-containing SiC fiber, which fabricated from electron beam anaerobic curing process, has been developed[9,10] (Hi-Nicalon).
This fiber is more heat resistant than conventional Nicalon fiber. In this work carbon- or boron nitride-coated Hi-Nicalon fiber reinforced silicon carbide and silicon nitride matrix composites were fabricated by the slurry route and structure and mechanical properties were characterized.

EXPERIMENTAL PROCEDURE

Two kind of matrix composites (SiC and Si_3N_4) were fabricated and characterized. A SiC fiber [Hi-Nicalon (Nippon Carbon Co.): $14\mu m\phi$; 500 fil./yarn; $=2.74 g/cm^3$; Si 63.7, C 35.8, O 0.5 % in weight), which was coated with either pyrolytic carbon (PyC: $0.07\mu m$) or boron nitride (BN: $0.5\mu m$), was used as a reinforcement . The PyC coating on the fiber was carried out with CVD by using propane gas as a source material. BN was coated on the fiber using following reaction at 1500°C, 0.05~0.1 Torr: $BCl_3 + NH_3 \rightarrow BN + 3HCl$. Fine powders, either β-SiC(Ibiceram Ultrafine, $0.3\mu m$: which was mixed with 6% AlB_2 and 1% TiC as sintering additives) or α-Si_3N_4(UBE COA: which contained 5% Al_2O_3 and 5% Y_2O_3 as sintering additives: s.s.a. $=10.6 m^2/g$), were used as a slurry element(filler). An uniaxial fiber prepreg was made with the filament winding method by using a slurry. The slurry consisted of the filler, an organo-silicon resin[SiC matrix: polycarbosilane(Nippon Carbon); Si_3N_4 matrix: polysilazane(Chisso NCP 201)] and toluene.
The prepreg was cut into segments of equal size to from a postform in a manner so that the fibers were arrayed in the same direction(1D fiber reinforcement). In order to make a composite the post form was pyrolyzed(SiC: 700°C in Ar; Si_3N_4: 850°C in N_2), followed by hot pressing(SiC matrix: 1750~1800°C in Ar; Si_3N_4: 1550~1600°C in N_2).
Structure of composite was characterized either an EPMA, an Auger electron spectroscopy an optical microscope.

The EPMA micrograph was taken by using JEOL JXA-8900L for a composite surface perpendicular to fiber axis and the Auger spectrum was taken by using JEOL JAMP-10S.

The flexural strength (σ_f) was measured by 3-point bending whereas the fracture toughness(K_{IC}) was measured by 4-point bending single edge notched beam(SENB) method. Both properties were measured at room and high temperatures in vacuo.

RESULTS AND DISCUSSION

The composite fabricated in the present work had fiber volume fraction (V_f) of 46± 6%(SiC matrix) and 48 ±6%(Si_3N_4 matrix), respectively. Table-1 shows some properties of Hi-Nicalon fiber reinforced SiC and Si_3N_4 matrix composites. SiC matrix composites had higher open porosities(lower bulk densities) than those of Si_3N_4 ones. In the case of SiC matrix composite reinforced with C-coated Hi-Nicalon fiber, the strength in 1600°C was much lower than that of room temperature(RT) though the fracture toughness in 1400°C was nearly the same as that of RT. This trend was maintained in the case of SiC composite reinforced with BN-coated fiber. Similarly in the case of Si_3N_4 matrix composites the high temperature strength and fracture toughness were much lower than those of RT.
This tendencies were remarkable in the Si_3N_4 matrix composite reinforced with C-coated fiber compared to that reinforced with BN-coated fiber. In order to observe a temperature dependence of the strength and fracture toughness of the composite(reinforced with the C-coated Hi-Nicalon fiber), they were measured at various temperature under vacuum(Fig.1). The strength and fracture toughness had nearly constant values from RT up to a crital temperature(SiC matrix: ~1300°C, Si_3N_4 matrix: ~1200°C), then they abruptly decreased beyond the crital temperature.

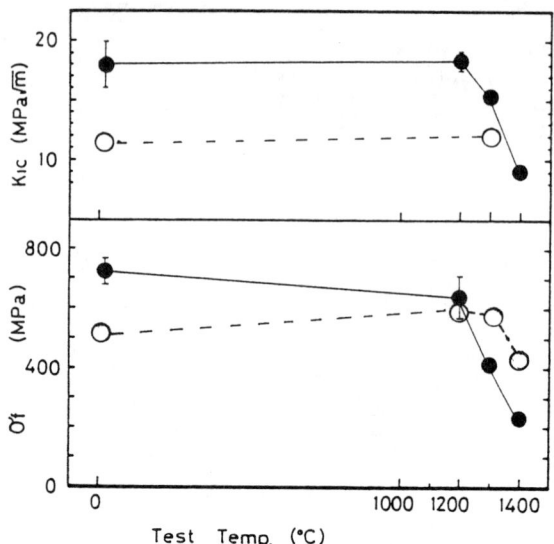

Figure 1 Test temperature vs. flexural strength(σ_f) and K_{IC} of C-coated Hi-Nicalon fiber reinforced composites.
○: SiC matrix(hot pressed at 1750°C); ●: Si_3N_4 matrix(hot pressed at 1600°C).

In the external appearance of specimens after bending test, RT and 1200°C tested ones showed no special features in the both matrix cases(Fig.2- (a), -(b), Fig.3-(a)). However, the specimen tested above the critical temperature showed either straight crack path(SiC matrix: tested at 1600°C: Fig.2-(c)) or zigzag crack path with pronounced delamination(Si_3N_4 matrix with C- and BN-coated fiber: tested at 1400°C: Fig.3-(b)). In the case of SiC matrix composite, bonding between fibers and matrix might be tight compared to that of Si_3N_4 composite, because the crack paths of SiC composite were rather straight than those of Si_3N_4 composite(cf. Fig.2 and Fig.3). Moreover, the Hi-Nicalon fibers which used as the reinforcement of SiC composite might have some deterioration because of high hot pressing temperature(1750~1800°C). These could be reasons that the fracture toughness and flexural strength of SiC matrix composite were lower than those of Si_3N_4 one.

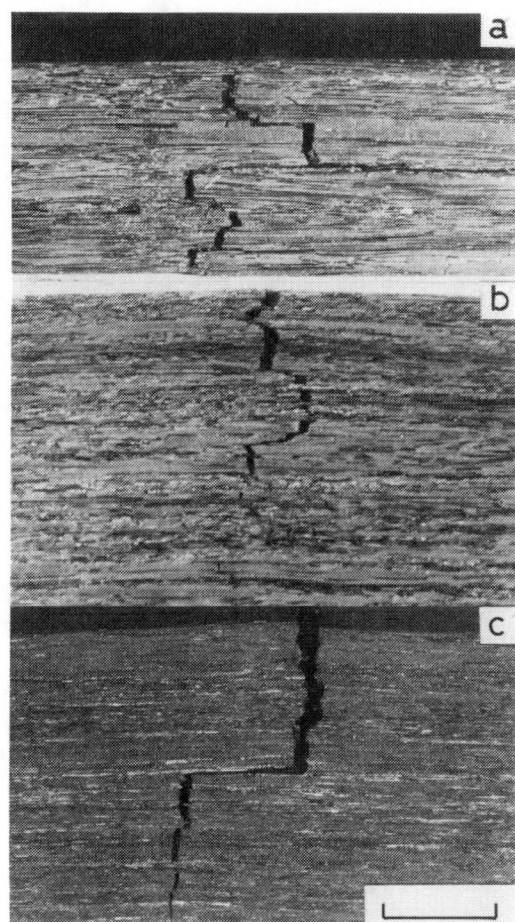

Figure 2 Crack paths on specimen of SiC matrix composite after bending test(bar scale: 1mm), a) reinforced with C-coated fibers, RT test; b) reinforced with BN-coated fibers, RT test; c) reinforced with BN-coated fibers, 1600°C test.

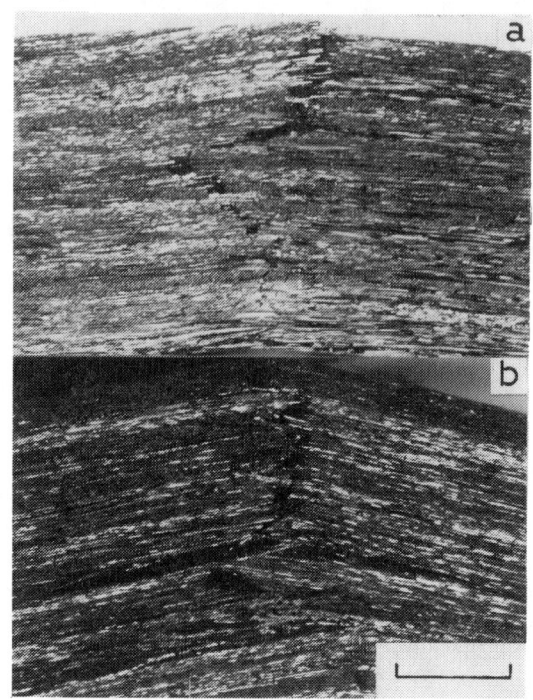

Figure 3 Crack paths on specimen of BN-coated fiber reinforced Si_3N_4 matrix composite after bending test (bar scale: 1mm), a) RT test, b) 1400°C test.

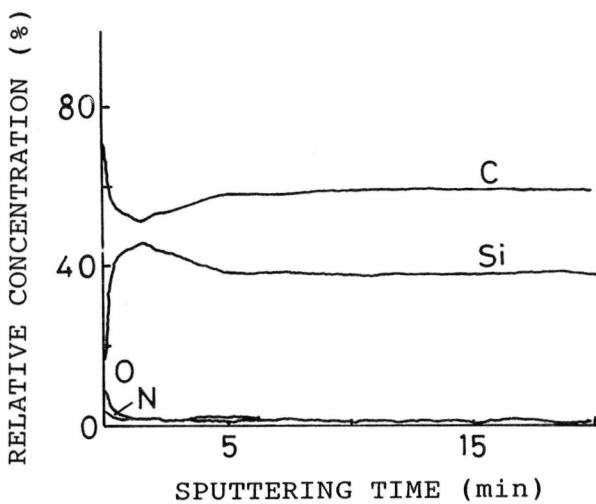

Figure 4 Surface analysis of Hi-Nicalon(SiC) fibers by Auger electron spectroscopy(etching rate: 12.1mm/min(SiO_2)).

Fig.4 shows surface analysis of a Hi-Nicalon fiber by Auger electron spectroscopy(AES). Unlike to a commercial Nicalon fiber [11,12], the Hi-Nicalon fibers had low concentration of oxygen even on a surface of the fiber. Fig.5 shows Auger spectrum from a fiber surface on a fracture surface of bending test specimen in Si_3N_4 composite reinforced with C-coated fiber [13].
Evidently the intensity from carbon atoms decreased with increase of the temperature in bending test, suggesting that the carbon element disappeared with the increase of test temperature. In a EPMA back-scattered electron micrograph of non-coated Hi-Nicalon fiber reinforced Si_3N_4 composite, a band zone, which had high content of Al and O, was observed periphery of each fiber(Fig.6-(a)). This zone may be formed from a reaction between the sintering additives(Al_2O_3, Y_2O_3) and the fiber.

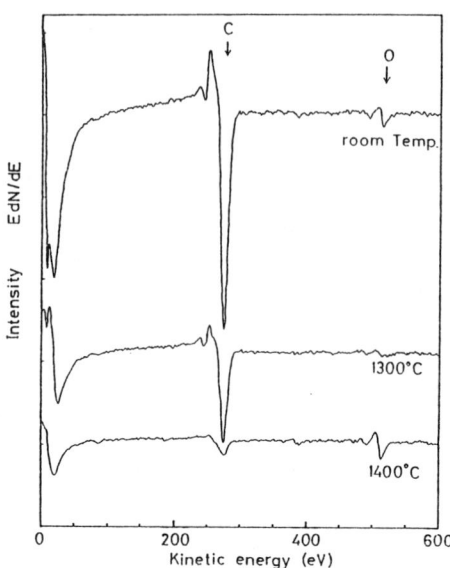

Figure 5 Auger spectrum from a fiber surface on a fracture surface of a bending test specimen in Si_3N_4 composite reinforced with C-coated fiber. The bending test was carried out RT, 1300 and 1400°C.

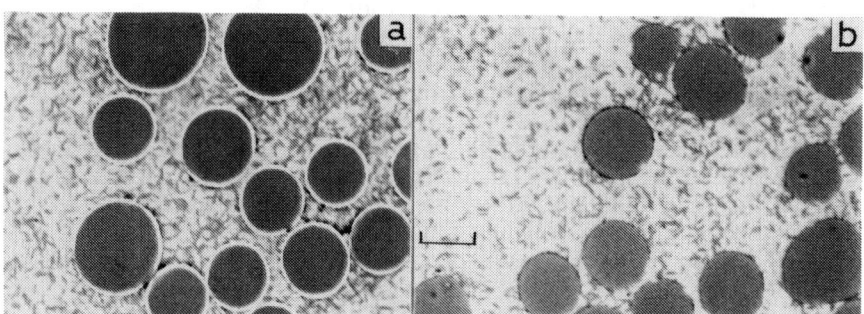

Figure 6 EPMA back-scattered electron image of Hi-Nicalon fiber reinforced Si_3N_4 composite(bar scale: $10 \mu m$), a) non-coated fibers, b) C-coated fibers.

However, in the C-coated Hi-Nicalon fiber reinforced Si_3N_4 composite, a gap space, which may be formed from dissipation of the coated carbon interphase, was observed around each fiber instead of the band zone(Fi.6-(b)).
The dissipation of carbon interphase may promote delamination of the C-coated fiber reinforced Si_3N_4 composite, which reflected to the decrease of the high temperature strength and fracture toughness of the composite(Table1, Fig.1).

Table 1 Some properties of Hi-Nicalon fiber reinforced composites

Matrix	Fiber	Hot pressing temperature (°C)	Open porosity (%)	Bulk density (g/cc)
SiC	C-coat	1750	11.4 (0.6)	2.56 (0.03)
	BN-coat		4.9 (0.2)	2.51 (0.02)
Si3N4	C-coat	1550	4.4 (0.4)	2.75 (0.01)
	BN-coat		2.9 (0.5)	2.81 (0.02)

Matrix	σf(MPa)			K_{IC}(MPa√m)		
	RT	1400°C	1600°C	RT	1400°C	1600°C
SiC	492 (57)		80	11.3 (0.9)	11.9 (0.8)	
	296 (46)		174 (20)	8.2 (2.0)		6 (2)
Si3N4	890 (10)	196 (14)		29 (8)	6.9 (0.40)	
	702 (60)	426 (10)		20.2 (1.7)	16.3 (0.6)	

number in a prenthesis show a standard deviation

Figs. 7 and 8 show EPMA back-scattered electron image and line profile of some element around a fiber in BN-coated Hi-Nicalon fiber reinforced SiC and Si_3N_4 matrix composites.

The BN interphase seems to loosely bond to the fiber in the both matrix cases.

B and N elements can be detected in the interphase in the both matrix cases.

Thickness of BN interphase around the fiber tend to decrease and in many fibers, the interphase seemed to disappear in the case of SiC matrix composite (Fig.9-(a)).

This may reflect to the tight bonding between the fiber and the matrix as mentioned before.

However in the case of Si_3N_4 matrix compsite BN interphase tend to be thicker than that of SiC one (Fig.9-(b)).

This interphase may act as a weak bonding between the fiber and Si_3N_4 matrix as well as protection of the fiber from chemical attack by the matrix.

Cracks straightly passed in the matrix and rounded about the interphase of fibers as seen in the Fig.9-(b).

These cacks were freguently observed prallel to fibber direction in the BN-coated Si_3N_4 matrix composite after high temperature bending test.

The cracks promoted delamination of the Si_3N_4 composite (Fig.3-(b)).

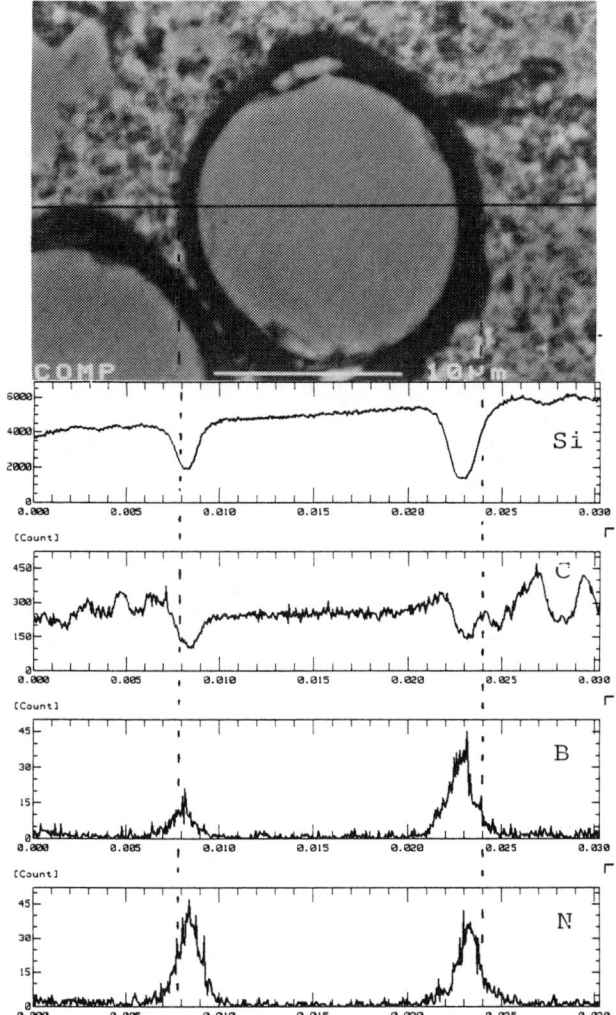

Figure 7 EPMA back-scattered electron image and line profiles of some elements around a fiber in BN-coated Hi-Nicalon fiber reinforced SiC matrix composite (bar scale: 10μm).

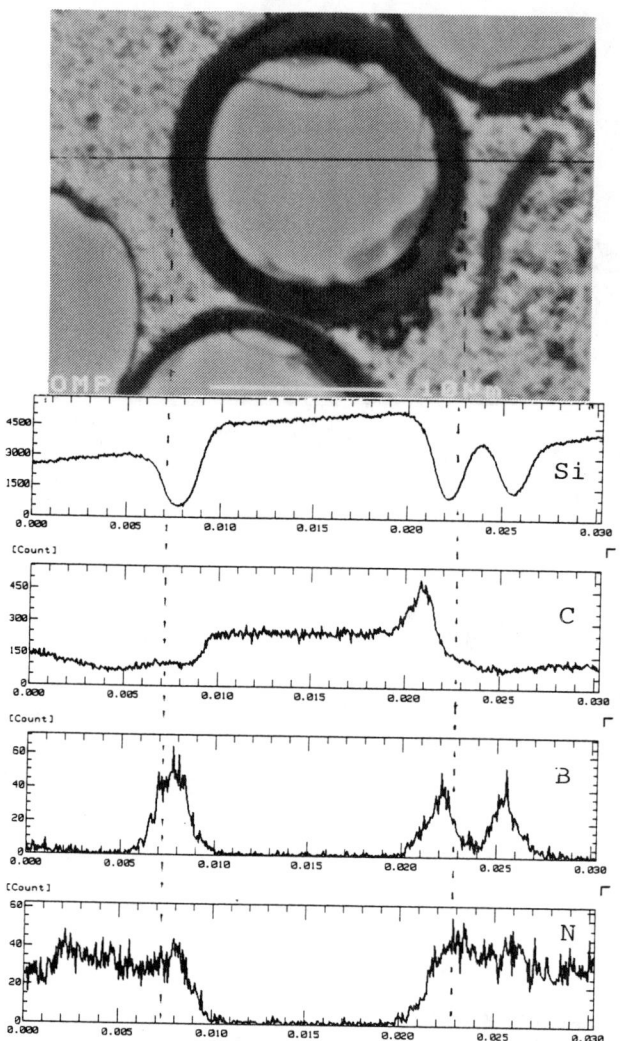

Figure 8 EPMA back-scattered electron image and line profiles of some elements around a fiber in BN-coated Si_3N_4 matrix composite (bar scale: $10\mu m$).

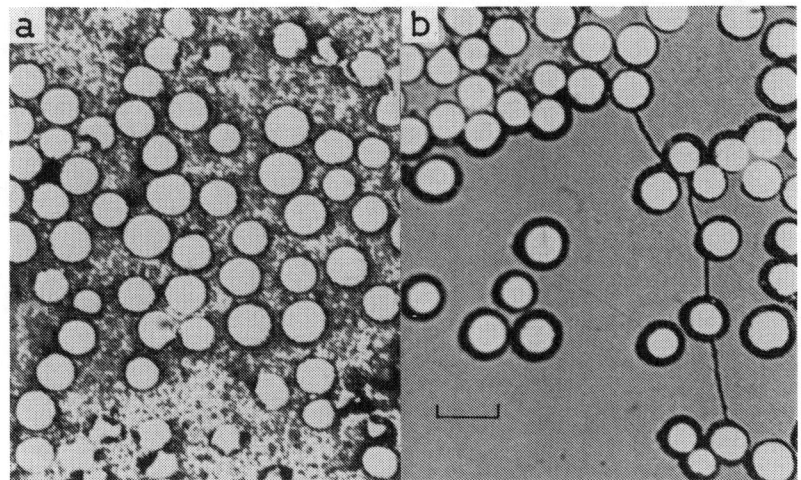

Figure 9 Optical micrograph of SiC(a) and Si3N4(b)matrix composites after bending test (SiC matrix :1600°C;Si3N4 matrix:1400°C).
Bar scale is 20 um.

SUMMARY

ID carbom- or BN-coated Hi-Nicalon fiber reinforced SiC and Si3N4 matrix composites were fabricated by the slurry route followed by hot pressing.
Mechanical and structural characterization of the composites have been conducted and the majar findings are summarized in the following:

1) Open porosities of the composites were less than 5% except carbon-coated fiber reinforced SiC matrix composite.
   Bulk densities of the composites showed the values between 2.51 and 2.81g/cm^3.

2) In the case of C-coated fiber reinforced composites the strength and fracture toughness were nearly constant up to a critical temperature, then they abruptly decreased abone the critical temperature. Room temperature strength and fracture toughness of the composites reinforced with BN-coated fiber were considerably higher than those of high temperature.

3) C- or BN-coated fiber reinforced Si_3N_4 matrix composite showed a remarkable delamination in the high temperature bending test.
Thickness of BN-interphase around the fiber tend to decrease, and in many fibers the interphase seemed to disappear in the case of SiC matrix composite.

REFERENCES

1) B.E. Walker, R.W.Rice, P.F.Becher, B.A.Bender and W.S. Coblentz, "Preparation and Properties of Monolithic and Composite Ceramics Produced by Polymer Pyrolysis", Ceramic Bulltin, 62 [8] 916-923 (1983).

2) E.Fitzer and R. Gadow, "Fiber-Reinforced Silicon Carbide", Am.Cer.Soc. Bull., 65 [2]326-335 (1986).

3) K. Nakano, A. Kamiya, T. Suganuma and I. Maki, "Fabrication of Fiber Reinforced Silicon Carbide Composites"; pp. 1350-1355 in *Sintering '87 vol.2,* Proceedings of the International Institute for the Science of Sintering Symposium (Tokyo, Japan, November, 1987). Edited by S. Somiya, M. Shimada, M. Yoshimura, and R. Watanabe. Elsevier Applied Science, 1988.

4) K. Nakano and H. Ogawa, "Carbon Fiber Reinforced Silicon Carbide Composites"; pp. 419-424 in *Developments in the Science and Technology of Composite Materials,* Proceedings of the European conference on composite materials (Stuttgart, F. R. G. ,September, 1990). Edited by J. Füller, G.Grüninger, K. Schulte, A. R. Bunsell, and A. Massiah. Elsevier Applied Science, 1990.

5) K. Nakano, A. Kamiya, "Carbon Fiber Reinforced Silicon Nitride-Sialon Matrix Composites"; pp. 697-702 in *Developments in the Science and Technology of Composite Materials,* Proceedings of the 5th European Conference on Composite Materials (Bordeaux, France, April, 1992). Edited by A. R. Bunsell, J. F. Jamet, A. Massiah. European Association for Composite Materials,1992.

6) H. Ichikawa, S. Mitsuno, Y. Imai, and T. Ishikawa, "Mechanical and Electrical Properties of SiC Fiber (Nicalon) and Their Composites; pp. 923-928 in Society for the Advancement of *Materials and Process Engineering,* Proceedings of the First Japan, International SAMPE Symposium and Exhibition(Chiba, JAPAN, November, 1989).

Edited by N. Igata, I. Kimpara, T. Kishi, E. Nakata,A. Okura, and T. Uryu. The Nikkan Kogyo Shinbun, Ltd.

7) K. Sato, T. Suzuki, O. Funayama, and T. Isoda, "Preparation of Carbon Fiber Reinforced Composite by Impregnation with Perhydropolysilazan followed by Pressureless Firing"; pp. 614-621 in *Ceramic Engineering and Science Proceedings,* Proceedings of 16th Annual Conference on Composites and Advanced Ceramic Materials (Cocoa Beach, FL, January, 1992).Edited by J. B. Watchman. The American Ceramic Society, Westerville, OH.

8) K. Nakano, A. Kamiya, Y. Nishino, T. Imura, and T. W. Chou, "Fabrication and Characterization of 3D Carbon Fiber Reinforced SiC and Si_3N_4 Composites", submitted for J. Amer.Ceram.Soc.

9) F. Shimoo, T. Hayatsu, M. Narusawa, M. Takeda, H. Ichikawa, T. Seguchi, and K. Okamura, "Mechanism of Ceramization of Electron-Irradiation Cured Polycarbosilane Fiber", J. Ceram. Soc. Japan 101 [1] 809-903 (1993).

10) F. Shimoo, T. Hayatsu, M. Narusawa, M. Takeda, H. Ichiikawa, T. Seguchi, and K. Okamura, "Pyrolysis of Low Oxygen SiC Fiber Prepared by Electron-Irradiation Curing Methed",J. Ceram. Soc. Japan 101 [12] 1379-1383 (1993).

11) R. Naslain, O. Dugne, A. Guette, J. Sevely, C. R. Brosse, J-P. Rocher, and J. Cotteret, "Boron Nitride Interphase in Ceramic-Matrix Composites" J. Amer. Ceram. Soc., 74 [10] 2482-2488 (1991).

12) M. Takeda, Y. Imai, H. Ichikawa, T. Ichikawa, N. Kasai, T. Seguchi, and K. Okamura, "Thermal Stability of the Low Oxygen Silicon Carbide Fibers Derived from Polycarbosilane"; pp.209-217 in *CeramicEngineering Science Proceedings,* Proceedings of 16the Annual Conference on Composites and Advanced Ceramic Materials (Cocoa Beach, FL, Janualy, 1992). Edited by J. B. Watchman. The American Ceramic Society, Westerville, OH.

13) A. Kamiya, K. Nakano, S. Moribe, T. Imura, and H. Ishikawa, "Mechanical Properties of Unidirectional, Hi-Nicalon Fiber-Reinforced Si_3N_4 Matrix Composites", J. Ceram. Soc. Japan, 102 [10] 957-960 (1994).

RAPID DENSIFICATION OF CARBON-CARBON COMPOSITES BY THERMAL-GRADIENT CHEMICAL VAPOR INFILTRATION

I. Golecki, R.C. Morris, D. Narasimhan and N. Clements, AlliedSignal Inc., Corporate Research and Technology, 101 Columbia Road, Morristown, NJ 07962.

ABSTRACT

Porous carbon-carbon preforms, 10.8 cm od x 4.4 cm id x 3.0 cm thick have been densified in a one-cycle, 26 h process. The disks are heated by induction, creating an inside-out thermal gradient, and are exposed to cyclopentane vapor in a water-cooled vacuum chamber. Rough-laminar carbon microstructure is obtained; a compressive strength of 268 MPa is measured at 1.79 g/cm^3 density. The densification rate is monitored in real time. The precursor utilization efficiency is 20-30%. Our patented process can be applied to other materials, has significant scale-up potential and is economically competitive.

INTRODUCTION

Composite materials, such as carbon-carbon (C-C), offer advantages of low density and excellent mechanical and thermal properties, especially at high temperatures, for a variety of aerospace and other applications, e.g., brake pads and uncooled engine parts[1]. One of the most common fabrication methods of composites is densification of a porous body of the desired shape by chemical vapor infiltration (CVI). In CVI, a precursor vapor flows over and around the part, which is kept at a temperature sufficient to decompose the precursor, resulting in deposition of the desired element or compound within the pores of the part, thus increasing its density. The deposition rate usually increases moderately with increasing precursor partial pressure and exponentially (with a ≈4 eV/molecule activation energy for carbon) with increasing substrate temperature. A common application of CVI involves densification of porous carbon, where a large number of substrates may be placed in an enclosure uniformly heated to a temperature of about 1000°C and exposed to e.g., methane. This approach is known as hot-wall CVI and its major drawback is an extremely long CVI time of 600-2000 h to achieve the desired density. Usually, the process must also be interrupted several times to grind the exterior surfaces of the substrates in order to open the pores and allow further infiltration. It is very desirable to shorten the processing time, but increasing the pressure and/or temperature beyond certain ranges may produce unwanted homogeneous nucleation of powders in the gas phase, premature surface crusting, and undesirable microstructure of the material. Also, the progress of the densification is not readily measurable.

Here, we describe a novel patented method[2] for densifying porous C-C preforms in as little as 26 h in a one-cycle process. The disks are heated by induction so as to create an inside-out thermal gradient, and are exposed to cyclopentane (C_5H_{10}) vapor in a water-cooled vacuum chamber. This approach allows significantly higher operating temperatures (by 200°C+) compared to isothermal CVI and produces the desirable rough-laminar carbon microstructure[1] with very high efficiency, very little tar and no soot.

EXPERIMENTAL APPARATUS

The thermal-gradient C-C CVI system is shown in Fig. 1. The main part of the stainless steel deposition chamber is 38 cm id x 43 cm high. A 38 cm high, hydraulically-operated conical flange allows access from the top. The chamber is pumped by a 606 m^3/h Roots blower backed by a PFPE oil lubricated, 78 m^3/h rotary forepump (Leybold-Heraeus). The pump outlet is connected to an oxidation furnace (Delatech), which converts the exhaust gases to CO_2 and water vapor. The C_5H_{10} is delivered to the chamber as a vapor through a calibrated, mass-flow controlled gas line maintained at 70°C. C_5H_{10} is a liquid at 25°C with a vapor pressure[3] of 321 Torr and a boiling point of 50°C. In volume quantities it costs[4] about $ 0.44/kg. The total pressure in the deposition chamber is measured using an MKS # 107B diaphragm pressure gauge and controlled by means of a # 253A butterfly throttle valve and a # 652B controller. Two stainless steel, optically dense traps contain triply-wound tubing cooled to -12°C, to minimize condensibles and particulates from reaching the pumps.

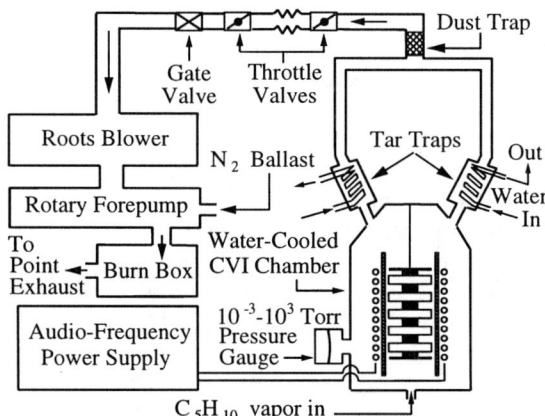

Figure 1. Simplified schematic diagram of AlliedSignal patented thermal-gradient CVI reactor.

The porous C-C preform disks are heated by induction inside a 15 cm id x 22 cm long, water-cooled copper coil, which is energized with a Pillar power supply. The preforms are made of non-woven polyacrylonitrile (PAN)[1] carbon fibers and have an initial density in the range 0.4-0.6 g/cm^3. Multiple C-C disks can be infiltrated simultaneously; the results described here were obtained with three disks per run, each 10.8 cm od x 4.4 cm id x 3.0 cm thick. The disks were placed around a Mo or Al_2O_3 mandrel, spaced about 1 cm apart, and hung from the top of the deposition chamber. An electrically conductive mandrel is not required, as the electrical conductivity of these preforms is sufficient to couple directly to the electromagnetic field. In order to reduce radiative heat losses, several 0.6 mm thick grafoil plates were placed above and below the mandrel. A 14.1 cm od x 13.5 cm id x 47 cm high quartz flow channeler tube was inserted between the C-C disks and the coil in some runs, while in other runs a much shorter, 11.4 cm high quartz tube was placed between the bottom of the chamber and the coil, the outer diameter surfaces of the C-C preforms then staring directly at the coil and chamber walls. The temperatures of the C-C disks were measured with 0.25 mm diameter, Pt-13% Rh/Pt thermocouples inserted half-way through the thickness of the disks. The power and frequency of the induction power supply (8.8-13.2 kW, 4.9-8.6 kHz), total pressure (20-100 Torr) and C_5H_{10} flow rate (170-540 sccm) were controlled during carbon CVI runs.

RESULTS AND MECHANISM OF DENSIFICATION

Fig. 2 demonstrates average carbon pick up rates per disk of 9.5 g/h or 10.6%/h. The whole-disk density increases from 0.41 to 1.541 g/cm^3 in just 26 h, representing an average rate of increase of 0.044 g/cm^{3}h. Whole-disk densities of 1.68 g/cm^3 were obtained, compared to 1.80 g/cm^3 for Poco graphite. The shape of the density vs. time curves is sigmoidal-linear rather than the much slower exponential approach to final density in isothermal CVI. Generally, the highest density was found in the middle disk in the stack, due primarily to lower temperatures at the outer surfaces of the two extremal disks. Improved end-insulation and optimized coil design will reduce such disk-to-disk variations. Also, in a scaled-up stack containing a larger number of disks, the two extremal disks represent only a very small fraction of the total number. The density uniformity within a disk, as measured from cored and sliced samples, was within ±(5-8)%. The overall utilization efficiency of the precursor, computed as the amount of carbon added to the disks divided by the amount of carbon flowing as C_5H_{10}, was 20-30%. Higher efficiencies were generally obtained with the long flow channeler.

The microstructure of the deposited carbon was determined by measuring the extinction angle of the Maltese cross in polarized-light microscopy[1]. Depending on processing conditions, rough- (19-23°), smooth→rough- (13-15°) or smooth-laminar (10-12°) microstructure was obtained; isotropic carbon (0°) was not found. The rough-laminar structure is generally desired for braking applications. The compressive strength measured for 6.6 mm diameter x 7.9 mm long samples (Fig. 3), increases steeply with increasing density. At 1.79 g/cm^3 the compressive strength is 268 MPa (39 ksi), considered very good.

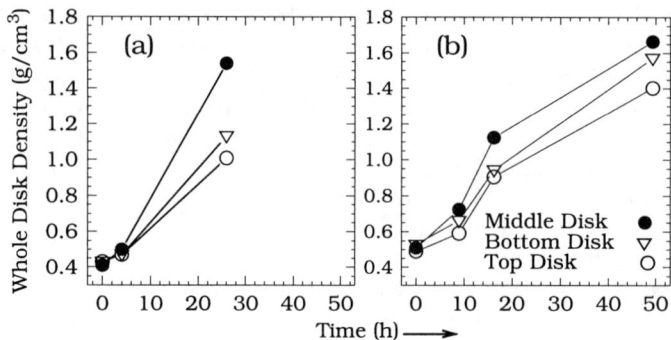

Figure 2. Carbon density vs. time in thermal-gradient CVI: (a) One set of disks without insulation, *with* long flow channeler (b) Three separate sets of disks with grafoil insulation, *without* long flow channeler. The lines are an aid to the eye.

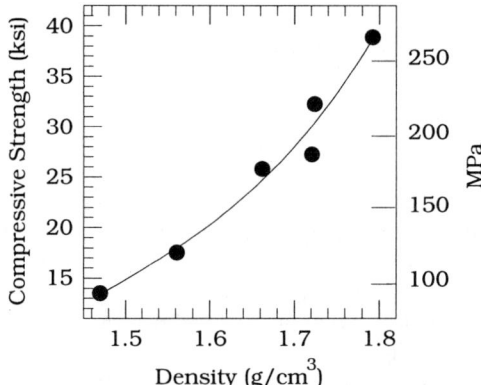

Figure 3. Compressive strength vs. density for C-C rapidly densified by thermal-gradient CVI.

Fig.4 illustrates the spatial and temporal temperature variation in the middle disk in a CVI run with the long flow channeler. Initially the id region of the C-C preform disk was hottest and all three temperatures increased as a function of time; the temperature difference between the id and od decreased as a function of time. Temperatures inside the disk reached almost 1200°C, significantly higher than in isothermal CVI. In this run, some surface crusting was noted and yellow vapors were observed in the chamber towards the end of the run. In runs which were stopped earlier, no surface crusting occurred and no yellow vapors were seen. The amount of liquid tar was significantly smaller in such runs, compared to the estimated 1-3% of the incoming C_5H_{10} for runs continued past surface crusting. No solid or powdery soot was found.

In this thermal-gradient CVI process, the substrates are Joule heated by circumferential induced currents flowing inside them. In our geometry, the induced power is initially highest near the od and diminishes to zero in approximately three skin depths[5], δ, where $\delta = 1581.2/(\sigma f)^{0.5}$, with δ in cm, the electrical conductivity σ in ohm^{-1}m^{-1} and the frequency f in kHz. For example, for a material with an initial uniform electrical conductivity $\sigma=10^4$ ohm^{-1}m^{-1}, $\delta = 5.4$ cm at 8.5 kHz. The substrate temperatures, which increase with power, are dominated by the radiation losses, ΔQ, to the water-cooled coil and walls, where $\Delta Q = \varepsilon\sigma_o(T^4_{od} - T^4_{wall})$, with ΔQ in J; ε is the emissivity (a number between 0 and 1), $\sigma_o=5.67\times10^{-8}$ J/K^{4}m^{2}s is the Stefan-Boltzmann constant and T is in K. The temperature will initially be highest in the interior regions of the preforms, lower at the top and bottom surfaces and lowest at the od. The gas-phase diffusivity of the precursor is high, ensuring that the initial densification rate will be highest in those hottest interior regions, consistent with our density distribution measurements (to be published). The presence of an electrically conducting Mo mandrel results in additional heating near the id of the preforms, but does not change this general picture. With the long quartz flow channeler, T_{wall} is in the range of 300-800°C, instead of 20-50°C. As densification progresses, the electrical and thermal conductivities of the preform increase. The improved electrical conductivity results in a shallower skin depth and improved coupling to the coil, i.e. higher induced current and higher temperatures near the exterior surfaces of the preform, other conditions being equal. Thus an "inside-out" densification front exists in the preforms. As the pores inside the disk become smaller and finally the surface starts crusting, the carbon deposition rate decreases, due to the much smaller density of nucleation sites, and the temperature distribution becomes more uniform.

SUMMARY

Our patented[2] thermal-gradient CVI technique reduces the total densification time of porous, 10.8 cm od x 3.0 cm thick C-C preforms with initial density of 0.4-0.6 g/cm^3 to a single-cycle, 26-50 h step, which is more than ten times faster than in isothermal CVI. This accomplishment is made possible by higher temperatures in the interior regions of the preforms and other innovations. Additional optimization could shorten the cycle time further. The densified substrates have the rough-laminar microstructure, which is desired for brake pads[1] and has excellent friction and wear properties. Densities as high as 1.84 g/cm^3 have been obtained in regions. The compressive strength is excellent. Real-time monitoring and control capabilities of the densification rate and the end-point of the process have been developed (to be published). The process is highly efficient even with modest chamber loading and results in only a very small amount of tar and no solid soot. No dipping in flammable liquids[6] is required and the pressure is adjustable; our system is safe and stable and does not require an explosion-proof room. No special fixtures or machining are required as in forced-flow CVI[7]. Our process has real scale-up potential, as no fundamental technical barriers exist to significantly increasing the substrate diameter and the number of substrates per run. This process is economically competitive, providing the ability of batch manufacturing of several dozen substrates with a turn-around time of less than one week, allowing potentially significant reductions in stock

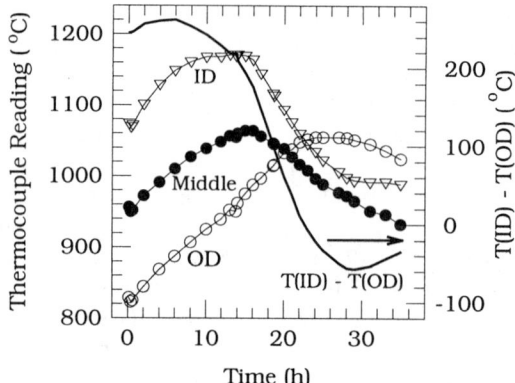

Figure 4. Time evolution of the temperatures in three radial locations inside the middle disk in a carbon thermal-gradient CVI run.

inventory and "just-in-time" operation. Materials other than C-C, singly or in combination, can be densified using this method, including metals, semi-metals, semiconductors and even room-temperature insulators; e.g. Al, B, TiN, SiC-C, SiC-SiC, and ZrO_2. To exploit the inherent speed of this process, the preform or the matrix needs to be sufficiently electrically conducting at the densification temperature to couple to the electromagnetic field. The frequency of the field can be adjusted to the material system and dimensions.

ACKNOWLEDGEMENT

The financial support of the AlliedSignal Aerospace Company, Aircraft Landing Systems, South Bend, IN (D. Hayes, J. Pigford, J. Hendricks and N. Murdie) is acknowledged.

REFERENCES

1. G. Savage, "Carbon-Carbon Composites" (Chapman and Hall: Cambridge, England, 1993).
2. I. Golecki, R.C. Morris and D. Narasimhan, U.S. Patent # 5,348,774, "Method of rapidly densifying a porous structure", 9-20-1994.
3. CRC Handbook, 55th ed., p.D-175 (1974).
4. E.g., Chemical Marketing Reporter.
5. G.H. Brown, C.N. Hoyler and R.A. Bierwirth, "Theory and Application of Radio-Frequency Heating" (Van Nostrand: New York, 1948).
6. Michel Houdayer, Jean Spitz, Danh Tran-Van, U.S. Patent # 4,472,454, "Process for the densification of a porous structure", 9-18-1984.
7. T.M. Besmann and J.C. McLaughlin, "Scale-up and modeling of forced-flow chemical vapor infiltration", Ceramic Eng. and Sci. Proc. <u>15</u>, No.5, 897-907 (1994).

CHARACTERIZATIONS OF CARBON-CARBON COMPOSITES ELABORATED BY A RAPID DENSIFICATION PROCESS

B. NARCY, F. GUILLET, F. RAVEL, P. DAVID
Commissariat à l'Energie Atomique, Centre d'études de Bruyères le Châtel
BP 12, 91680 Bruyères le Châtel, FRANCE

Carbon-Carbon composites were produced by a rapid densification process. Composites of 1.4/1.8 densities were obtained in only a few hours. The influence of elaboration temperature on densification rates was studied. Optical microscopy, X-ray diffraction and Transmission Electron Microscopy were used to characterize the synthesized materials and showed the graphitizability of the carbon matrices.

1 INTRODUCTION

The densification process used to produce the C/C composites studied in this paper presents many advantages as compared to other classical routes: it is fast, not so expensive and does not require heavy equipment. The process itself has been detailed elsewhere [1,2]; this study is only concerned with the characterization of the produced materials. The porous substrates used are felts, 2D or 3D preforms. The process makes it possible to obtain composites of various shapes in only a few hours time. The density after densification ranges from 1.4 to 1.8, depending on the substrate used.

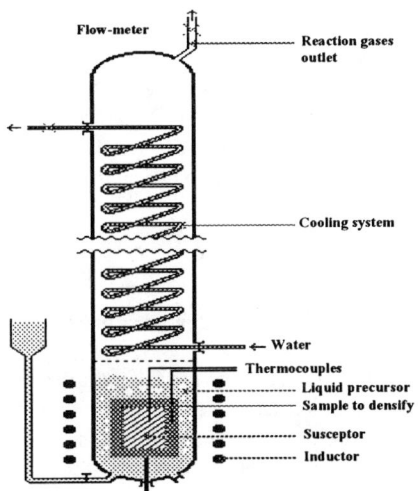

Figure 1: Experimental device

The principle of the process consists of heating by induction a graphite susceptor, directly submerged in a liquid hydrocarbon precursor, which is surrounded by the porous substrate to be densified (Fig. 1). The vapors of the precursor are decomposed and cracked in the preform hot zone (near the susceptor at the beginning), leading to the deposition of carbon at the fiber surface. During the deposition, the hot zone moves to the outside of the preform, inducing the densification of the whole porous substrate. This leads to very strong convections in the preform and to a densification front moving from the inside to the outside of the preform; there are no external porosity plugging.

The uncracked vapors are condensed in a cooling system above the reactor and the reaction gazes are released at the top of the installation.

The different parameters to be taken into account are the nature of the substrate to be densified, the nature of the precursor, the elaboration temperature and the HTT applied to the obtained composites. The substrate and precursor considered in this study are identical for all samples. The effects of these two parameters are to be characterized later on..

2 EXPERIMENTAL PARAMETERS

The substrates consist in small tubes (h = 65 mm, $\Phi_{int} \approx 16$ mm, $\Phi_{ext} \approx 22$ mm) of carbon felt (RVC 2000 type from the society "Le Carbone Lorraine", with an initial density of about 0.1 g/cm^3). Cyclohexane was used as precursor. The elaboration temperature influence on densification kinetics was studied in the range 900-1200°C and on nanostructure in the range 900-1120°C. The HTT applied were 2h at 2450°C/Ar or at 2700°C/Ar.

The different matrices were studied using optical microscopy under polarized light, X-ray diffraction and TEM (Transmission Electron Microscopy) with a Philips CM 12 (120kV). These studies were carried out before and after High Temperature Treatments (HTT) in order to characterize the graphitizability of the matrices.

3 RESULTS

3.1 Influence of elaboration temperature on densification rates

The mass increase of the felt samples has been measured as a function of cracking time and deposition temperature. For each temperature, the mass increase is linear with respect to cracking time (and cracking surface) and allowed us to determine the kinetics of densification. By plotting the logarithm of the kinetics with respect to the inverse of the deposition temperature (Fig. 2), an apparent activation energy (Arrhenius form: $v = A \exp(-E_a/RT)$) of about 53 kcal/mol was obtained, in good agreement with previous work [3].

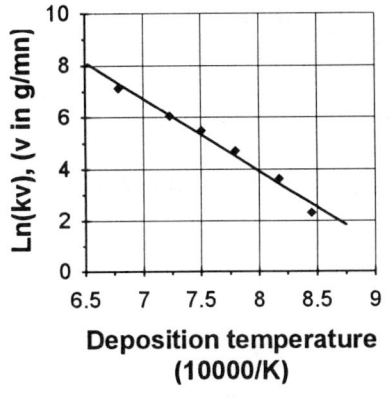

Figure 2: Logarithm of kinetics with respect to the inverse of synthesis temperature

Figure 3: Polarized light optical micrography of a C felt/C composite synthesized at 1020°C

3.2 Optical microscopy

Polarized light optical microscopy was used to determine the nature of the deposited pyrocarbon (rough or smooth laminar, isotropic, etc.) and to visually assess the nature of the matrix. Figure 3 shows large conical growth areas, distorted Maltese crosses, high optical contrasts. According to the classification [4,5], the deposited pyrocarbons are rough laminar (and thus should be graphitizable) in the temperature range 900-1200°C.

3.3 X-ray diffraction

An X-ray powder diffraction study of C felt/C composites has been carried out in order to show the influence of the synthesis temperature and of the HTT. The studied parameters are d_{002} (interlayer spacing along the c-direction), Lc (crystallite size along the c-direction) and the evolution of)11(band with HTT. All data reported here were taken at room temperature with a type F Siemens diffractometer (CuKα_1 + Kα_2 radiation). Because of the low fiber volume fraction (< 6 %), the contribution of the fibers (which are not graphitizable) is negligible. The results were obtained using a NBS standard silicon powder so as to correct 2θ values and instrumental broadening.

The classical Scherrer equation [6,7] (equation 1) was used to calculate Lc for the as prepared composites. The equation (equation 2) proposed by Ergun [8], Thrower and Nagle [9] and Seehra and Pavlovic [10], which is more complete and takes into account the effect of strain on the line broadening (equation 2 is for a Gaussian strain distribution), was used for C matrices after HTT.

$$L_c = \frac{k\lambda}{\beta \cos(\theta)} \quad (1) \qquad \beta \cos(\theta) = \frac{\lambda}{L_c} + \frac{2\pi^2 \lambda \sigma^2}{c} l^2 \quad (2)$$

where k is a constant (one used k = 1)
  λ is the X-ray wavelength
  β is the full width at half maximum (FWHM) of (00l) lines on classical 2θ scans
  σ is the strain parameter

$$\beta = B\left(1 - \frac{b^2}{B^2}\right) \quad (3)$$

Equation 3 was used to correct for the instrumental broadening (where b is the linewidth of silicon at nearly the same 2θ as the experimental line of the sample of linewidth B).

d_{002} is calculated using the corrected value of θ and the (002) line (1st method) and by plotting $d = n\lambda/(2\sin\theta)$ against $x = [(\cos^2\theta/\sin\theta)+(\cos^2\theta/\theta)]/2$ (for the uncorrected value of θ) and determining the limit for d for $x = 0$ (2nd method) [14]. These two methods lead to nearly the same values of d_{002}. By plotting $\beta \cos(\theta)$ versus l^2 for l = 2, 4, 6 (Fig. 4), it is possible to calculate Lc and σ. Figure 5 shows the influence of synthesis temperature on Lc after a 2450°C/Ar (2h) HTT. The different results are presented in Table 1.

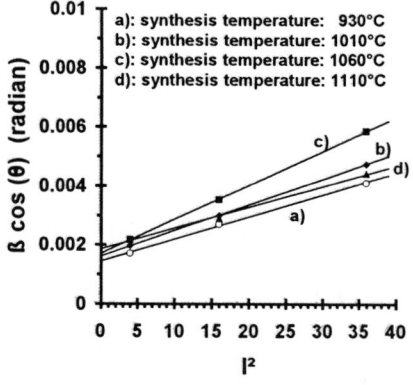

Figure 4: $\beta\cos(\theta)$ vs l^2 for (samples treated 2h at 2450°C/Ar)

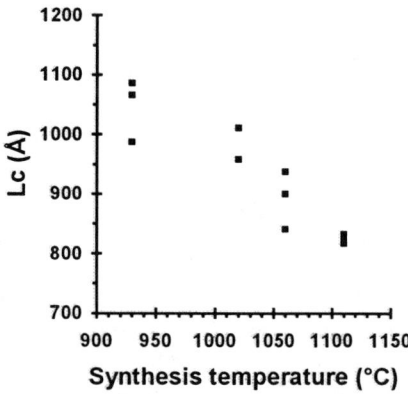

Figure 5: Lc vs synthesis temperature (samples treated 2h at 2450°C/Ar)

Table 1: various results obtained by X-ray diffraction

T synthesis (°C)	930	1010	1060	1110
As prepared				
d_{002} (1st method) (Å)	3.477	3.480	3.472	3.474
Lc (equation 1) (Å)	20	20	20	30
After 2h at 2450°C/Ar				
d_{002} (1st method) (Å)	3.367_5	3.367_6	3.369_5	3.367_4
d_{002} (2nd method) (Å)	3.367_7	3.368_0	3.370_4	3.367_3
Lc (equation 1) (Å)	900	790	700	700
Lc (equation 2) (Å)	1070	960	900	830
σ (10^{-3})	4.1	3.2	5.0	4.0
After 2h at 2700°C/Ar				
d_{002} (1st method) (Å)	3.356_7	3.357_7	3.358_4	3.357_2
Lc (equation 1) (Å)	> 1500	> 1500	> 1500	> 1500

The major result is that all the synthesized matrices are graphitizable in the temperature range 930-1120°C [11]:
. d_{002} decrease from about 3.47/3.48 Å to 3.357 Å after 2h at 2700°C/Ar
. Lc increase from 20/30 Å up to 1500 Å after 2h at 2700°C/Ar
. The study of the evolution of)11(band with HTT showed clearly the split of the band in the 2 peaks corresponding with the (110) and (112) reflections.

Lc results seem to be strongly depending on diffractometer setting: by intercalating Solers slits, Lc values increase from about 500 Å up to the values indicated in Table 1. This phenomenon can be easily explained by the difference of absorption between carbon and silicium. As a matter of fact, the low absorption of carbon induces an asymetrical 002 peak with a tail toward low angles. This broadening does not occur when considering the silicium reference and the carbon peak thus appears more broadened. This effect is considerably decreased when using Solers slits.

An other result is that the low temperature (930°C) matrices are more likely to produce ''graphite like structure'' at lower temperatures (Fig. 5). The Basic Stuctural Units (BSU) of C are probably less disoriented when deposited at lower temperatures and so need less energy to transform into the graphite structure.

Because of the preferred orientation effects of the powder during sample mounting, the intensities of the peaks corresponding to (hk) reflections are very low as regard to (00l) intensities and did not allow us to determine La by measuring the FWHM of the (112) peak.

These results have to be completed by La calculations and by the study of the matrices obtained at higher temperatures (> 1120°C).

3.4 TEM observations

Bright Field (BF) TEM observations of C felt/C composites synthesized at 1060°C were carried out before (Figure 6) and after (Figure 7) HTT (2h at 2450°C/Ar) in order to characterize the carbon matrix graphitizability.

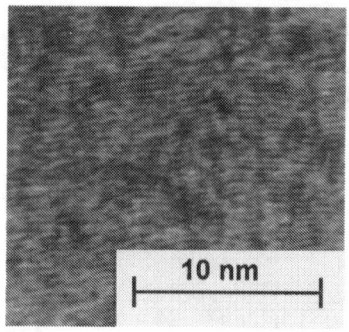 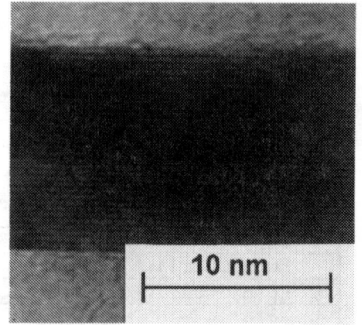

Figure 6: BF image of transverse section of C matrix (as-prepared)

Figure 7: BF image of transverse section of C matrix (after 2h at 2450°C/Ar)

The as-prepared carbon matrix (Fig. 6) shows large areas of preferentially oriented carbon layer planes. This structure changes after HTT (2h at 2450°C) in large areas of well organized carbon planes in the 3 directions (Fig. 7). Carbon matrix/carbon fiber interfaces have not yet been characterized. This study has to be completed; these first results show that the C matrices (synthesized in the 930-1120°C range) are graphitizable.

4 CONCLUSIONS

The process we used to produce C/C composites is very quick, the C matrices synthesized in the temperature range 930-1120°C are graphitizable. The next step will be the elaboration of composites at higher temperatures. A more accurate TEM study has to be carried out in order to characterize the fiber/matrix interface.

REFERENCES
1 M. HOUDAYER, J. SPITZ, D. TRAN VAN, French Patent n° 81 22163-26/11/81
2 B NARCY, P. DAVID, F. RAVEL, to be published in *Silicates Industriels* (mid 95)
3 P.A. ETSNER, A.I. ECHEISTOVA, *Doklady Akad. Nauk SSSR*, 87, n° 6, pp 1029-1031 (1952)
4 P. LOLL, P. DELHAES, A. PACAULT, A. PIERRE, *Carbon*, Vol. 15, N° 6, p 383 (1977)
5 B. GRANOFF, *Carbon*, Vol. 12, pp 405-416 (1974)
6 W. RULAND, In *Chemisty and Physics of Carbon*, Vol. 4, pp 1-84, Marcel Dekker, New York (1971)
7 R. PERRET and W. RULAND, *J. Appl. Cryst.* 1, 257 (1968)
8 S. ERGUN, *Carbon*, Vol. 14, p 139 (1976)
9 P.A. THROWER AND D.C. NAGLE, *Carbon*, Vol. 11, pp 663-664 (1973)
10 M.S. SEEHRA and A.S. PAVLOVIC, *Carbon*, Vol. 31, N° 4, pp557-564 (1993)
11 A. OBERLIN, *Carbon*, Vol. 22, N° 6, p 521-541 (1984)

OPTIMIZATION OF THE FUGITIVE COATING THICKNESS IN PRESSURE INFILTRATED MULLITE-ALUMINA COMPOSITES

E. H. Moore
WL / MLLN
Wright Laboratory Materials Directorate
Wright-Patterson Air Force Base OH 45433-7817

S. Shamasundar
UES Inc.
4401 Dayton-Xenia Road
Dayton OH 45432

J. L. Kroupa
University of Dayton Research Institute (UDRI)
300 College Park Avenue
Dayton OH 45469-0128

INTRODUCTION

There is an increasing interest in oxide-oxide refractory composites and ceramic matrix composites (CMC) because of their high strength, high creep resistance and resistance to crack propagation in high-temperature structural and non-structural applications. High strength and low modulus oxide fibers are introduced into ceramic oxide matrices in order to resist crack growth (i.e., increase the composite's strain to failure or "toughness"). Nevertheless, the introduction of a 2-D fibrous matte or 3-D fibrous preform into a ceramic matrix constrains the densification of the composite. (As a result, as prepared composite's typically will have about 20 percent residual porosity.) Although higher densification is possible by free or pressure-less sintering, degradation of the mechanical properties of the fibers at elevated temperatures (e.g., normally above 1100°C for mullite fibers) prevents the application of high-temperature processing.

An oxide-oxide composite composed of a high-purity alumina matrix and mullite fibrous reinforcement has been used in this study. A fugitive carbon coating has been applied to 2-D fibrous mattes and 3-D preforms by chemical vapor deposition (CVD) and by polymer pyrolysis of a polymeric based resin system. This paper will only discuss the processing and applicable analysis of the CMC prepared with the applied polymeric pyrolyzed carbon coating.

To the extent authorized under the laws of the United States of America, all copyright interests in this publication are the property of The American Ceramic Society. Any duplication, reproduction, or republication of this publication or any part thereof, without the express written consent of The American Ceramic Society or fee paid to the Copyright Clearance Center, is prohibited.

The carbon coated fibrous preform is initially processed by pressure infiltration (PI) with an alumina slurry. After appropriate freezing, ambient drying and thermal treatments (furnace drying), the composite is vacuum impregnated (VI) two times with an alumina sol to reduce the composite's porosity. The composite is then thermally treated to remove the "fugitive" carbon coating and to consolidate the CMC. The "fugitive" coating is thermally removed during this stage to leave a micro-gap between the matrix and fiber. The general processing methods used in this work have been described in a previous publication by Moore, Mah and Keller [1]. Other investigators have prepared CMC in a similar fashion with 20-27 percent open porosity [2].

An important issue in the design of these composites is the thickness of the micro-gap resulting from the "fugitive" coating. A higher value for the fugitive coating thickness means a larger volume fraction of gap in the composite. A lower value for the fugitive coating thickness means the possibility of contact and micro-gap closure between the fiber and the matrix is increased. In certain composite systems, such contacts can also lead to reactions between the constituent phases.

This paper is written in two sections. The first section concerns CMC processing. The second section concerns modeling of the residual stresses in the CMC, using a general purpose finite element method (FEM) code. The purpose of the FEM analysis here is to quantify the effects of "fugitive" coating thickness (i.e., for reducing residual stresses) for a given composite system under operating conditions.

PART-I. EXPERIMENTAL (PROCESSING)

Materials And Reagents

Nextel 440™ mullite/α-alumina composites were prepared using fractionated Sumitomo alumina powders (AKP™-15 and AKP™-53) with a median particle diameter of about 0.7 and 0.2 µm, respectively. The reinforcement phase consisted of a 3-D angle interlock woven fabric, manufactured by the multi-warp weaving method, that was about 6.5 mm thick. The mullite fibrous 3-D preform makes up 40 volume percent of the resultant processed CMC. The preforms contain a proprietary sizing that is removable at 600°C. An approximately five volume percent Nyacol™ alumina sol, with ≤0.05 µm diameter alumina, was used for VI. Colloidal or dispersed slurries were prepared for PI using Sumitomo AKP™ alumina powder's, deionized water and nitric acid (to adjust pH between 3.9 to 4.1). Carbon coatings were applied with an Ashland Aerocarb 240™ pitch.

Sample Preparation And Composite Processing

The 3-D preforms were carbon coated using an Ashland™ Aerocarb 240 pitch. The pitch and toluene solvent were mixed with a six weight percent pitch in the resin solution. The preforms were saturated with the resin based solution, ambient dried and heated at 800°C for one hour in a Brew-188 furnace. Before the furnace was heated to 800°C, a vacuum of less than one micron was pulled and the furnace was back filled with argon.

The CMC was fabricated by placing a 127 mm (five inch) diameter piece of 3-D fabric into the PI device onto nylon cloth with 0.22 μm sized openings. The 3-D fibrous preform covered over 95 percent of the surface of the nylon filtering cloth. The PI device was filled with an appropriate amount of dispersed alumina slurry, closed and a pressure of about 0.62 MPa (90 psi) of air was applied to the inlet as a stimulus to remove the liquid and consolidate the fibrous preform with ceramic powder. A vacuum pump was used in combination with a liquid nitrogen cold trap, and with the applied air pressure, to pull a vacuum on the outlet of the PI device. The alumina slurry infiltrated sample was consolidated in less than two days, removed from the PI device, allowed to partially dry and then totally immersed into liquid nitrogen at -197°C for one hour to freeze the liquid in the pores of the composite. The sample was removed, allowed to ambient dry in a hood and then furnace dried at 400°C for two hours in an air atmosphere. The CMC was trimmed, returned to the PI device and PI again with an alumina slurry (≤0.35 μm diameter). Once again, the CMC was consolidated in less than two days, air dried and furnace dried for two hours at 400°C in air. The CMC was then cut into appropriate sizes for compact-tension (C-T) and four-point bend (flexure) test. The CMC was then VI with the fine Nyacol™ alumina sol, ambient air dried, furnace dried for two hours in air at 400°C, VI with the fine Nyacol™ alumina sol a second time, ambient air dried and furnace dried for two hours in air at 400°C. A C-T and a flexure specimen were heated for four hours in an inert atmosphere (nitrogen) at 950°C, a second set of specimens at 1100°C and a third set of specimens at 1250°C. All specimens were additionally heated in a circulating air furnace at 650°C for 24 hours to insure that the "fugitive" interface was removed prior to final preparation for mechanical testing. Samples were cut for flexure and C-T test, and core drilled for C-T test. Processed CMC were examined using scanning electron microscopy for analysis of the microstructure, by optical microscopy for analysis of fiber pull-out and examination of macrostructure. Figure 1 shows CMCs prepared by this processing route.

PART-II. OPTIMIZATION OF THE FUGITIVE COATING THICKNESS

In this paper, a FEM simulation procedure is used to study the effect of the thickness of "fugitive" coating on the residual stress state of the composite. Processing temperature and coating thickness are optimized by adopting a unit cell model. It is well known in composite processing that a mismatch in the coefficient of thermal expansion (CTE) between the fiber and matrix material results in a residual stress state in the composite [3, 4]. In conventional composites, the usual objective of the post consolidation thermo-elastic residual stress analysis is to predict the mechanical properties of the composite (which is significantly altered due to the presence of a residual stress state) and to predict the failure mechanisms in certain cases. However, in the present work, residual stress analysis is carried out to determine the effect of a range of coating thickness for a set of processing conditions.

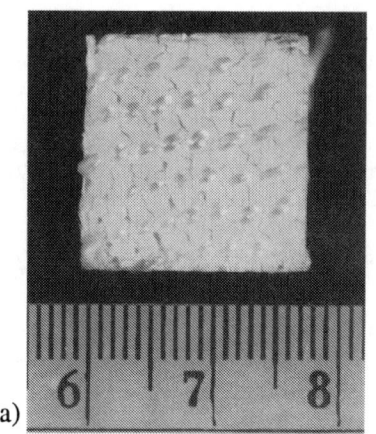 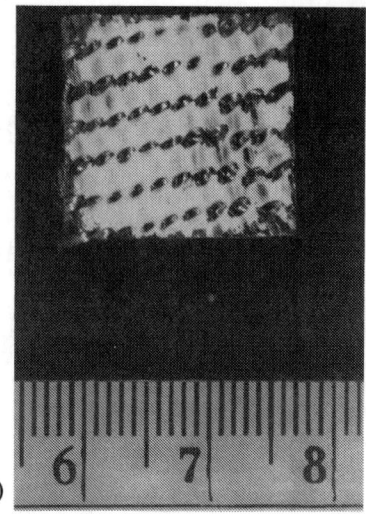

Figure 1. Mullite-Alumina CMCs prepared in this study: (a) heated at 800°C for four hours in air, and (b) as-prepared and unheated.

An idealized unit cell of fiber-interface-matrix is shown in Figure 2. Generalized plane strain and symmetry conditions are imposed on appropriate boundary nodes. The material property data for alumina and mullite were taken from the CRC Materials Science and Engineering Handbook [5]. Three different temperatures (i.e., 950°C, 1100°C and 1250°C) were considered for thermal treatment. Once the "fugitive" coating has been removed, the interface region is removed from the computational domain. The "fugitive" carbon coating thickness is treated as an initial gap (g) in the finite element analysis (FEA) and the gap was kept between 0.02 to 0.6 μm. In order to effectively tackle the above, the FEM is used to optimize the coating thickness for given operating conditions. ABAQUS, a general purpose finite element code is used for analysis.

Figure 3 shows the radial, hoop and the longitudinal stresses at the end of the cooling cycle with a thermal treatment temperature of 925ºC. As expected, as the gap is reduced, the residual stresses increase. Figure 4 shows a summary of the simulation for various temperatures and gap sizes, where the maximum residual stress is plotted as a function of the gap thickness for various temperatures. It is seen that the FEA provides a means for the quantitative prescription of the coating thickness.

CONCLUSIONS

Experiments show that composites can be prepared with a "fugitive" coating. More experiments are required to demonstrate the reduction of residual stresses in

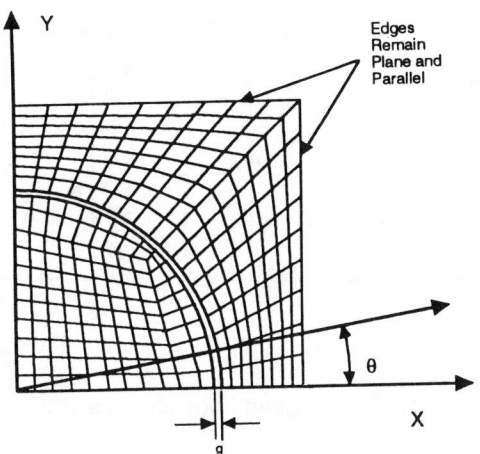

Figure 2. An idealized unit cell showing the fiber-gap-matrix.

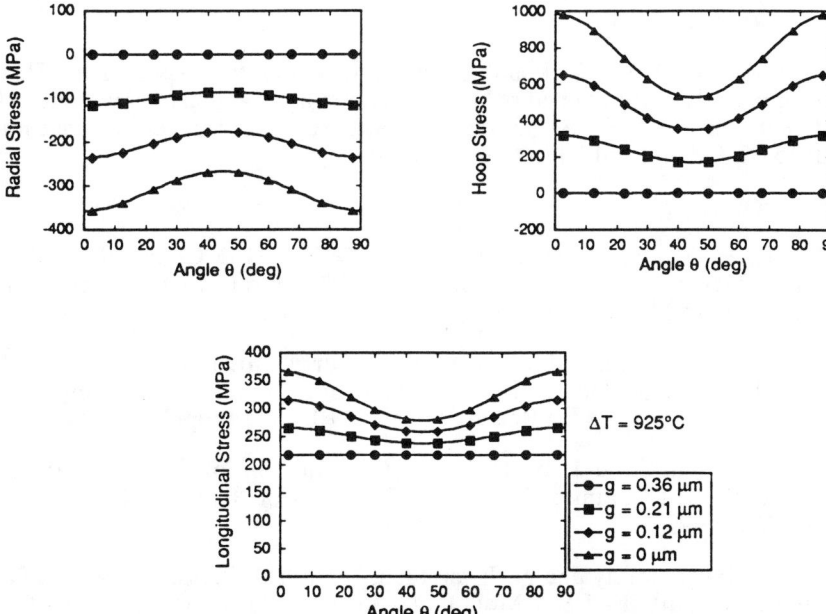

Figure 3. Residual stress distributions (i.e., radial, hoop and longitudinal) at the end of the cooling cycle in a Mullite-Alumina CMC processed at 925°C, with various gap sizes.

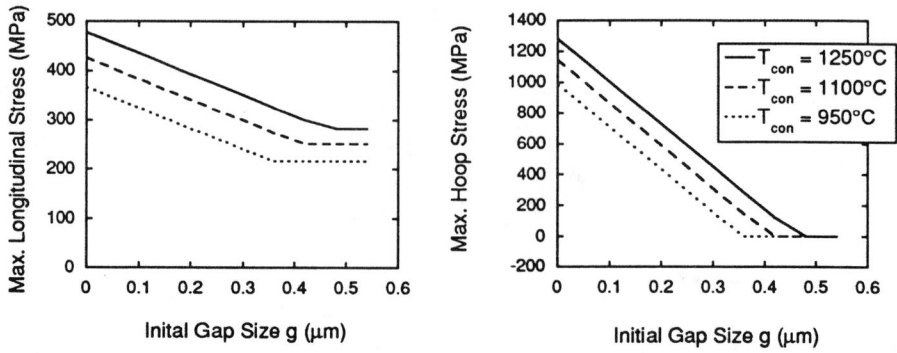

Figure 4. Maximum longitudinal and hoop stresses as a function of the gap size.

actual composites for operational temperatures of interest. Numerical analysis indicate that residual stresses can be tailored with the introduction of a gap of appropriate thickness via "fugitive" interfaces.

ACKNOWLEDGMENTS

The authors thank M. Cook and C. Pierce for laboratory assistance. This work was supported and conducted at Wright Laboratory Materials Directorate, Wright-Patterson Air Force Base OH; the work was supported under contract numbers F33615-92-C-5900 and F33615-94-C-5200.

References

1. E. H. Moore, T. Mah and K. A. Keller, "3-D Composite Fabrication Through Matrix Slurry Pressure Infiltration," Ceramic Engineering & Science Proceedings, 15 [4] 113-120 (1994).

2. S.-M. Sim and R. J. Kerans, "Slurry Infiltration of 3-D Woven Composites," Presented at the 16th Annual Conference on Composites and Advanced Ceramics, Cocoa Beach FL, Jan. 7-10, 1992 (unpublished).

3. Y. H. Park and J. W. Holmes, "FE Modeling of Creep Deformation in Fiber Reinforced Ceramic Composites," Journal of Materials Science, 27 [23] 6351-6361 (1992).

4. P. Rangaswamy and N. Jayaraman, "Residual Stresses in SCS-6/Ti-24-11 Composites: Part II - FEM Modeling," Journal of Composite Technology and Research, 16, 54-67 (1994).

5. CRC Materials Science and Engineering Handbook, 2nd Ed, Edited by J. F. Shackelford, W. Alexander and J. S. Park, CRC Press, Boca Raton FL, 1994.

EFFECTS OF CARBON COATINGS ON SiC(Nicalon)-Pyrex COMPOSITES

B. MOUHAMATH - N. TAKEDA

Research Center for Advanced Science and Technology - University of Tokyo
4-6-1 Komaba, Meguro-ku, Tokyo 153 - JAPAN

ABSTRACT

The behaviour at elevated temperatures of unidirectional Nicalon SiC/Pyrex borosilicate glass composite was investigated. Studies were carried out on three types of materials fabricated with uncoated, thin and thick carbon coated fibres. Flexural tests were performed at temperature ranging from room temperature to 700°C. The response of the materials has been found depending on the morphology of the composite and the behaviour of the individual constituents of material.

1. INTRODUCTION

In last few decades, efforts have been made to improve the reliability of fibre-reinforced ceramic matrix composite materials for use in high temperature structural applications. There is a great deal of interest in producing ceramic matrix composites (CMCs) which exhibit good mechanical properties to face the undesirable phenomena such as the oxidation at elevated temperatures. It is well know that the interface is a key feature of the ceramic composite performance. The fibre-matrix interface has a great influence on the fracture behaviour and mechanical properties of ceramic composites. The composite should be strong, tough and sufficiently stable that can survive in severe environments while the interface fibre-matrix is required to be relatively weak for matrix deflection. If the interface is too strong, crack propagation cannot be deflected by the interface and the composite fails in a weak and brittle manner, if it is too weak it may cause the fibre-matrix debonding and the low interfacial shear strength results in diminution of strength characteristics of material. Good control of the fibre-matrix interface is then requisite to obtain the desired properties. One of the promising approach is to utilise the coating fibres. A coating layer can modify the interfacial bond character, cause or prevent fibre debonding or sliding. The final properties of composite is critically dependent upon two parameters of coating. First, the nature of coating must be adequately chosen to be oxidatively stable and second, it thickness must be adjusted to prevent the interdiffusion fibre-matrix phenomenon. So far, carbon has been one of the most widely used as coating [1-2].

This paper describes a study of carbon coating used to modify the mechanical behaviour of Nicalon fibre-reinforced borosilicate glass composite. The oxidation inhibition properties of carbon coating were examined. Flexural test at the temperature ranging from room temperature to 700°C were performed to assess the effectiveness of the coating.

To the extent authorized under the laws of the United States of America, all copyright interests in this publication are the property of The American Ceramic Society. Any duplication, reproduction, or republication of this publication or any part thereof, without the express written consent of The American Ceramic Society or fee paid to the Copyright Clearance Center, is prohibited.

2. EXPERIMENTAL PROCEDURE

2.1. Composite material

The materials contained continuous, aligned Nicalon SiC fibres were manufactured by a hot pressing process. The glass matrix was a borosilicate supplied by Corning Glass Works (commercially known as "Pyrex 7740"). In spite of the limited temperature capability of the borosilicate is about 500°C, the attractive feature of this glass matrix is its coefficient of thermal expansion which is slightly similar than that of Nicalon fibres. It means that there are no residual stresses in the SiC/Pyrex system. Three different types of materials, named CG, A and B were investigated. They corresponded respectively to the material systems that contained uncoated, thin and thick carbon coated fibres. The thickness of the coating was 20-30 nm for the material system A and 140 nm for the system B. The carbon coating was performed by a chemical vapour deposition (CVD) process. The 100 mm square composite plates were fabricated by Nippon Carbon Co., Ltd. (NCK) and supplied to the University of Tokyo for testing.

2.2. Characterisation and Testing

Test bars were cut from the composite plate using a diamond blade, the low speed saw was chosen to minimise the cutting defect. No matrix microcracking was observed in the specimens after cut. The average dimensions of the specimen were 1.8 x 5 x 50 mm. Flexural tests were carried out using a three-point bending configuration with a span of 30 mm. Tests were performed at temperature ranging from room temperature to 700°C in air with a crosshead speed of 0.5 mm/min. Two or three samples were tested at each of testing temperature. The number of tests is insufficient to offer any statistical analysis, our attempt was only to observe the dispersion of results. Load-crosshead displacement curves were recorded and were used for comparison between three types of material.

3. RESULTS AND DISCUSSION

The flexural strengths are summarised as a function of temperature up to 700°C (Table 1) for three different types of material CG, A and B. The difficulties in interpreting composite flexural test result are recognised, hence the numerical data are reported here in an attempt only to compare the behaviour of three materials tested.

From this data, it can be concluded that the composite prepared from coated Nicalon fibres (types A and B) exhibit higher strengths than those fabricated with uncoated fibres (type CG). Figure 1 shows a plot of the flexural strengths of the materials as a function of test temperature. It is interesting to note that the flexural strengths were found affect differently from one material to an other. Flexural strength of composite A were decreased continually from the room temperature to 700°C, while for the composite CG and B, they were found dependent on the temperature differently. It has been found that they were affected by temperature up to 400°C and an abrupt increase at about 500°C and decreased above this temperature. Similar behaviour of the Type CG and B materials, has been observed in fibre reinforced borosilicate glass matrix [3-4]. However, caution should be exercised when one defines the strength from flexural tests at elevated temperatures since it occurs that no fracture was observed and therefore the stress attained when specimen touch the support is not the ultimate strength. Attempt to explain the performance of composite at temperatures should also remain, of course, the possibility that, despite of the important deflection, specimen fracture may be incomplete.

Table 1 - Flexural strengths of Nicalon SiC/borosilicate matrix (in MPa).

Temp. (°C)	MATERIAL		
	CG	A	B
RT	493.76	670.28	772.57
200	482.95	622.30	758.03
360	-	-	710.81
400	564.61	-	-
500	-	561.10	811.90
707	587.90	516.80	679.90

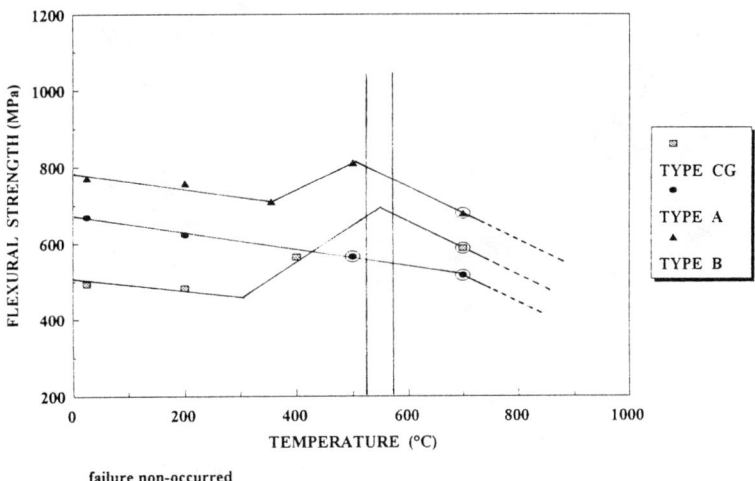

failure non-occurred

Figure 1 - Three-point flexural strength of Nicalon SiC/Pyrex borosilicate glass matrix composites as a function of test temperature.

An explanation for the behaviour at elevated temperature of the Nicalon SiC/Pyrex materials may be offered by considering the morphology of the composite and the behaviour of the individual constituents of material. The morphology analysis of materials was based on the work performed at United Technologies Research Center (UTRC) [5] which has shown through the replica and thin foil TEM micrographs the fibre-matrix region of three types of materials CG, A and B.

It has been found for the Type A material, that the matrix elements may diffuse through the coating into the Nicalon fibres and formed an amorphous phase (figure 2). This observation suggested that the difference in flexural strength between the Type A material and those as Type CG and B was related to the presence of the new amorphous phase. When taking account this observation in accordance with the data of three-point bending tests, it has been thought that the presence of such amorphous phase influenced strongly the performance of composite at elevated temperatures. Therefore it was reasonable to conclude that the thermal expansion coefficient of the amorphous in the present material is quite different compared to those of the Nicalon SiC fibre and the borosilicate matrix.

Hence, it may be argued that the difference in the flexural strengths between the Type A material and the Type CG and B materials are related to the presence of an amorphous phase in the microstructure. The flexural strength evolution at elevated temperatures of the type CG and B materials can be correlated with the behaviour of the material constituents *e.g.*, fibre and matrix. It should be noted the full potential of the Nicalon SiC fibre was not utilised in the Nicalon SiC/Pyrex system because the glass Pyrex began to soften and flow below the fibre degradation.

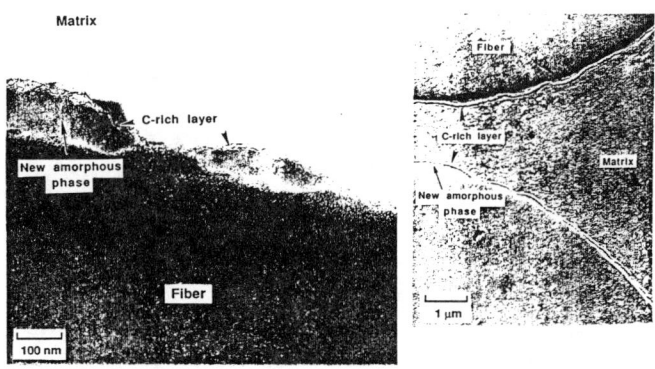

Figure 2 - TEM micrographs of the fibre-matrix interfacial region in the Type A material [5].

Figure 3 shows the characteristic temperatures of the borosilicate glass as a function of its viscosity [6]. From this curve, one should note that the region around 500°C is a critical area. Two main important temperatures, the strain point and the annealing point are located in this "critical zone". For the glass 7740, the strain point is measured at 510°C and the annealing point at 560°C. It means the borosilicate glass cannot be efficient beyond 510°C and above 560°C, it can deform without any applied stress. Therefore, an evolution in flexural strength can be explained by considering the characteristics of the matrix. When the test temperature was around 500°C, the decrease in viscosity of borosilicate glass was observed, which increased the matrix deformation capability, and permitting a stress redistribution in the material. As a result, it led to an increase in flexural strength. On the other hand, the observation of the load-deflection curves at 500°C suggested that the load transfer governed by the interface was different for the three types of materials (figure 4). The load-deflection curve of the Type A material presented many drops that corresponded probably to the debonding whilst the curve of the material B was slightly smooth. Beyond the critical zone, the plasticity of the matrix increased permitting to the fibres to play a role in the fracture of the material. When a fibre breaks, it caused the fibre-matrix debonding and the low interfacial shear strength resulted in diminution in flexural strength of material.

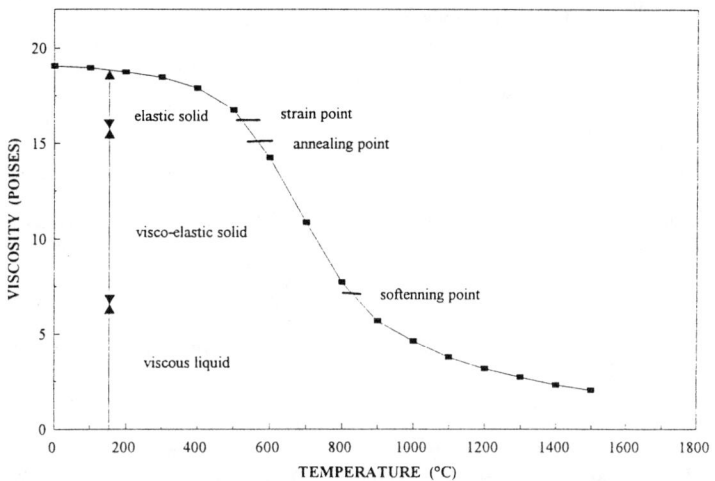

Figure 3 - Characteristic temperatures of borosilicate glass as a function of its viscosity [6].

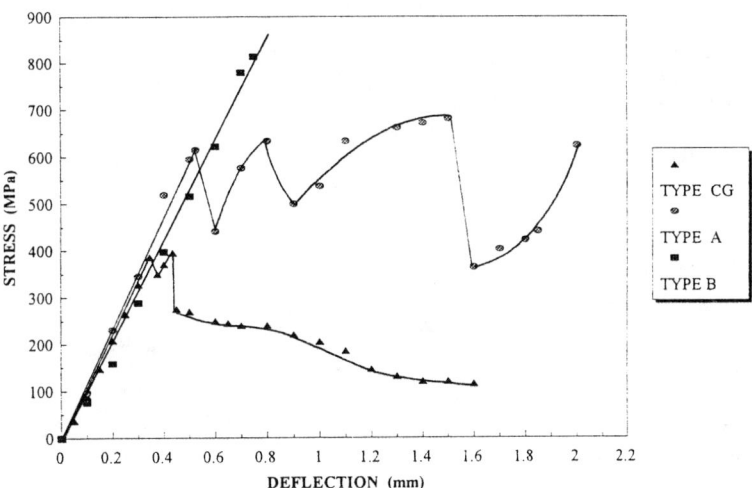

Figure 4 - Load-deflection curve of Nicalon SiC/Pyrex glass at 500°C.

4. CONCLUSION

The flexural strengths at elevated temperatures of three types of Nicalon SiC/Pyrex glass composite were discussed in relation to the morphology of the composite and the behaviour of the individual constituents of material. It has been noted that the response of the materials to the test temperature was conditioned by their microstructure via the thickness of the C-coating layer. The thin C-coating on the Nicalon fibre used in the Type A material was not thick enough to prevent the interdiffusion phenomenon. It led to the formation of an amorphous phase which distinguished its behaviour to that of the Type CG and B materials. The evolution flexural strengths of the uncoated and thick C-coated materials could be explained by the behaviour of the matrix whether it was viscous or presented a plastic property.

5. REFERENCES

[1] Caputo, A. J., Stinton, D. P., Lowden, R. A. and Besmann, T. M., "Fiber-Reinforced SiC Composites with Improved Mechanical properties," *Am. Ceram. Soc. Bull.,* **66** [2], 1987, pp. 268-72.
[2] Brennan, J. J., "Interfacial Characteristics of Glass-Ceramic Matrix/SiC Fiber Composites," *J. Phys., Colloq.,* C5 Suppl., **10** [49], 1988, pp. 791-809.
[3] Prewo, K. M. and Brennan, J. J., "Silicon Carbide Yarn Reinforced Glass Matrix Composites," *J. of Mater. Sci.,* **17**, 1982, pp. 1201-6.
[4] Ramakrishnan, V. and Jayaraman, N., "Tensile Behaviour of Borosilicate Glass Matrix-Nicalon (silicon carbide) Fibre Composite," *J. of Mater. Sci.,* **27**, 1992, pp. 2423-8.
[5] Tredway, W. T. and Prewo, K. M., "Fiber-Matrix Interface Characteristics in Nicalon Fiber Reinforced Borosilicate Glass Matrix Composites," in : *Proc. of the 3rd Japan International SAMPE Symposium,* JISSE-3, Chiba, 1993, pp. 531-6.
[6] "7740 High Reliability Sheet Glass for Liquid Cristal Displays," Product Information - Corning Glass Works, Corning - New York 14831.

ACKNOWLEDGMENTS

B. Mouhamath appreciates the financial support of the Japan Society for the Promotion of Science under the auspices of the Foreign Researchers in Japan program. Both authors would like to thank the Nippon Carbon Co. for provision of SiC/Pyrex composites. We also thank Mr. T. Nose of Nippon Steel Corporation for assistance in conducting the experimental tests.

EFFECT OF POLYMER ADDITION ON THE STRENGTH OF CARBON FIBER REINFORCED REACTION-BONDED SiC MATRIX COMPOSITES

Eiji TANI and Kazuhisa SHOBU
Kyushu National Industrial Research Institute
Shuku, Tosu, SAGA 841, JAPAN

ABSTRACT

Unidirectional carbon fiber reinforced SiC composites were fabricated with polymer pyrolyzed and reaction-bonded SiC matrix. In spite of a low density of about 1.85 Mg/m^3, the flexural strength of the composites was about 530 MPa. Addition of polymer decreased the open porosity and increased the elastic modulus of the composites.

INTRODUCTION

We[1] had reported that the unidirectional fiber reinforced SiC composites could be fabricated by the reaction bonding method with phenolic resin and silicon powder. However, the fracture behavior showed a brittle manner and the flexural strength was as low as about 150 MPa. The reason of such a low strength of the composites would be the low volume fraction of fiber due to the large particle size of silicon (< 74 μm) and the inadequate interphase mechanical properties between matrix and fiber. The composites which were made from polymer pyrolysis without recycling impregnation showed nonlinear fracture and did not show the sharp fiber-bundle failure but very low strength[2]. The purpose of the present study was to investigate the effect of the addition of PSS (polysilastyrene) and the refinement of silicon powder by ball milling on the mechanical properties of the carbon fiber reinforced SiC composites.

EXPERIMENTAL

The carbon fiber used in this study was CARBONIC-HM60-6K (Petoca Co., Ltd., Japan). The bulk density and average diameter of the fiber were 2.17 Mg/m^3 and about 10 μm. The average tensile strength, elastic modulus, and elongation of

---

To the extent authorized under the laws of the United States of America, all copyright interests in this publication are the property of The American Ceramic Society. Any duplication, reproduction, or republication of this publication or any part thereof, without the express written consent of The American Ceramic Society or fee paid to the Copyright Clearance Center, is prohibited.

the fiber were 2.9 GPa, 590 GPa, and 0.50 %, respectively. The slurry with phenolic resin (Novolac type; Hitachi Chemical Co., Ltd., Japan), silicon powder (< 74 μm in size; Sumitomo Sitix Co., Ltd., Japan), polysilastyrene (PSS-400; Nippon Soda Co., Ltd., Japan) and solvent was ball milled for 1 or 3 days. The ratio of silicon and phenolic resin was fixed at Si/C=1 after the pyrolysis of phenolic resin at 1000°C. The amount of PSS was defined as PSS/(PSS+Si). The carbon fiber reinforced SiC composites were fabricated by passing the carbon yarn through the slurry onto a rotation drum, drying the resultant tape at about 70°C for 12 h and then cutting the tapes into the appropriate length. After the tapes were cut to length, sufficient tape was pressed in a steel die under a pressure of about 50 MPa at about 140°C for 20 min and then sintered at 1450°C for 1h in Ar.

Flexural strength samples (about 2 mm thick by 10 mm wide with 40 mm long) were ground with a 200 grit diamond wheel. Flexural strength measurements were performed at room temperature using a three-point loading device with a span of 30 mm and a crosshead speed of 0.5 mm/min. The elastic modulus of the composites was calculated using beam theory.

RESULTS AND DISCUSSION

The effect of ball milling on the particle size of silicon powder was examined by SEM observation. Some large particles of 20 μm were observed in the raw powder. The particle size of silicon decreased from 10 μm to < 5 μm by ball milling from 1 to 3 days.

The bulk density of the composites ball milling for 1 day was 1.6~1.7 Mg/m^3 (Fig. 1). However, the bulk density of the composites ball milling for 3 days was higher by about 0.1 Mg/m^3, except at the composition of PSS=80 wt%. Because the smaller the particle size of silicon was, the closer the carbon fiber packed. The addition of PSS=20~60 wt% increased the bulk density of the composites. This increase of the bulk density might be due to the increase of the volume fraction of fiber and/or the decrease of the porosity. Because the amount of PSS is inversely proportional to the amount of silicon powder. However, the decrease of the bulk density at PSS > 60 wt% might be due to the lower volume fraction of fiber. Because the only composite at PSS=100 wt% expanded slightly after the heat treatment.

The flexural strength of the composites ball milling for 1 day without PSS was 140~200 MPa (Fig. 2). The composites with PSS=20~80 wt% showed a little higher flexural strength of 170~280 MPa. However, the composites with PSS=100 wt% (without silicon and phenol resin) showed low strength of 50 MPa and low density of 1.3 Mg/m^3. However, the flexural strength of the composites

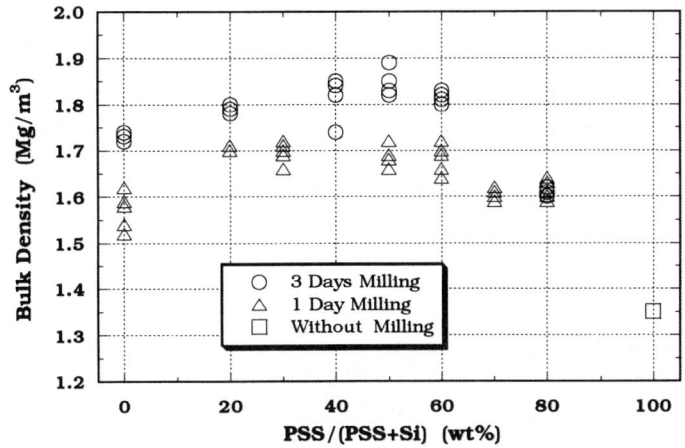

Fig. 1 Effect of the amount of polysilastyrene addition on the bulk density of the composites.

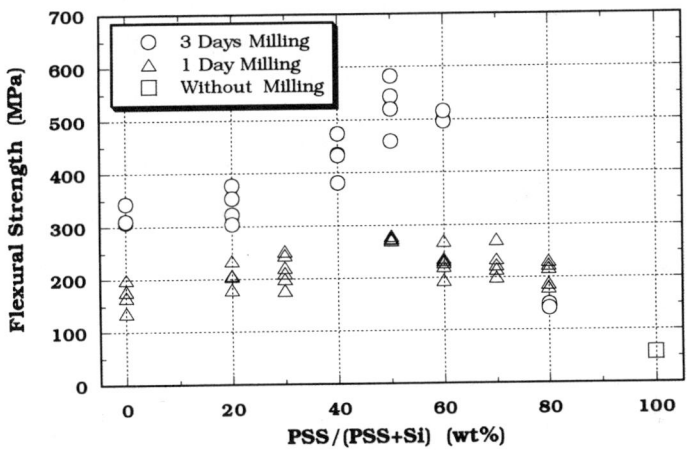

Fig. 2 Effect of the amount of polysilastyrene addition on the flexural strength of the composites

ball milling for 3 days increased from 320 to 530 MPa with increasing PSS addition from 0 to 50 wt%. This result means that the effect of the addition of PSS on the mechanical properties of the composites is more clear in the composites with smaller particle size of silicon. This strength of the composites ball milling for 3 days with PSS=40~60 wt% is almost the same value with powder & hot-pressing method[3] and polymer pyrolysis & recycling impregnation method[4]. However, the density of those samples needs more than 2.3 Mg/m^3.

The variation of the elastic modulus of the composites with PSS addition was almost the same as that of the flexural strength. The elastic modulus of the composites ball milling for 1 day varied from 80 to 180 GPa. However, the elastic modulus of the composites ball milling for 3 days increased from 150 to 250 GPa with increasing PSS addition from 0 to 50 wt%, and it abruptly decreased to 70 GPa at the composition of PSS=80 wt%. The strain of the composites with PSS=0~60 wt% at the maximum stress was almost constant of 0.2~0.3 %. Therefore, the increase of the strength of the composites was attributed to the increased of the elastic modulus.

The open porosity of the composites is shown in Fig. 3. The open porosity of the composites without PSS ball milling for 1 day was quite high value of about 38 vol%. It decreased to about 30 vol% for the composites with PSS=20~80 wt%. The open porosity of the composites ball milling for 3 days decreased from 30 to 20 vol% with increasing of the PSS addition from 0 to 50 wt%. The decrease of the open porosity seemed to be related with the increase of the elastic modulus and flexural strength of the composites.

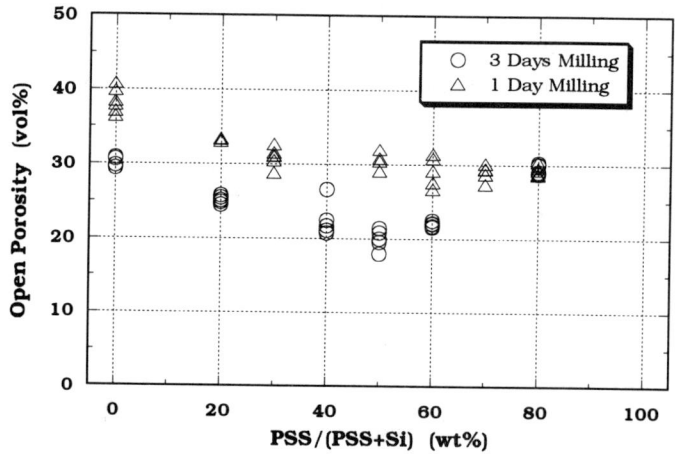

Fig. 3   Effect of the amount of polysilastyrene addition on the open porosity of the composites.

The fracture behavior of the composites ball milling for 3 days shows from linear-elastic, brittle to nonlinear manner with addition of PSS (Fig. 4). This figure might clear show that the interfacial mechanical properties of the matrix between fiber were different between reaction-bonded SiC and polymer pylolyzed SiC. The brittle stress-strain behavior is attributed to the strong interfacial bonding caused by reaction-bonded SiC. The composites with PSS=50 wt% exhibited initially a linear stress-strain response, followed by nonlinear behavior.

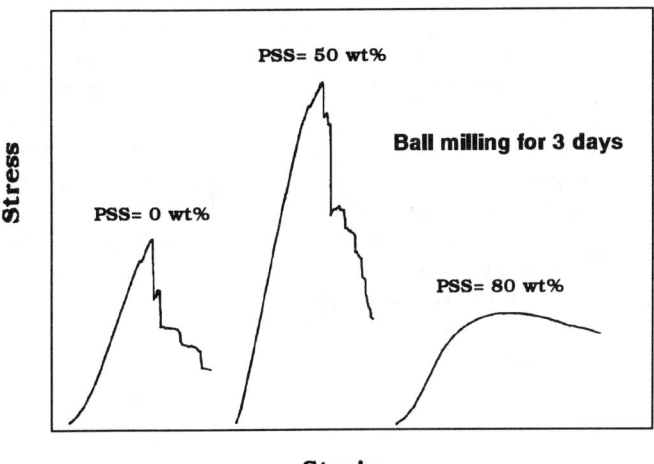

Fig. 4 Effect of the amount of polysilastyrene addition on the fracture behavior of the composites ball milling for 3 days.

The fracture behavior of the composites with PSS=80 wt% showed typical nonlinear manner, and did not show the abrupt drop of the stress. It might be considered that the reaction-bonded SiC preferentially adhered on the surface of the fiber, however, polymer pyrolyzed SiC did not consolidate on the surface of fiber but between fiber. Addition of PSS made dense matrix network between fibers and decreased the open porosity, then it increased the elastic modulus but not failure strain.

SUMMARY

Unidirectional carbon fiber reinforced SiC composites were fabricated by pyrolyzed PSS and reaction-bonded SiC from phenolic resin and silicon at 1450°C. Refinement of silicon particle size to < 5 μm and PSS addition decreased the open porosity of the composites, and increased elastic modulus and flexural strength of the composites. The PSS addition changed the fracture behavior of the composites from linear-elastic to nonlinear. In spite of the low density of about 1.85 Mg/m^3, the elastic modulus and the flexural strength of the composites were about 250 GPa and about 530 MPa, respectively.

REFERENCES

1) E.Tani, K.Shobu and T.Watanabe, "Carbon Fiber-Reinforced SiC/C Composites Produced by Reaction-Bonding" pp. 207-13 in *High Temperature Ceramic Matrix Composites*. Edited by R.Naslain, J.Lamon and D.Doumeingts, Woodhead Publishing Ltd., Cambridge, England, (1993).

2) H.Yoshida, N.Miyata, M.Sagawa, S.Ishikawa, K.Naito, N.Enomoto and C.Yamagishi, "Preparation of Unidirectionally Reinforced Carbon-SiC Composite by Repeated Infiltration of Polycarbosilane" *J.Ceram. Soc. Japan*, **100** [4] 454-58 (1992).
3) K.Nakano, A.Kamiya, H.Ogawa and Y.Nishino, "Fabrication and Mechanical Properties of Carbon Fiber Reinforced Silicon Carbide Composites" *J.Ceram.Soc. Japan*, **100** [4] 472-75 (1992).
4) K.Sato, T.Suzuki, O.Funayama and T.Isoda, "Fabrication of Silicon Nitride Based Composites by Impregnation with Perhydropoly-silazane" *J.Ceram.Soc. Japan*, **100** [4] 444-47(1992).

NOVEL ORGANOSILICON PRECURSORS FOR Si-C CERAMICS AND CERAMIC MATRIX COMPOSITES

E. A. Chernyshev, S. A. Bashkirova, T. V. Tikchonovich, V. V. Ivanov

The State Scientific Research Institute of Chemistry and Technology of Organoelement Compounds (GNIIChTEOS), 38, sh. Entuziastov, 111123 Moscow (Russia)

ABSTRACT

New organosilicon polymers with functional groups Si-H, C=C and C≡C have been prepared by hydrosilylation. The developed polymers were used both as impregnating compositions for reinforcement of ceramic articles produced from silicon carbide and as binders for powder ceramic reinforcers for manufacturing of highly strong thermally stable products. Advantages of the novel pre-ceramic polymers in ceramic production in comparison with conventional procedure are shown.

INTRODUCTION

Some organosilicon polymer classes are efficient sources of non-oxide silicon carbide ceramics. They include polycarbosilanes (PCS), first obtained by Yajima and co-authors [1]. PCS contain $-CH_3(H)SiCH_2-$ as the basic repeated unit. Their pyrolysis at $1200^0 C$ results in ceramic residue with 55-60% yields. However, for many preceramic polymer applications higher ceramic yields under pyrolysis are desirable. Investigations have been devoted recently both to chemical modification of commercial PCS by means of functional groups introduction, ethynyl [2], vinyl [3] ones, for example, and

---

To the extent authorized under the laws of the United States of America, all copyright interests in this publication are the property of The American Ceramic Society. Any duplication, reproduction, or republication of this publication or any part thereof, without the express written consent of The American Ceramic Society or fee paid to the Copyright Clearance Center, is prohibited.

search for oxygen-free cross-linking by e-beam irradiation [4] or thermal cross-linking with unsaturated compound vapors [5]. But all these methods have the same drawbacks as the Yajima process, that can be summarized in the following way: 1) prerceramic PCS is produced in two steps: (i) synthesis of polydimethylsilane from $(CH_3)_2SiCl_2$ in presence of sodium and (ii) its thermal rearrangement; 2) PCS structure is not definite, it is difficult to reproduce from one synthesis to another one.

Recently there was a number of publications about synthesis of polyacetylene-polysilane [6] and silylene-diacetylene [7] polymers, that could be efficient ceramic precursors as at 1000^0C they lost only 20% of weight. However these polymer syntheses are rather complicated and require low temperatures (-78^0C) and organolithium compounds.

EXPEREMENTAL

^1H NMR spectrum was recorded on a Bruker AM 360 spectrometer. IR and TGA data were obtained respectively on a Bruker IFS 113 V and Du Pont 990 Analyzer.
The number average molecular weight (M_n) of synthesized polymers with polystyrene standard was measured with a liquid chromatograph and GPC column.
Pyrolysis was conducted in a graphite furnace under flowing argon atmosphere.
Mechanical strength was determined on "INSTRON 1122" testing machine.

RESULTS AND DISCUSSION

Our target was the one-stage synthesis of new efficient organosilicon precursors of Si-C ceramics by catalytic process. In this work we obtained polymer (I) with functional groups Si-H, C=C and C≡C by bis(dimethylsilyl)acetylene hydrosilylation in presence of H_2PtCl_6 at $60-90^0C$:

$$HMe_2SiC≡CSiMe_2H \xrightarrow[60-90^0C]{H_2PtCl_6} HMe_2SiC≡CSiMe_2-[C=CHSiMe_2]_n-H$$
$$\underset{HSiMe_2}{|}$$

where n= 9-22 (I)

Polymerization was carried out in hexane or chlorobenzene. Chain length can be varied according to the solvent. Polymer (I) was characterized by ^{1}H NMR, IR and Raman spectra. ^{1}H NMR (CDCl$_3$, 20^{0}C) δ 0.01-0.16 (s, Si-Me), 3.72-4.82 (br s, Si-H), 6.30-6.56 (m, =CH); IR (film, cm^{-1}) 2955, 2895, 2081 (Si-H), 1620 (C=C), 1408, 1248, 835, 752; Raman: 2089 cm^{-1} (C≡C). Elemental analysis: C, 51.22; H, 9.96; Si, 38.81; calcd: C, 50.70; H, 9.87; Si, 39.43%.

Reactive functional groups presence allows to use these polymers (I) with various chain lengths as preceramic precursors as well as to conduct their oxygen-free cross-linking in an inert atmosphere.

Polymer samples heating at the rate of 10^{0}C/min in N$_2$ flow to 300^{0}C provided complete polymer cross-linking. Further pyrolysis resulted in ceramic residue in the range of 80-85 wt.%. TGA curves for polymer (I) samples with n=9-15 and n=17-22 are shown in Fig.1.

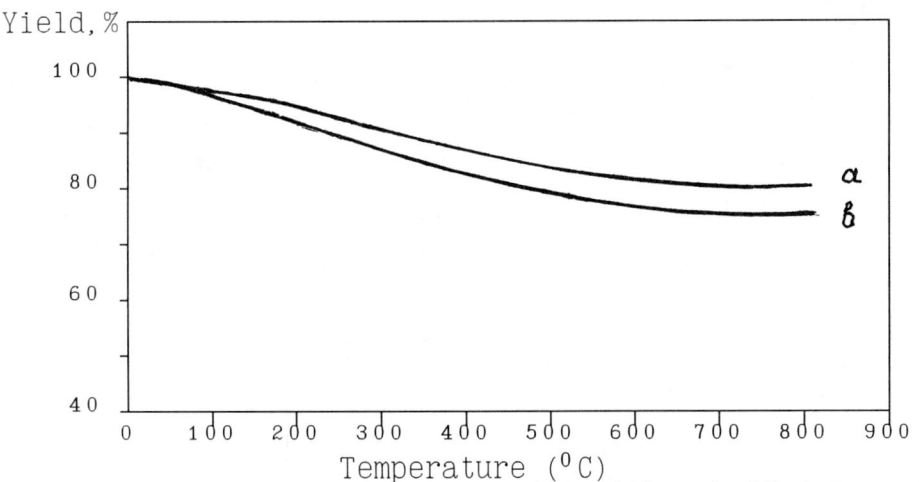

Fig.1 TGA analysis of polymer (I) with n=9-15 (a) and n=17-22 (b)

There is a good perspective for functional polymers impregnation as 10-40 wt.% solutions for ceramic and pyrographite products [8].

Porous ceramic articles (pipes-shells for thermocouples, dosing apparatuses) with the porosity of 30.5% and strength of 25.9 MPa were produced from SiC-ceramic by means of slip casting method without any additions.

The first impregnation of the articles by polymer (I, n=9-22) solution (20%) and the following pyrolysis in

an inert atmosphere at $1400^0 C$ resulted in the porosity of the articles of 20.1-25.6% and strength - 118.1-234.6 MPa (see Table 1).
As a result of thrice-repeated impregnation of the articles by polymer (I) solution with the following pyrolysis in an inert atmosphere at $1400^0 C$ these articles porosity was 10.2% and compression strength 550 MPa (see Table I, test No.6).

Table I. Test results of refractory silicon carbide articles treatment by impregnating compositions

No. samples	Properties before impregnation			Properties after impregnation		
	Porosity, %	Apparent density, g/cm^3	Compr. strength, MPa	Porosity, %	Apparent density, g/cm^3	Compr. strength MPa
1	30.5	2.18	25.9	20.1	2.36	178.9
2	--	--	--	20.6	2.33	171.2
3	--	--	--	22.9	2.26	125.3
4	--	--	--	23.2	2.29	234.6
5	--	--	--	25.6	2.23	118.1
6*	--	--	--	10.2	2.91	550.0

* These data are the result of thrice-repeated impregnation of silicon carbide articles

So our technology of ceramic strengthening by polymer (I) impregnation allows to use low-grade ceramics as the base. This ceremics obtains required properties after reinforcement. The developed ceramics is successfully used for the production of protective shells for temperature control of non-ferrous metal melts up to $1400^0 C$ and of gas phases up to $1600^0 C$, as well as for dosing apparatuses of non-ferrous casting.
X-ray diffraction analysis after pyrolysis at $1400^0 C$ shows that silicon carbide is present as flat thin layer on the surface and inside silicon carbide products.
Ceramics chemical composition is the following: C, 44.95; H, 0.15; Si, 54.80%, $SiC_{1.76}$.
The possibility of production of ceramic matrix composites with "Nicalon"-type fibers obtained from PCS (SiC-0-C) and polymer(I)-based binder was demonstrated. The composite remains thermally stable up to $1500^0 C$.

This could not be achieved before, due to low composite strength.
The following results were obtained: σ_{bend} = 50.5 MPa; σ_{compr} = 58.0 MPa
The developed polymers (I) were also used as binders for powder ceramic reinforcers for manufacturing of highly strong thermally stable construction-grade products. SiC and Si_3N_4 were used both separately and in various combinations as ceramic powder reinforcements. Binders quantity was varied in the range of 15-20%. Compression strength of new binder based ceramic samples raised to 50 GPa after firing at $800^0 C$, but the samples can undergo mechanical treatment (drilling of holes etc) with relative ease. After that they are thermally treated at $1000-1500^0 C$.

CONCLUSION

Novel polymers with functional groups (Si-H, C=C, C≡C) have been prepared. Their pyrolysis in an inert gas results in the formation of silicon carbide with a yield higher than 80 wt.%. The compression strength of refractory silicon carbide articles impregnated with these novel polymers can be increased about 21 times in comparison with initial materials. They remain thermally stable up to $1500^0 C$.
The possibility of the production of highly strong ceramic matrix composites from Nicalon-type fibers and a binder based on the novel polymers with thermal stability up to $1500^0 C$ has been demonstrated.

ACKNOWLEDGMENTS

We are very grateful to the employees of Ucr. GNIIO Fedoruc R.M. and Degtyareva L.M. for their assistance in the fabrication and analysis of the composites.

REFERENCES

1. S.Yajima, K.Okamura, I.Hayashi and M.Omori, "Synthesis of continuous SiC fibers with high tensile strength", J.Am. Ceram.Soc., **59**, 7, 324-327 (1976)
2. K.J.Thorne, S.E.Johnson, H.Zheng, J.D.Mackenzie and M.F.Hawthorne, "Chemically designed, UV curable polycarbosilane polymer for the production of silicon carbide", Chem.Mater., **6**, 110-115 (1994)

3. E. Chernyshev, S. Bashkirova, I. Raisin, T. Tikchonovich, L. Ul'yanova, "Polycarbosilanes - perspective precursors for silicon carbide composites", pp 194-199, Moscow Intern. Composites Conf., (Elsev. appl. sci.), (1990)
4. M. Takeda, Y. Imai, H. Ichikawa, T. Ishikawa, T. Seguchi and K. Okamura, "Properties of the low oxygen content silicon carbide fiber after high temperature heat treatment", Ceram. Eng. Sci. Proc., **12**, 1007-1018 (1991)
5. Y. Hasegawa, "Synthesis of thermally stable Si-C fibre", pp 59-65, Proc. 6-th Eur. Conf. on composite mater., HT-CMC-1, (R. Naslain et al. ed.), Woodhead Publ. Limited, (1993)
6. T. J. Barton, "Polyacetylene-polysilane and polyacetylene-polysilylene ceramic precursors", Pat. 4 940 767, US (1990)
7. S. Ijadi-Maghsoodi, T. J. Barton, "Synthesis and study of silylene-diacetylene", Macromolecules, **23**, 4485-4486 (1990)
8. S. Bashkirova, T. Tikhonovitch, L. Ui'yanova, E. Chernyshev "Polyorganosilane: Si-C pyrolysis and ceramic application", Proc. 6-th Eur. Conf. on composite mater., HT-CMC-1, (R. Naslain et al. ed.), Woodhead Publ. Ltd., (1993)

ANTIOXIDATIVE PROTECTIVE COATINGS FOR CARBON MATERIALS

G.A.Kravetskii, V.I.Kostikov, A.V.Demin, V.V.Rodionova
NIIGrafit, 2, Electrodnaya St., Moscow, 111524 Russia

A widespread use of carbon-carbon and carbon-ceramic materials (CCM) in the aerospace industry, metallurgy (crucibles for melting metals) and electrical engineering is limited because of the need for protecting CCM parts against oxidation at service temperatures above 500 to 700 oC.
At temperatures up to 1300-1400 oC, the problem can be solved by volume siliconizing CCM parts, impregnating C-C substrates with organosilicon compounds or gas-phase depositing (CVD process) silicon-containing compounds (SiC or Si_3N_4) [1, 2].
For CCM parts to be used at temperatures above 1500 oC in oxidative environments (space structures; aircraft gas-turbine engine components in contact with a hot gas; crucibles for melting metals), the following techniques are being devised to apply protective coatings, as evident from a patent literature analysis:

* Application of SiC coatings onto the surface of graphite or C-C parts by the CVD or CVR methods; such coatings can be quite efficient for parts operating short time at temperatures up to 2000 oC, for example in rocket engines [3];

* SiC coatings applied onto the surface of large-sized or intricately-shaped parts frequently experience cracking; this necessitates the application of multilayered or multicomponent coatings (by subsequent impregnation with various silicate compositions, covering with glass or glass-like compositions to "heal" cracks; applying surface oxide or silicate coatings) [4, 5];

* Application on the surface of CCM parts of refractory, self-healing-in use coatings containing refractory borides and silicides; to this end the CVD method and plasma spraying in controlled atmospheres are employed [6, 7].
Given below are results of the investigations conducted at NIIGRAFIT in the above-mentioned directions with the use of the slip-casting technology.

To the extent authorized under the laws of the United States of America, all copyright interests in this publication are the property of The American Ceramic Society. Any duplication, reproduction, or republication of this publication or any part thereof, without the express written consent of The American Ceramic Society or fee paid to the Copyright Clearance Center, is prohibited.

APPLICATION OF SiC COATINGS USING THE SLIP-CASTING PROCESS

We have developed SiC coatings involving a two-step process:
- Application onto the carbon part surface to be protected of a slip based on a fine-disperse carbon powder of increased reactivity;
- Heat treatment of the coated carbon part in the presence of gaseous Si (CVD process).

The heat treatment results in penetrating gaseous Si into a substrate material and coating followed by its interaction with the carbon to form SiC. The microstructure of such coatings is shown in Fig.1. The coating consists essentially of SiC with inclusions of free Si and C.

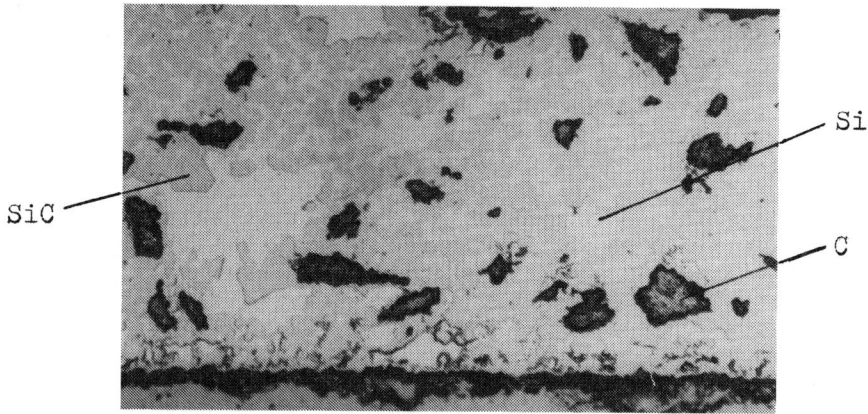

Figure 1. Microstructure of a SiC-based coating obtained by the slip technology.

Free silicon serves as the source for forming a sufficiently dense oxide film due to its interaction with oxidizing components under service conditions, while carbon inclusions serve to increase the crack resistance of the coating by fixing cracks initiated in the SiC film.

GLASS-SILICIDE SLIP-TYPE COATINGS

With the aim to improve the reliability in service and heat resistance of siliconized carbon materials that form a low-density and not continuous carbide film during their production, we in collaboration with the Institute of Silicate Chemistry have developed a slip-type coating that provides a peculiar kind of healing discrete siliconized or borosiliconized surface areas at service temperatures above 1500 oC.

On the basis of the investigation results, a composition based on borosilicate glass with a MoSi2 additive has been proposed to protect borosiliconized CCM against oxidation at temperatures up to 1500-1600 oC. The borosilicate glass is SiO2 doped with B and Al oxides. Chemical composition of the glass allows a vitreous gas-tight coating to be produced due to the diffusion under high temperature treatment. Moreover, B and Al oxides serve to decrease the SiO2 ability to crystallization and impart more stability to the coating performance, especially at high temperatures. MoSi2 exhibits a high heat resistance up to 1700 oC due to the presence of a gas-tight bonded vitreous silica film.

The principal advantage of such glass silicide coatings is their high resistance to thermal shocks. The reason is that, in addition to the good CTE match between the substrate and coating materials, at temperatures above 1200 oC the latter is in a plastic state, and this facilitates the relaxation of thermal stresses which may arise. Moreover, the plastic state of the coating material provides, to some extent, self-healing of small defects produced under service conditions.

Such coatings feature a low thermal conductivity [1.71 W/(m*K) at 20 oC] and good electrical isolation properties ($\rho = 10^4$ Ohm*m). Water absorption in the coated samples after a 3-week exposure was 10 times lower than that in the samples without coating (0.043 and 0.33 %, respectively). Strength characteristics of both samples, with and without coating, are practically alike.

Metallographic, X-ray, and chemical analyses have shown that the coating is essentially an amorphous SiO2 phase with inclusions of crystalline MoSi2 phase. In the course of the coating formation at 1350 oC, a weak diffusive-chemical interaction between molten glass and MoSi2 occurs to form MoB. On long heating (40 h) at 1550 oC in air, extra MoB2 and Mo2B3 phases (in small quantities) are found in the coating. Molybdenum borides surround MoSi2 inclusions.

The developed coating exhibits a relatively high oxidation resistance both in static tests in air (1000 h at 1500 oC) and in a dissociated air flow under low vacuum (0.01 MPa, a few hours). In the latter case, the mass loss in test samples of coated (MoSi2 + borosilicate glass) borosiliconized CCM at 0.01 MPa gas flow pressure was practically independent on the gas flow temperature within the temperature range at the sample surface T = 1000-1500 oC and came to 10^{-5} kg/(m^2*s) (Fig. 2). A rise in the gas flow temperature up to 1500-1520 oC leads to a sharp noncontrolled temperature increase at the sample surface up to 2000 oC. After a 2-3 min vigorous gas release through the molten coating material, the temperature decreases to 1500 oC and the gas release ceases. This results in increasing the average mass loss rate up to 10^{-4} kg/(m^2*s) after a 60 min exposure, with 80 % of the mass loss occurring in the first 2-3 min.

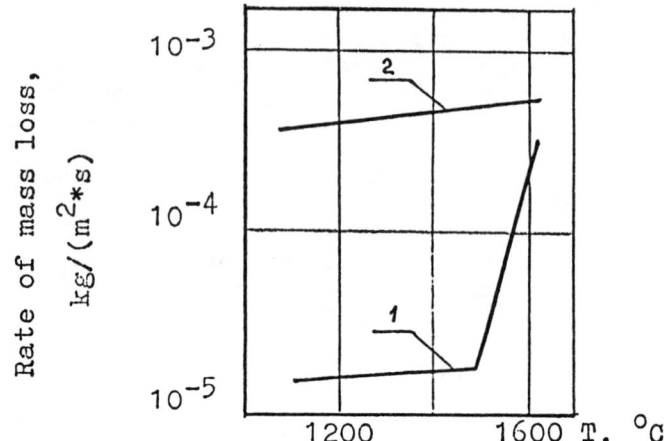

Figure 2. Oxidation resistance of samples in a dissociated air flow with a glass-silicide coating (1) and without such (2).

Samples of uncoated borosiliconized CCM when tested under the same conditions for 10 min at T = 1480 oC showed a mass loss rate higher by an order of magnitude than that of the coated material.
Spectral analysis of the boundary layer in the samples as well as metallographic examinations allowed a conclusion on the interaction mechanism between an air plasma flow and borosiliconized CCM protected with a glass silicide coating. At relatively low service temperatures (1000-1300 oC), the breakdown of the coating occurs through evaporation of Si and B oxides. As temperature rises, the coating material viscosity decreases, resulting in the release of those gaseous inclusions that are present beneath the coating in a carbon substrate. As temperature further increases, an exothermic $MoSi_2$ oxidation with atomic oxygen occurs accompanied by a momentary temperature increase at the sample surface up to 2000 oC. The temperature at which gas release starts depends both on the substrate pore structure and melt viscosity whose value, in turn, is much dependent on the temperature and boron oxide content in the glass.
On the basis of the oxidation mechanism proposed, the following ways for improving heat resistance of protective coatings onto CCMs have been directed:
* Decrease or replacement with other heat-resistant materials of the $MoSi_2$ phase in the coating composition;

* Surface densification of or porosity decrease in the substrate;
* Decrease of the boron oxide content in the coating composition;
* Substrate vacuum outgasing during coating formation.

The realization of the above directions has enabled us to create a modified glass-silicide coating (M55VA) that provides an efficient operation of parts up to 1600 oC. Thus, the major advantages of glass-silicide coatings are as follows:
* Service temperatures as high as 1500-1600 oC;
* Stress relaxation due to the plastic state at high temperatures
* Good CTE match between coating and substrate;
* Self-healing of small defects under service conditions;
* Reduced (by a factor of 10) water absorption in coated samples;
* Simple technology;
* The possibility of applying coatings onto large-sized parts (up to 2000 mm) of intricate configuration.

SLIP-TYPE COATINGS BASED ON HIGH-MELTING BORIDES

Coatings based on Si oxides are not suitable to protect graphite and CCM parts in oxidazing environments at temperatures as high as 1700-2000 oC. The investigations conducted show that boride-carbide coatings of the (Hf, Zr, Ti) system with additions of high-melting components are most promising for these purposes as they provide formation of protective films in the coated parts under service conditions.
The selected slip-type coating composition offers resistance to thermal shocks and self-healing of small defects in coatings. Fig. 3 shows changes in the service temperature as a function of the boride type for boride-protected parts. As can be seen, the highest service temperature is offered by the HfB_2-based coating. It should be noted that the cost of borides increases in the order TiB_2 ---> ZrB_2 ---> HfB_2.
In Fig. 4 are presented experimental results for siliconized graphite samples with and without coating tested in static air at 1700 oC. The uncoated sample failed in 30 min. In the coated sample, some mass gain due to oxidation of the coating constituents was observed when it was tested for 3 h.
The HfB_2-based coatings were tested on samples and models simulating service conditions of aerospace parts and structures. Slip-type boride coatings have the following features:
   - Service temperatures of above 1700 oC;
   - Self-healing of small defects under service conditions;
   - The possibility of combining the coating formation process and the volume siliconizing;
   - The possibility of applying coatings onto large-sized parts.

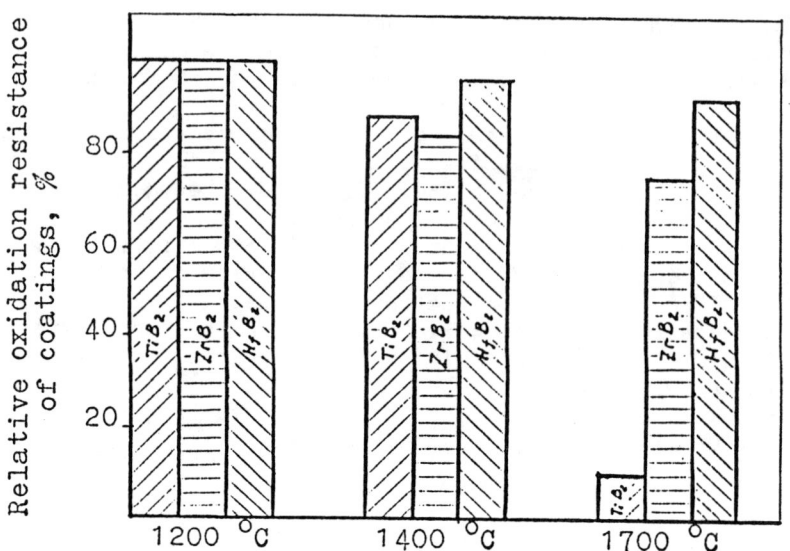

Figure 3. Comparative resistance of coatings based on titanium, zirconium, and hafnium borides.

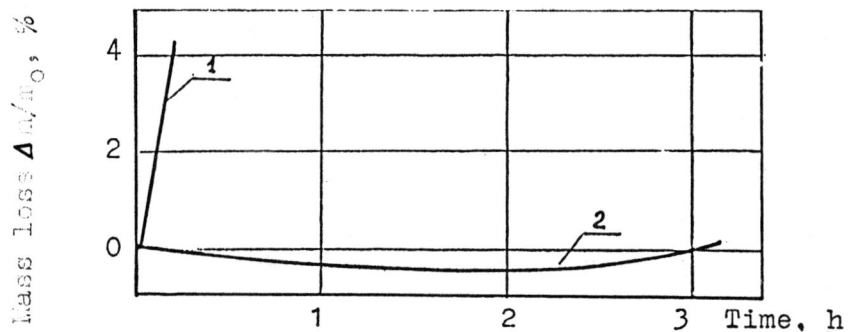

Figure 4. Oxidation resistance of siliconized graphite without coating (1) and with HfB_2-based coating (2) in air at 1700 °C.

ACKNOWLEDGEMENT

The authors wish to thank Dr. M.Yakushin and Dr. A.Gordeev of the Institute of Problems in Mechanics, Russian Academy of Sciences, for their technical contribution and support in implementing the dissociated air flow tests.

REFERENCES

1. Hide Itiro. Process for producing a barrier layer on a carbon material surface/Patent No. 59-171334, Japan, Publ. 10.03.1986.
2. R.D.Weltri, F.S.Gallosso. SiC/Si3N4 coating on graphite-carbon materials/Patent No. 8,417,030, UK, Publ. 08.10.1986.
3. E.V.Gellhorn, V.Gruber, H.Leis. SiC coated C-C composite:modern material for rocket systems/STAR, V. 27, No. 16, P. 2672, 1989.
4. D.M.Carry,J.A.Cunningham,J.R.Frank. Space Shuttle Orbiter leading edge structural subsystem thermal performance/AIAA Paper No 82-0004.
5. V.I.Kostikov, G.A.Kravetskii,A.V.Kuznetsov, V.V.Rodionova.Heat-resistant antioxidative protective coating for C-C, C-SiC, and graphite materials/1st Int. Aerospace Conf. Perspectives of Mastering Outer Space, 28 Sept.-2 Oct.,1992, Proc., V. 5, p. 249-254, Moscow, 1995.
6. Improved oxidation-resistant CVD coating for high temperatures/ Advanced Materials, V. 11, No. 3, p. 3, 1988.
7. Process for Applying a HfB2 + 20 % SiC coating in low-pressure plasma/United Technologies Corp. Catalogue, 1988.

The Nanoscale Microstructure of 2D C/C-SiC Ceramic Composites Processed via Silicon Capillary Impregnation

W. Braue[*], R. Pleger[*] and R. Weiss[#]

(*) German Aerospace Research Establishment (DLR), Materials Research Institute, D-51147 Cologne, Germany
(#) Schunk Kohlenstofftechnik GmbH, D-35452 Heuchelheim, Germany

ABSTRACT

Silicon capillary impregnation offers a cost-effective siliconization route for C/C preforms resulting in a damage tolerant C/C-SiC structural material with reasonable internal oxidation protection and thus improved lifetime at high temperatures. The nanoscale microstructures of both the C/C preform and the converted C/C-SiC materials employing different heat treatments prior to silicon impregnation are investigated via HREM and AEM in order to characterize the different carbon structural variants and evaluate the effects of carbon fiber impurity sources during siliconization.

INTRODUCTION

C/C-SiC ceramic composites can be fabricated via silicon capillary impregnation of C/C preforms with subsequent conversion to SiC [1]. This process offers a cost-efficient alternative compared to i) chemical vapor infiltration of a fibrous preform or ii) infiltration and pyrolysis of silicon-polymer precursors [see ref. 2 for review].
C/C-SiC composites exhibit an excellent strength-to-density ratio and reasonable short-term oxidation resistance in air which can be significantly improved through an additional surface oxidation protection [3]. Depending on the conversion rate, siliconization results in a trade-off between the oxidation rate and the retained ductility. Fully converted C/C-SiC composites exhibit brittle fracture, but nevertheless offer promising tribological applications.
We present an HREM/AEM investigation of a phenolic resin-based 2D C/C-SiC composite focusing on (i) the characterization of the dominant carbon structural variants from intrabundle regions in the C/C preform, (ii) nucleation and growth of SiC polycrystals during silicon impregnation and (iii) segregation effects due to inherent fiber impurity sources.

To the extent authorized under the laws of the United States of America, all copyright interests in this publication are the property of The American Ceramic Society. Any duplication, reproduction, or republication of this publication or any part thereof, without the express written consent of The American Ceramic Society or fee paid to the Copyright Clearance Center, is prohibited.

EXPERIMENTAL PROCEDURE

Commercially available 2D C/C preforms consisting of PAN-based HT-carbon fibers embedded in a carbonaceous matrix derived from a phenolic resin were employed in this study. Evaluation of microstructure is focused on two C/C-SiC batches derived from two C/C materials experiencing different heat-treatment temperatures (HTT) prior to silicon impregnation: (i) a low carbonization HTT of 1000°C, without additional graphitization and (ii) a carbonization HTT of 1100°C followed by a graphitization treatment at 2100°C/1h. While the former material was prepared with particular emphasis on the effects of fiber impurity sources during siliconization, the latter reflects the significance of stress-oriented graphite at the fiber/matrix interface. The silicon impregnation process was supported by capillary treatments and performed at 1700°C in vacuum. Silicon uptake was of the order of 25% leaving very few residual silicon grains dispersed in the microstructure. Thin TEM foils were prepared from C/C and C/C-SiC materials via grinding, dimpling and argon-ion beam milling. The TEM studies were performed in a Philips EM 430 TEM/STEM microscope operating at 300 kV.

RESULTS AND DISCUSSION

In the graphitized 2D C/C preform typically employed for silicon capillary impregnation in this research, the intrabundle microstructure consists of a complex matrix carbon architecture involving different carbon structural variants [4,5]. Vitreous carbon is the most abundant matrix carbon (Fig. 1a).

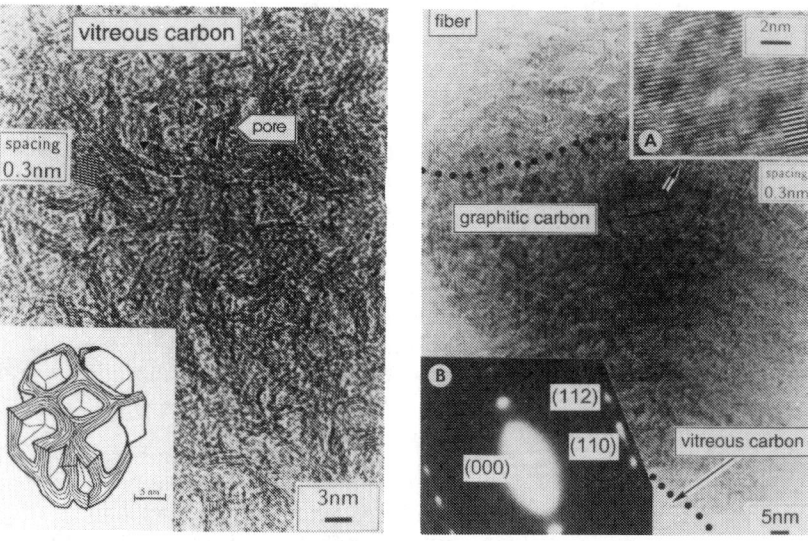

Fig. 1 (a) HREM image of vitreous carbon in graphitized 2D C/C, (b) HREM image of the fiber/matrix interface in graphitized 2D C/C, emphasizing the formation of stress-oriented graphite

This nongraphitizing constituent is characterized by an isotropic closed-foam structure with cell walls defined by stacks of turbostratic graphite basal planes, as shown in the HREM image and the structural model (see inset in Fig. 1a, after [6]) respectively. During preform processing, the high shrinkage of the thermoset-derived matrix in the pyrolysis step gives rise to multiaxial deformation of the carbonaceous matrix at the fiber/matrix interface [7]. This complex process is characterized by circumferentially (with respect to the fiber periphery) acting tensile stresses which are sufficiently high to cause preorientation of matrix molecules and the formation of stress-oriented graphite (Fig. 1b) after prolonged heat-treatment at 2100°C. Its degree of orientation varies from a turbostratic to a graphitic structure exactly at the interface, as indicated by the ({hkl}, with h, k ≠ 0, l ≠ 0)-type Bragg reflections in the SAD pattern (see inset B in Fig. 1b). In terms of intrabundle failure, the implication of matrix orientation at the fiber/matrix interface is emphasized by the common observation that the far majority of cracks propagates along the graphite (001) cleavage planes with only few cracks bound to the isotropic structure of vitreous carbon [8].

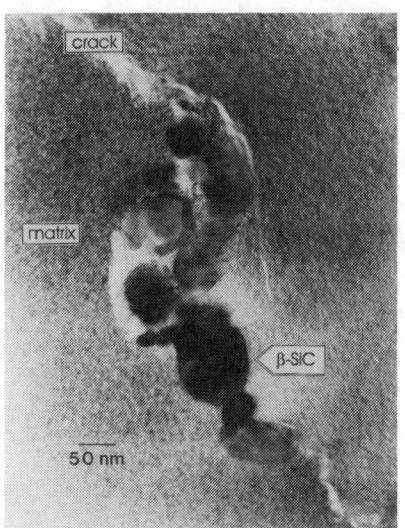 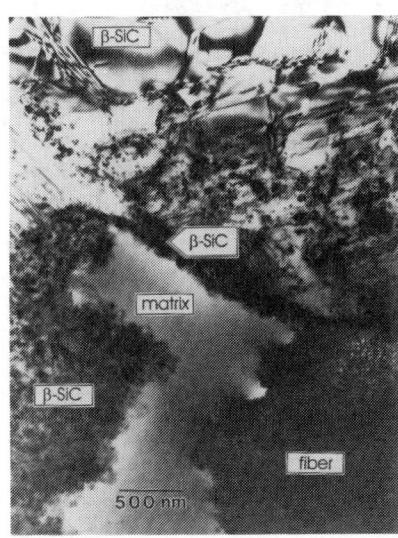

Fig.2 (a) Early stages of ß-SiC formation along a microcrack in C/C-SiC, (b) SiC conversion front in C/C-SiC (prepared from graphitized C/C preform) leaving residual carbonaceous matrix regions

Propagation of the SiC conversion front in C/C-SiC is supported by capillary forces along macroscopic as well as microscopic (Fig. 2a) cracks due to the high shrinkage stresses introduced during preform processing. Although different conversion kinetics have been reported for different single-phase fiber and matrix carbons in separate infiltration experiments [9], the main difference in conversion rate between fiber and matrix carbons in C/C-SiC is related to the higher crack density of the matrix which provides superior diffusion paths in the early stage of silicon impregnation until the conversion rate becomes limited by silicon diffusion through the SiC layer following a parabolic rate law.

The siliconized regions in C/C-SiC (Figs. 2b, 3a) typically consist of a fine-grained twinned ß-SiC layer followed by a coarser grain structure (see SAD insets A, B in Fig. 3a). This sequence monitors the multiple SiC nucleation sites and grain coarsening processes in the wake of the liquid silicon/graphite reaction front as siliconization proceeds.

The typical C/C-SiC intrabundle microstructure derived from the graphitized C/C preform is displayed in Fig. 2b revealing no evidence for interfacial reaction products.

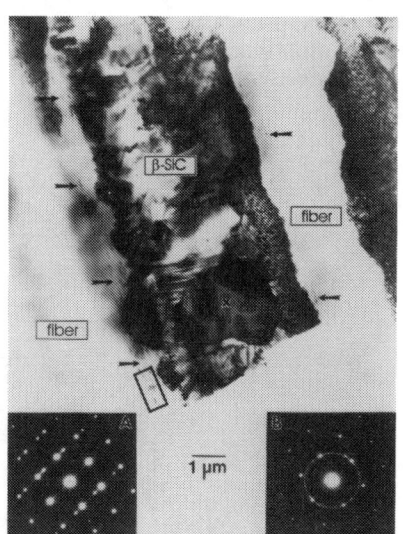 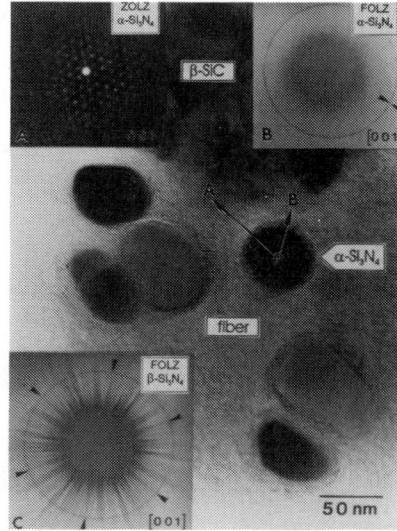

Fig. 3 (a) SiC conversion front in C/C-SiC material (low carbonization HTT), note needle-shaped precipitates as indicated by arrows and boxed region, (b) α-Si_3N_4 precipitates in C/C-SiC, as identified via convergent beam electron diffraction (CBED)

In the C/C material derived from the low carbonization HTT, the matrix carbon exhibits a turbostratic structure. During siliconization, inherent fiber impurity sources, such as nitrogen, can be activated, giving rise to a high amount of filament-like precipitates, later identified as α-Si_3N_4, as shown in Figs. 3a,b. They are enriched in the C/C-SiC fiber/matrix contact zone and have been recently reported for the first time [10]. Because of their small grain size, the true crystallographic nature of the precipitates could only be identified via convergent-beam electron diffraction (Figs. 3b, 4b). The α-Si_3N_4 polymorph can easily be distinguished from the β-phase because of its different HOLZ-ring radii (compare insets B, C in Fig. 3b) in appropriate orientations. Due to the low carbonization

HTT, the PAN-derived HT carbon fiber employed in this study still contains an estimated amount of 5 wt.% nitrogen [11]. During the impregnation process highly reactive nitrogen bearing species (e.g. heterocycles) can interact with the advancing silicon-rich vapor phase ahead of the SiC reaction front to form α-Si_3N_4. This mechanism exhibits noticeable analogies to the growth conditions of crystalline α-Si_3N_4 whisker and filaments reported from other systems [12,13].The high amount of α-phase precipitates compared to the few transformed into ß-phase is in good agreement with the common observation that (i) impurity-assisted CVD reactions lead to the formation of the α-Si_3N_4 polymorph and that (ii) kinetics of the α/ß-Si_3N_4 transformation promoted by thermal activation only are very sluggish in the absence of a liquid phase [14].

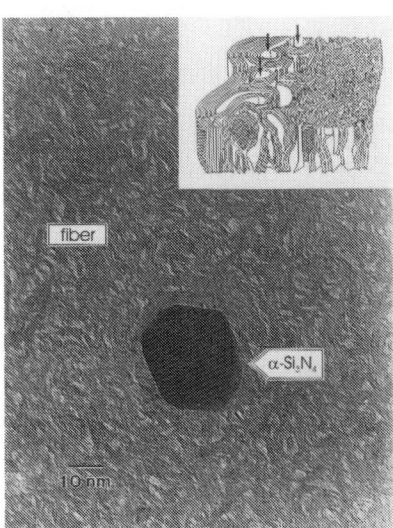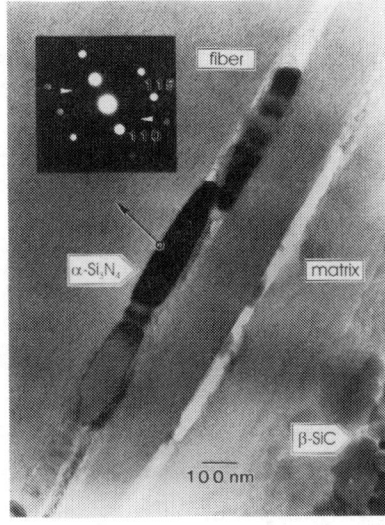

Fig. 4 (a)HREM image of faceted α-Si_3N_4 particle embedded in turbostratic fiber graphite (cross section perpendicular to fiber axis), α-Si_3N_4 filament embedded in continuous fiber pore channel (cross section parallel to the fiber axis).

A comparison of HREM/BF images taken from C/C-SiC sections both perpendicular and parallel to the fiber axis (Figs. 4a,b) clearly indicates that the continuous pore channels aligned with the fiber growth axis (compare with structural model [15] in Fig. 4a) furnish ideal nucleation sites for α-Si_3N_4 precipitation thus promoting the fibrous nature of the α-Si_3N_4 precipitates. It is anticipated that during cyclic loading of C/C-SiC such precipitation phenomena confined to the carbon fiber periphery may contribute to the weakening of the fiber structure, e.g. via a stress concentration effect. This interpretation however is tentative and will be addressed in future research.

CONCLUSIONS

The microstructural development of C/C-SiC during silicon capillary impregnation is controlled by the defect density (cracks, open porosity) of the carbon structural variants preexisting in the C/C preform. The intrabundle matrix regions of the graphitized C/C preform consist of vitreous carbon and stress-oriented graphite. In case of a rather low carbonization HTT prior to silicon impregnation, the PAN-derived carbon fiber act as an internal nitrogen source giving rise to α-Si_3N_4 precipitation in C/C-SiC along the continuous pore channels of the fiber. High HTT prior to silicon impregnation or the employment of pitch-based instead of PAN-based carbon fibers are recommended as alternatives choices in order to avoid impurity segregation at the fiber/matrix interface.

REFERENCES

[1] E. Fitzer and R. Gadow, Am. Ceram. Soc. Bull. **65**, 326 (1986)

[2] W. Krenkel, in AGARD Report 795 on "Introduction of Ceramics into Aerospace Structural Composites", AGARD (1993)

[3] U. Papenburg, K. K. O. Baer, R. Heidenreich and R. Weiß, Fortschrittsberichte der Deutschen Keramischen Gesellschaft **9**, 141, (1994)

[4] L. H. Peebles, R. A. Meyer and J. Jortner, in Interfaces in Polymer, Ceramic and Metal Matrix Composites, (H. Ishida, editor), Elsevier, 1 (1988)

[5] R. Pleger and W. Braue, Proc. 5th International Carbon Conference, June 22-26, 1992, Essen/Germany, Deutsche Keramische Gesellschaft, 653 (1992)

[6] A. Yoshida, Y. Kaburagi and Y. Hishiyama, Carbon **29**, 1107 (1991)

[7] G. S. Rellick, D. J. Chang and R. J. Zaldivar, J. Mater. Res. **7**, 2798 (1995)

[8] T. Fend and J. Goering, Ceramic Transactions **46**, 165, The American Ceramic Society (1995)

[9] R. Gadow, Ph.D. Thesis, University of Karlsruhe, Germany, 1986

[10] R. Pleger and W. Braue, Mat. Res. Soc. Sym. Proc. **365**, The Materials Research Society, Pittsburgh/PA, 1995, in press

[11] E. Fitzer and M. Heine, in Fibre Reinforcements For Composite Materials, edited by A. R. Bunsell, Elsevier, 73 (1988)

[12] U. Vogt, H. Hofmann and V. Krämer, Key Engineering Materials **89-91**, 29, Trans Tech Publications, Switzerland, USA, (1994)

[13] Y. C. Zhou, X. Chang, J. Zhou and F. Xia, J. Mater. Sci. **26**, 3914 (1991)

[14] A. J. Moulson, J. Mater. Sci. **14**, 1017 (1979)

[15] S. C. Bennet and D. J. Johnson, Carbon **17**, 25 (1979)

SILICON NITRIDE FIBER SYNTHESIS FROM POLYCARBOSILANE FIBER BY RADIATION CURING AND PYROLYSIS UNDER AMMONIA

Seiji KAMIMURA, Kiyoshi WATANABE
Hitachi Cable Ltd, Hitachi, Ibaraki 319-14 Japan
Noboru KASAI*, Tadao SEGUCHI*
*JAERI Takasaki, Takasaki, Gunma 370-12 Japan
Kiyohito OKAMURA**
** University of Osaka Prefecture, Sakai, Osaka 593 Japan

ABSTRACT

Silicon nitride(Si-N) fiber was synthesized by pyrolysis of radiation cured polycarbosilane(PCS) fiber in ammonia(NH_3) gas flow. The properties of Si-N fiber were very dependent on oxygen content included in the fiber during and after pyrolysis. The tensile strength of Si-N fiber was 2.5 GPa for a small content of 3wt% oxygen, and the electric resistivity was very high. The heat resistance in tensile strength and in electric resistivity was estimated to be 1300 °C by ageing test in air for 1 hour.

1. INTRODUCTION

In recent years, the development of ceramic fiber synthesis from organometallic polymers has been progressed, for example silicon carbide(SiC) fiber was synthesized from polycarbosilane(PCS)[1]. The silicon nitride(Si-N) fiber synthesis from polysilazane was developed[2]. A new process of Si-N synthesis was invented by Okamura et al.[3,4], which process was the nitridation of PCS fiber by the pyrolysis in NH_3 gas. The PCS fiber was cured by ionizing irradiation in the absence of oxygen. In the fundamental research, the quantity of Si-N fiber synthesis was less than 1g, and the tensile strength of Si-N mono-filament was 1.8 GPa at maximum.

The target of this work is the scale up of Si-N fiber synthesis from PCS fiber. The first step was the development of pyrolysis system for 100g order PCS fiber. We studied the pyrolysis condition to get the uniform long fiber, and the properties of the Si-N fibers.

To the extent authorized under the laws of the United States of America, all copyright interests in this publication are the property of The American Ceramic Society. Any duplication, reproduction, or republication of this publication or any part thereof, without the express written consent of The American Ceramic Society or fee paid to the Copyright Clearance Center, is prohibited.

2. EXPERIMENTAL
2.1 Preparation of Si-N fibers

Figure 1 shows the process of Si-N fibers synthesis. PCS as the starting material was melt-spun into fiber. The PCS fiber was supplied from Nippon Carbon Co. Ltd. (NCK). The fiber was cured by electron beam(EB) irradiation in Helium(He) gas atmosphere. The cured PCS fiber of 20 to 100g was set in a ceramic furnace and heated up to 1000°C in NH_3 gas flow, and followed by a heat treatment in nitrogen(N_2) gas flow at various temperatures from 1000 to 1500°C.

2.2 Properties of Si-N fiber

The mechanical properties of Si-N fiber were determined by tensile tests on the so-obtained mono-filament at room temperature on 25 mm gauge length. The chemical composition of the fiber was determined by fluorescent X-rays analysis. The chemical structure and morphology were observed respectively by FT-IR spectrometer and by scanning electron microscopy (SEM).

3. RESULTS AND DISCUSSION
3.1 Radiation curing

For the PCS fiber curing, JAERI developed the process of EB irradiation in He gas atmosphere[5]. The He gas flow was applied for the exclusion of oxygen and the cooling of PCS fiber by EB heating during irradiation. After EB irradiation the PCS fiber was heat-treated for a short period to reduce the active species trapped in PCS fiber by EB irradiation. The cured PCS fiber contains oxygen less than 0.5wt%, and is rather stable in air at room temperature. The tensile strength is increased several times compared with the PCS fiber.

3.2 Si-N fiber synthesis

The nitridation reaction proceeds well by pyrolysis in NH_3 gas. The degree of nitridation was monitored from the amount of residual carbon content by the elemental analysis as shown in Fig. 2. It shows that the reaction starts at around 500°C and terminates at about 700°C. After heat-up to 1000°C in NH_3, the atmosphere is changed to N_2 gas at that temperature, then heat-treated up to various temperatures. The obtained Si-N fiber is white and its surface is very smooth.

The Si-N fiber after heat-treatment at 1000°C in N_2 gas for 1 hour has a low tensile strength of about 0.5GPa. It was found that the fiber contained much oxygen as listed in Table 1. The oxygen content changed with time when the storage was carried out under air. Figure 3 shows the FT-IR spectrum of Si-N fiber with 20wt% oxygen. It indicates the presence of Si-O($1070 cm^{-1}$), Si-N($850-900 cm^{-1}$), -OH($1600 cm^{-1}$), and $SiNH_2$($3400 cm^{-1}$)[6].

It is considered that the $Si-NH_2$ and Si-O- are formed by the following reactions.

$$\equiv\text{Si-NH-Si-} + \text{H}_2\text{O} \longrightarrow \equiv\text{Si-NH}_2 + \equiv\text{Si-OH} \quad (1)$$
$$\equiv\text{Si-NH}_2 + \text{H}_2\text{O} \longrightarrow \equiv\text{Si-OH} + \text{NH}_3 \quad (2)$$
$$\equiv\text{Si-OH} + \equiv\text{Si-OH} \longrightarrow \equiv\text{Si-O-Si}\equiv + \text{H}_2\text{O} \quad (3)$$

Until the heat treatment of 1000℃, a considerable hydrogen content is remained under the form of $\equiv$Si-NH-Si$\equiv$(silazane) in the fiber as shown in Table 2. The silazane would react gradually with H_2O in air after the pyrolysis, and cause an oxygen uptake in the fiber.

In the fiber, silazane content seems to be reduced by heat treatment at higher temperature in N_2 gas atmosphere. The elemental analysis of Si-N fiber after heat treatment is shown in Table 1. The oxygen content decreases with the increasing temperature. The following reaction should proceed in the process of heat treatment after nitridation.

$$\equiv\text{Si-NH-} \longrightarrow \equiv\text{Si-N-} + 1/2\,\text{H}_2 \quad (4)$$

The tensile strength of Si-N fiber after heat treatment in N_2 gas is plotted versus heat treatment temperature in Fig.4. The strength reaches a maximum with increase of heat treatment temperature up to 1300°C, but turns to decrease above 1300℃. The morphology of Si-N fiber will be convert to crystal form. For the fiber with 9.5wt% oxgen, the maximum strength is 1.4GPa. The oxygen is also brought into the fiber as impurity of cured PCS and NH_3 gas. By decreasing the total oxygen content in Si-N fiber down to 3 wt%, the tensile strength can be improved to 2.5 GPa.

At present, long size Si-N fibers of about 40g are well produced in a one time reaction, and the conversion is 76-80% by weight.

3.3 Properties of Si-N fiber

Figure 5 shows the SEM photograph of SiN mono-filament. The diameter is $15\mu m$ and the surface is very smooth. The physical properties of high tensile strength Si-N fiber synthesized from PCS fiber are shown in table 3. The fiber consists of 58% Si, 35% N, 3% O, and 4% C by weight. The properties of other fibers(commercial grade) is also listed in the same table as the reference of Si-N fiber. The Si-N fiber shows rather low density, $2.3 g/cm^3$, high tensile strength, and also the highest electrical resistivity at room temperature respect to the other fibers.

The variation of the tensile strength and of the volume resistivity was investigated at room temperature after ageing in air for 1 hour up to 1300°C, and the data are plotted in Fig.6. The tensile strength tends to decrease with increase of temperature, but the volume resistivity is almost constant, 10^{13} Ω-cm. The other fibers as alumina fiber and quartz fiber decrease in tensile strength above 1000℃ as seen in Fig.6. Figure 7 is the photograph of tape woven from Si-N fiber, which is a 0.4mm thick plain weave.

4. CONCLUSION

The flexible long size Si-N fiber was synthesized from radiation cured PCS fiber by pyrolysis in NH3 gas flow. The properties of Si-N fiber were much influenced by nitridation condition and also the heat treatment temperature after nitridation. Especially, the tensile strength decreased with increasing the content of oxygen which was introduced into the fiber after pyrolysis. The maximum tensile strength was 2.5GPa when the heat treatment was carried out up to 1300°C in N_2 atmosphere after pyrolysis in NH3 and the oxygen content decreased to 3wt%. The Si-N fiber shows a high electrical resistivity of $10^{13}\Omega$-cm and nearly retains its room temperature tensile strength after a heat-treatment up to 1300°C in air.

5. REFERENCES

[1] S.YAJIMA,K.OKAMURA,J.HAYASHI,andM.OMORI, J.Am.Ceram. Soc., 59 (78) 324 (1976)
[2] T.ISODA. Proc. of the Third Int.Conf. on Composite Interfaces (ICCI-III), May, 21-24, 1990, Cleveland, Ohio 255.
[3] K.OKAMURA,M.SATO,and Y.HASEGAWA, Ceramics International, Vol.1355 (1987)
[4] M.SATO,K.OKAMURA,S.KAWANISHI,and T.SEGUCHI, MRS Int'l. Mtg. on Adv. Mats.Vol.4, 197 (1989)
[5] T.SEGUCHI,N.KASAI,and K.OKAMURA,Proceeding Int.Cont.on Evolution in Beam Application, Takasaki. Japan ,Nov.5-7 (1991) P702
[6] Y.KANNO,K.SUZUKI,N.ISHIZUKA,and Y.KUWAHARA,The Chem. Soc. of Jap.,6, 808 (1984)

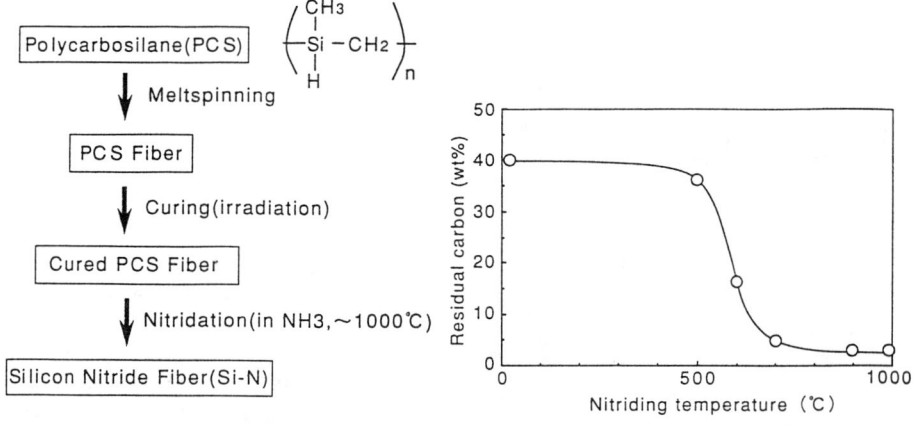

Fig.1 Preparation of Si-N fiber from PCS

Fig.2 The influence of temperature on nitridation

Table1. Chemical composition of Si-N fibers obtained by heat-treatment in N_2 at different heat-treatment temperature after pyrolysis of PCS fibers under NH_3

Heat treatment temperature (°C)	Element (wt%)			
	Si	N	O	C
1000	47.5	24.5	23.0	5.0
1200	51.7	28.7	15.4	4.2
1300	54.7	32.1	9.5	3.7
1400	55.1	32.5	7.7	4.7

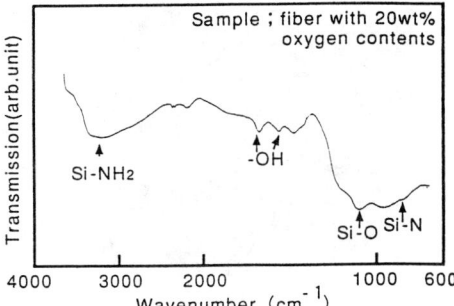

Fig.3 FT-IR spectrum

Table2. Supposed mechanism of nitridation of PCS fiber

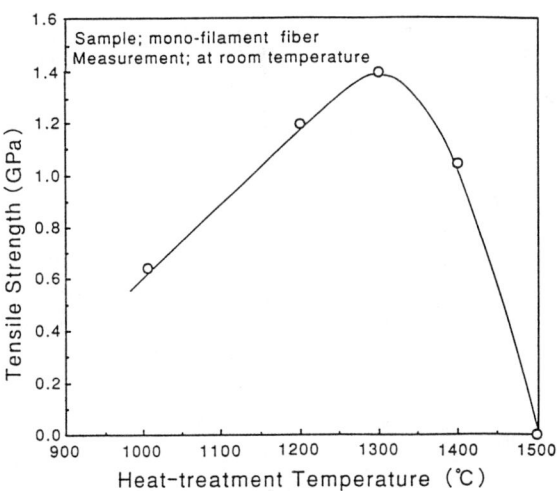

Fig.4 Effect of heat-treatment temperature on the tensile strength

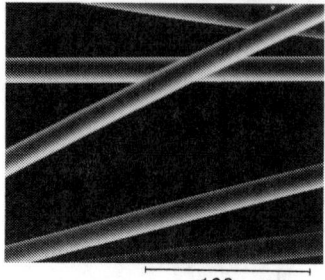

Fig.5 SEM photograph of Si-N fiber

Table3. Physical properties of Si-N fibers and other fibers

Item	Silicon Nitride	Alumina ***	Silicon Carbide ***	Quartz ***
State	amorphous	crystal	amorphous	crystal
Filament diameter (μm)	15	10	15	11
Density (g/cm^3)	2.3	3.6	2.3	2.3
Tensile strength * (GPa)	2.5	1.0	2.8	1.3
Volume resistivity ** (Ωcm)	10^{13}	10^{10}	10^{8}	10^{11}

* ; Maximum value, ** ; the data of woven tape
*** ; commercial

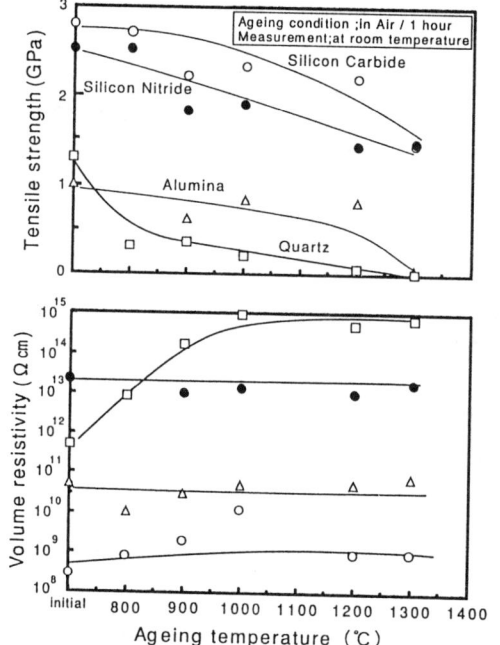

Fig.6 Heat resistance of Si-N fiber and other fibers

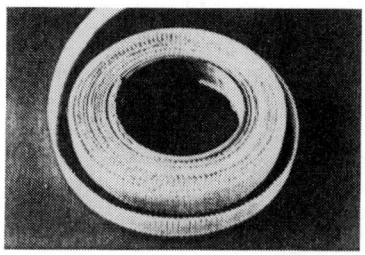

Fig.7 Photograph of tape woven from Si-N fiber

EFFECT OF RAPID HEAT TREATMENTS ON ELECTRICAL PROPERTIES OF POLYMER DERIVED CERAMIC FIBERS

Masaki Narisawa, Koji Nakashiba, and Kiyohito Okamura
Department of Metallurgy and Materials Science, College of Engineering, University of Osaka Prefecture
1-1, Gakuencho, Sakai, Osaka 593, JAPAN

ABSTRACT

The electrical resistivity of Si-C-O fibers and Si-C fibers were measured after rapid heat treatment. In the fiber containing oxygen, the resistivity of the fiber was highly influenced by the gas evolution during the heat treatment. The effect of the composition change on the absolute resistivity, temperature dependence of conductivity and Intensity-Voltage (I-V) characteristics of the fibers was analyzed and discussed in terms of Si/C atomic ratio.

INTRODUCTION

Processing SiC ceramic fibers from polymer precursors has been developed and has attracted much attention. These fibers maintain high strength up to high temperature and change in the mechanical properties during heat treatment has been widely investigated.[1,2] Not only the mechanical properties but also the electrical properties of these fibers are attracting attention in recent years.[3-5] The SiC fibers from polymer precursors are n-type semiconductor with the resistivity between 10^0-10^6 $\Omega \cdot m$.[5] In this study, the electric properties of silicon carbide fibers were investigated after rapid heat treatment. The Si-C-O fiber releases SiO and CO gases during the heat treatment,[6] and electric properties should reflect the change in composition and structure in the fiber during the heat treatment. The properties of various composition of oxygen free Si-C fibers were also investigated. The influence of Si/C atomic ratio on the electrical properties is expected to provide some clues on the microstructure in the silicon carbide fibers.

EXPERIMENTAL

Si-C-O fiber (Commercial name: NL 202) commercialized by Nippon Carbon Co. Ltd. (NCK) and various composition of Si-C fibers synthesized in NCK with modified "Hi-Nicalon" process were prepared.[7] The composition of the Si-C-O fiber is $SiC_{1.20}O_{0.41}$. Si/C atomic ratios in the Si-C fibers have been controlled between 0.8 and 1.58. 0.5g of the fiber in a graphite crucible was put into graphite furnace which had been preheated at appointed temperatures under a stream of Ar gas. An automatic balance was attached to the graphite furnace in order to measure the weight loss of the fibers. The resistivity of the monofilaments was measured at room temperature by two-probe direct current method with the voltage maintained at 10V and the terminal distance of 1.5mm. Temperature dependence of the conductivity was analyzed in the range of 320-480K. I-V characteristics were also analyzed for the same samples in the range of $10^{-2}-10^{1}$ V.

RESULTS AND DISCUSSION

Figure 1 shows the electrical resistivity of Si-C-O fibers after rapid heat treatment. The resistivity of the fiber decreases up to 1600K as heat treatment temperature increases. The resistivity, however, turns to increase beyond 1600K. In case of 3.6ks of heat treatment, resistivity comes to maximum at 1723K. In case of 18ks of heat treatment, the resistivity comes to maximum at 1623K. Beyond these temperatures, the resistivity suddenly decreases, and reaches stable value.

During the heat treatment, precipitation of β-SiC microcrystals is known to proceed as shown in the following equation.[6]

$$SiC_xO_y \rightarrow \beta\text{-SiC} + SiO(g) + CO(g) \quad (1)$$

This reaction is accompanied by gas evolution from the fiber. Figure 2 represents the weight loss of the Si-C-O fiber during heat treatment at high temperatures. The weight of the fiber begins to decrease at 1573K which is consistent with the temperature where the resistivity turns to increase. At finishing points of the weight loss, the resistivity also becomes stable. The observed resistivity maxima are found to be located between the beginning and the finishing point of the weight loss.

Figure 3 shows the temperature dependence of the conductivity in the heat treated fiber. Conductivity of as-received fiber shows sudden increase beyond 423K. The point of abrupt increase in conductivity, however, shifts to higher temperature and finally disappears as heat treatment temperature of the fiber increases.[3,4]

The resistivity and conductivity obtained in above experiments were measured under a voltage of 10V with the terminal distance of 1.5mm. The Si-C-O fibers, however, showed slight nonlinear I-V characteristics after the rapid heat treatment. Such non-linearity was analyzed with the following empirical equation.[8]

$$I = (V/C)^\alpha \quad (2)$$

where I is a current flowing through the fiber and V is an applied voltage. C is dependent on the diameter of the fiber and the terminal distance. The value of α of the heat-treated fibers changed into 1.2 from 1.0 with the heat treatment beyond 1873K.

The effect of the gas evolution on the resistivity suggests that the composition change during heat treatment would be the major factor dominating the electrical properties of the ceramic fibers. Figure 4 shows the influence of Si/C ratios on the resistivity of Si-C fibers which contains no oxygen. Near stoichiometric fiber has highest resistivity, and the fibers which contain excess silicon or excess carbon have lower resistivity. Figure 5 represents the temperature dependence of conductivity in Si-C fibers. The near stoichiometric fiber shows highest temperature dependence. The fiber containing carbon excess shows particularly low temperature dependence as compared with other fibers. In this fiber, the carbon excess is considered to form a conductive path which has a quasi-metallic character.

The observed change in electrical properties of Si-C-O fibers and Si-C fibers could be explained in terms of the carbon excess of the fibers. During heat treatment, two types of paths are considered to develop in the fiber matrix. One is the semiconductive path made from SiC micro-crystals, and another path is the quasi-metallic path made from carbon excess. The conductivity of silicon carbide fibers is reported to be dominated by basic structure unit (BSU) of carbon formed on SiC micro-crystals.[3,4] In our system, however, carbon is suspected to be eliminated as CO gas during the rapid heat treatment,[9] and rapid growth of SiC crystal may prevent contacts between BSU of carbon. In this case, SiC micro-crystals should dominate the electric properties of the fibers. As heat treatment temperature increases, the continuous carbon layer is considered to be formed on the SiC crystals again, and the resistivity begins to decrease. The temperature dependence of conductivity of these fibers also suggests the existence of two types of path in the fibers.

SUMMARY

A marked change in the electrical resistivity of Si-C-O fibers was observed after their rapid heat treatment under argon in the range 1600-1800K. This temperature range corresponds to gas evolution and composition change of the fibers. The electrical properties of oxygen free Si-C fibers were also investigated, and the effect of Si/C ratio on the resistivity was confirmed. The effect of composition change and the temperature dependence of conductivity of these fibers suggests the existence of two types of path which has different electrical characters.

Acknowledgements: We wish to thank Dr. Y. Imai, Mr. M. Takeda, Dr. H. Ichikawa of NCK and Dr. T. Seguchi of Japan Atomic Research Institute for providing Si-C fibers.

References

[1] S. Yajima, K. Okamura, T. Matsuzawa, Y. Hasegawa and T. Shishido, "Anomalous Characteristics of the Microcrystalline State of SiC Fibres," Nature, **279**, 706-707 (1979)

[2] S. Yajima, "Special Heat-Resisting Materials from Organometallic Polymers," Ceramic Bulletin, **62**, 893-904 (1983)

[3] E. Bouillon, F. Langlais, R. Pailler, R. Naslain, F. Cruege, P. V. Huong, J. C. Sarthou, A. Delpuech, C. Laffon, P. Lagarde, M. Monthioux and A. Oberlin, "Conversion Mechanisms of a Polycarbosilane Precursor into an SiC-based Ceramic Material," J. Mater. Sci., **26**, 1333-1345 (1991).

[4] E. Bouillon, D. Mocaer, J. F. Villeneuve, R. Pailler, R. Naslain, M. Monthioux, A. Oberlin, C. Guimon and G. Pfister, "Composition-Microstructure-Property Relationships in Ceramic Monofilaments Resulting from the Pyrolysis of a Polycarbosilane Precursor at 800 to 1400°C," J. Mater. Sci., **26**, 1517-1530 (1991)

[5] N. Muto, M. Miyayama, H. Yanagida, T. Kajiwara, N. Mori, H. Ichikawa and H. Harada, "Infared Detection by Si-Ti-C-O Fibers," J. Am. Ceram. Soc., **73**, 443-445 (1990)

[6] T. Shimoo, H. Chen and K. Okamura, "Pyrolysis of Si-C-O Fibers (Nicalon) at Temperatures from 1473K to 1673K," J. Ceram. Soc. Jpn., **100**, 48-53 (1992).

[7] M. Takeda, J. Sakamoto, Y. Imai, H. Ichikawa and T. Ishikawa, "Properties of Stoichiometric Silicon Carbide Fiber Derived from Polycarbosilane," Ceram. Eng. Sci. Proc., July-Auguest **15**, 133-141 (1994)

[8] T. Masuyama, "Electrical Non-Linearity of SiC Aggregate," Yogyo-Kyokai-shi, **77**, 323-328 (1969)
[9] Y. Sasaki, Y. Nishina, M. Sato and K. Okamura, "Raman Study of SiC Fibres Made from Polycarbosilane," J. Mater. Sci., **22**, 443-448 (1987).

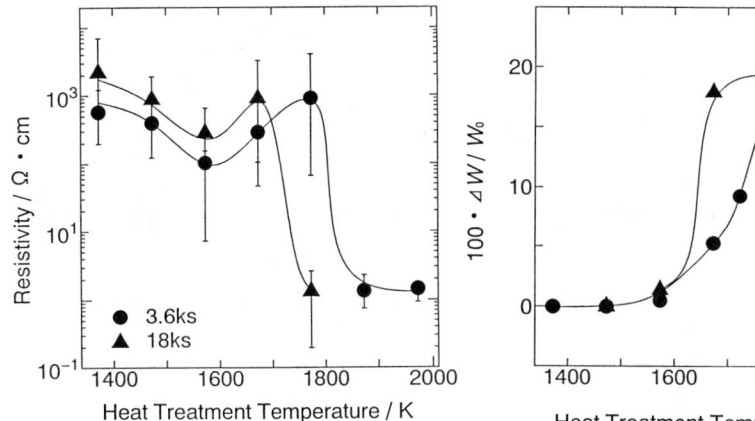

Fig. 1 Change in resistivity of the Si-C-O fiber during heat treatment

Fig. 2 Weight loss of the Si-C-O fiber during heat treatment

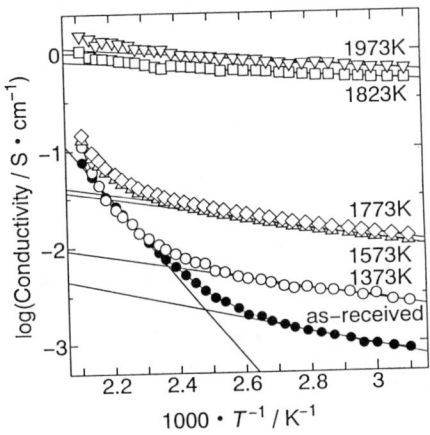

Fig. 3 Temperature dependence of conductivity in heat-treated Si-C-O fibers

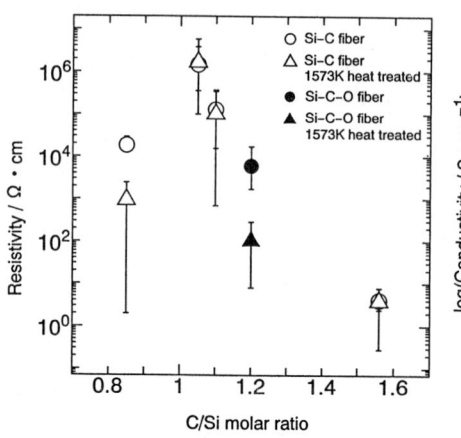 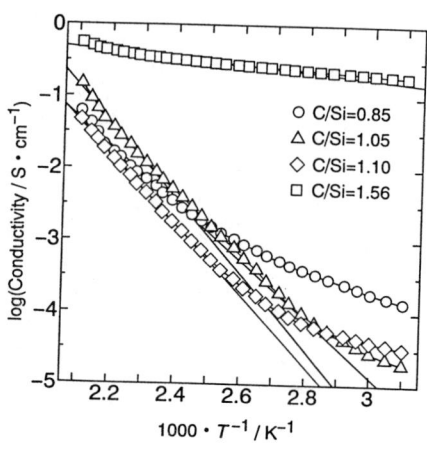

Fig. 4 Relationship between resistivity and C/Si molar ratio in Si-C fiber

Fig. 5 Temperature dependence of conductivity in Si-C fibers

REACTION MECHANISMS OF SiC FIBER SYNTHESIS FROM RADIATION CURED POLYCARBOSILANE FIBER

Masaki Sugimoto, Kiyohito Okamura*, and Tadao Seguchi
Japan Atomic Energy Research Institute, Takasaki Radiation Chemistry Research Establishment
1233 Watanuki-machi, Takasaki, Gunma 370-12, Japan
*Department of Metallurgical Engineering, College of Engineering, University of Osaka Prefecture
1-1 Gakuen-cho, Sakai, Osaka 593, Japan

ABSTRACT

Silicon carbide (SiC) fiber was synthesized from polycarbosilane (PCS) fiber by heat treatment after the electron beam irradiation curing. The reaction mechanisms for SiC fiber synthesis during heat treatment were investigated by gas analysis, free radical detection, and mechanical properties measurements. There was two steps in the major reaction; first step was 800-1200 K, where the chemical reactions related to Si atoms proceeded as the cleavage of Si-CH_3 and Si-H, and second step was 1000-1600 K, where the reactions related to C atoms proceeded.

INTRODUCTION

A flexible silicon carbide (SiC) fiber with high tensile strength and thermally resistant is obtained from polycarbosilane (PCS) precursor fiber. The fiber is potentially useful as a reinforcing fiber for composite materials with metals and ceramics for high heat resistant materials [1]. The SiC fiber derived from PCS fiber has been manufactured by Nippon Carbon Co. Ltd. as a name "Nicalon" [2]. The "Nicalon" contains oxygen of about 10 wt% which is introduced in the process of PCS curing by thermal oxidation. It is known that the oxygen in SiC fiber induces the decomposition at high temperatures [3,4].

In order to reduce the oxygen in SiC fiber, the radiation curing of PCS is developed and succeeded to decrease the oxygen content until 0.4 wt%. Then, the thermal resistance is improved to around 2000 K [5,6]. As the another application of radiation curing of PCS fiber, it is possible to increase the oxygen content in SiC fiber in a wide range of up to 30 wt% [5].

The pyrolysis reaction during the heat treatment process from the cured PCS fiber to SiC fiber should be related closely to the curing process. For the improvement of SiC fiber, the suitable heat treatment condition must be selected with knowledge of the reaction mechanisms. In this work, the cured PCS fibers with various oxygen contents were prepared by electron beam irradiation, and the mechanisms were investigated by the analysis of gas evolution, free radical detection during heat treatment, and the measurements of mechanical properties for the obtained SiC fiber.

EXPERIMENTAL

Curing of PCS fiber

The PCS fibers were obtained by melt spinning of PCS at 506 K, and the diameter was 20 μm. The PCS fibers were cured by electron beam(EB) irradiation in He and in air using the conditions shown in Table I. For EB curing, the PCS fibers were irradiated by 2MeV electron beams from an electron accelerator.

In a case of curing without oxygen, He gas was flowed in the quartz tube during irradiation. After irradiation the sample was heated up to 673 K for 1h to decay the survival radicals in PCS fiber formed by irradiation. Because, the survival radicals react preferably with oxygen at room temperature. For the EB oxidation curing, the PCS fiber was irradiated in air instead of He using the same apparatus. The oxygen content in the PCS fiber increases with the dose.

Table I. Curing of PCS Fiber, and the Oxygen Content in Cured PCS Fiber

Method	Curing condition	O2 content (wt%)
EB Crosslinking	1.3 kGy/s, 12 MGy in He	1.1
EB Oxidation	1.9 kGy/s, 2.5 MGy in Air	12
EB Oxidation	1.9 kGy/s, 3.5 MGy in Air	17

Gas analysis

The cured PCS fiber of 0.5 g was put in a sample tube, and the tube was evacuated. It was heated to a temperature (500 K) and held at the temperature for 2×10^3 sec. Then, it was cooled down to room temperature. The gases accumulated in the sample tube by the heat treatment were analyzed by gas chromatography. After the gas analysis, the sample tube was evacuated and heated again to the higher temperature by 100 K than the first heat treatment. The gases evolved by increasing the temperature by 100 K were analyzed and these procedures were repeated from 500 to 2000 K.

Free radical detection

Heat treatment of the cured PCS fiber was carried out using an electric furnace in Ar gas atmosphere. The fibers were heated to a temperature ranging from 500 to 2000 K and held for 2×10^3 sec, then cooled rapidly down to room temperature. Electron spin resonance (ESR) spectra were obtained at room temperature in air using a JEOL-FE3X X-band spectrometer. The yield of free radicals was determined from the ESR spectral intensity, and corrected by the comparison with the material containing a known free radical concentration, 2,2-diphenyl-1-picrylhydrazyl (DPPH, 1.46×10^{23} spin/g) and 2,2,6,6-tetramethyl-1-piperidinyl-oxide (TEMPO, 3.78×10^{23} spin/g).

Mechanical properties of SiC fiber

The tensile tests were carried out for SiC fibers obtained by heat treatment at various temperatures. The mono-filament of SiC fiber was tested at room temperature, and the strength and modulus were obtained.

RESULTS
Gas evolution during heat treatment

The gas evolution by heat treatment of the cured PCS fibers containing different oxygen content is plotted in Fig.1. The gases are mainly methane(CH_4) and hydrogen(H_2). Both gases start to evolve at around 800 K, and have a peak at about 1000 K. The CH_4 has only one peak, but H_2 has another peak at around 1300 K. The H_2 at 1000 K is decreases with increasing oxygen content. Above 1800 K, carbon monoxide(CO) evolves, and the yield increases with oxygen content in cured PCS fiber. As the minor gases are water(H_2O), carbon dioxide(CO_2) and ethane(C_2H_6), these gases are less yield than H_2 or CH_4 by two order. The H_2O and CO_2 evolve around at 600 K, and the C_2H_6 and CO_2 around at 1000 K.

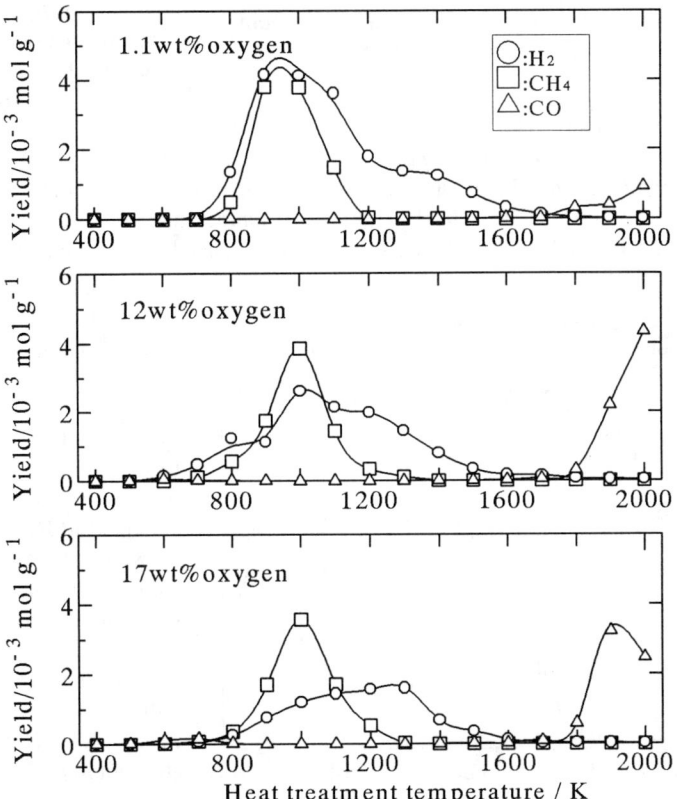

Fig. 1. Gas evolution during heat treatment of PCS Fiber cured by EB irradiation. (1.1 wt% oxygen : in He, 12 and 17 wt% oxygen : in air)

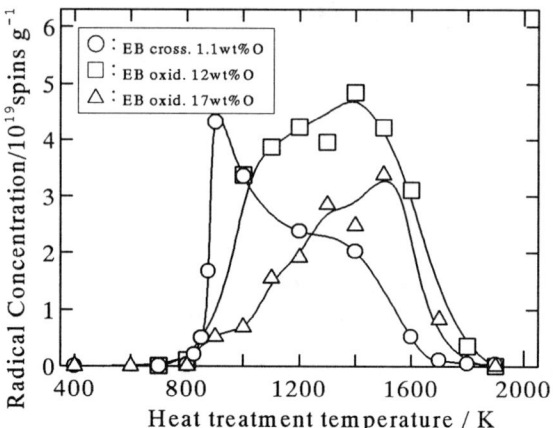

Fig. 2. Concentration of Radicals trapped at room temperature in SiC fiber by heat treatment of cured PCS fiber.

Free radical behavior during heat treatment

Free radicals are formed with the decomposition of organic materials of PCS by heat treatment above 800 K, and some of the radicals are trapped in the fiber after cooled down to room temperature. The radical concentration at room temperature is plotted against heat treatment temperature for PCS fiber with different oxygen content in Fig.2. The radical yield in the heat treatment is very different by the oxygen content in the fiber. The ESR spectrum was much changed by heat treatment temperature; its line width was 10 gauss at 800-900 K, and 2 gauss at 1200-1700 K. It is determined that the radicals are sited on Si atom for lower temperature, and on C atom for higher temperature. The radical yield against temperature is similar to the H_2 evolution in Fig.1. The radicals on Si and C atoms are induced with the cleavage of Si-H or C-H bonds.

Mechanical properties of SiC fiber

The tensile strength increases by the proceeding of ceramic SiC from organic PCS as shown in Fig.3. For PCS with 17 wt% oxygen, the synthesized SiC fiber reaches to maximum strength at 1200 K, and the strength decreases above 1400 K. For PCS with 1.1wt% oxygen content, the strength increases with the heat treatment temperature until 1800 K. For the SiC fiber formed from PCS fiber with very low oxygen less than 0.3 wt% maintains the strength until 2000 K [6,7].

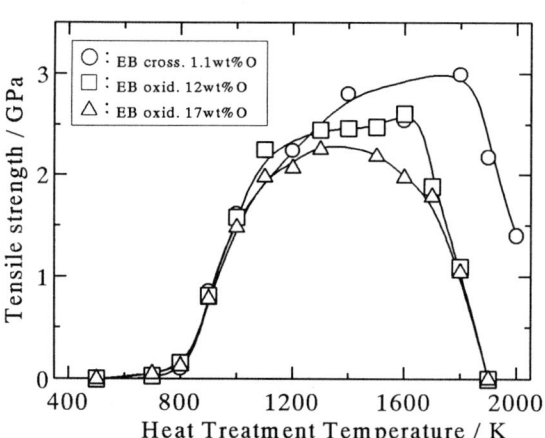

Fig.3. Tensile strength of SiC fiber formed from EB cured PCS fiber after heat treatment in Ar.

DISCUSSION

The decomposition of cured PCS fiber by heat treatment starts at 800 K and evolves CH_4 and H_2 gases, and reaches to maximum at around 1000 K. The CH_4 evolution ends completely at 1200 K. This reaction is the first step of ceramic formation, and the chemical reaction proceeds mainly at Si atoms in PCS; that is the cleavage of $Si-CH_3$ and Si-H bonds and the related bonding reactions. The second step reaction starts at around 1000 K and the rate is maximum at 1300 K, where H_2 gas is evolved and continues up to 1700 K. The second step reaction would take place on C atoms; cleavage of C-H bond and the related reactions. The temperature difference in the reaction at Si and C atoms would be the difference of bond energy between Si-H(295 kJ/mol) or $Si-CH_3$(290 kJ/mol) and C-H (413 kJ/mol) [8].

The first step reaction is much affected by the curing of PCS fiber as oxygen content. The oxidation in the PCS curing takes place mainly in the Si-H bonds, then the H_2 evolution at around 1000 K decreased to the quantity of remaining Si-H in PCS fiber after radiation oxidation curing.

During the chemical reactions from organic PCS to ceramic SiC, the free radicals are much formed. The radical species are sited on Si and C atoms which are produced by the cleavage of Si-H and C-H bonding. A part of radicals are trapped in the fiber by cooling to lower temperature. Therefore, the observed radical yield reflects the radical concentration during the pyrolysis at high temperature.

The reaction mechanisms of SiC fiber formation from radiation cured PCS fiber are analyzed by the radical reaction as demonstrated in Fig.4.

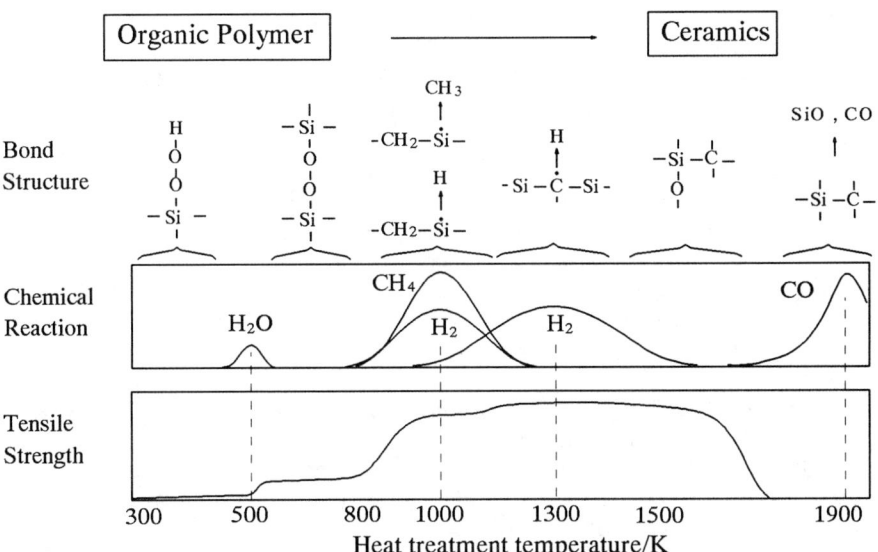

Fig.4. Reaction mechanisms of SiC fiber formation from radiation cured PCS fiber.

CONCLUSION

It was found that there were two steps in the pyrolysis reaction through free radical mechanisms for SiC fiber formation from cured PCS fiber, which was considered to be the difference of the chemical bond energy related to Si and C atoms in PCS main chains.

The pyrolysis process can be monitored by the gas evolution and free radical detection by ESR. The radical yield and the ESR spectrum indicate the state of the chemical reaction during pyrolysis. An understanding of reaction mechanisms and the kinetics is very useful to control of the pyrolysis process and the improvement of SiC fiber.

Acknowledgments

The authors wish to thank Mrs. Noboru Kasai and Takao Kanagawa of JAERI Takasaki for the help of EB curing.

References

[1] S. Yajima, K. Okamura, J. Hayashi, and M. Omori, "Synthesis of Continuous SiC Fibers with High Tensile Strength," *J. Am. Ceram. Soc.*, **59** [7-8] 324-27 (1976).
[2] T. Ishikawa and T. Nagaoki, "Recent Carbon Technology Including Carbon and SiC Fibers"; p. 348. JEC Press, Cleveland, Ohio, 1983.
[3] K. Okamura, M. Sato, T. Matsuzawa, and Y. Hasegawa, "Effect of Oxygen on Tensile Strength of SiC Fibers"; *Polym. Prepr.*, *Am. Chem. Soc.*, Div. Polym. Chem., **25** [1] 6-7 (1984).
[4] S. Yajima, "Silicon Carbide Fiber"; pp. 201-37 in *Handbook of Composites*, Vol.1, *Strong Fibers*. Edited by W. Watt and B. V. Pero. Elsevier Science Publishers B. V., New York, 1985.
[5] K. Okamura, T. Seguchi, "Application of Radiation Curing in the Preparation of Polycarbosilane-Derived SiC Fibers," *J. Inorg. Organomet. Polym.*, **2** [1] 171-79 (1992).
[6] T. Seguchi, N. Kasai, and K. Okamura, "Preparation of Heat-Resistant Silicon Carbide Fiber from Polycarbosilane Fiber Cured by Electron Beam Irradiation"; pp. 702-707 in Proceedings of the International Conference on Evolution in Beam Applications, Takasaki, Japan, November 5-8, 1991.
[7] T.Seguchi, M.Sugimoto and K.Okamura, "Heat Resistant SiC-Fiber Synthesis and Reaction Mechanisms from Radiation-Cured Polycarbosilane Fiber"; pp.51-57 in Proceedings of the High Temperature Ceramic Matrix Composites '93 (HT-CMC1), France Bordeaux, September 20-24, 1993.
[8] "Chemical Bond Energy"; p. 975 in *Chemical Handbook*, Vol. 2. Edited by S. Seki, Maruzen Co.,Ltd., Tokyo Japan, 1975.

STRUCTURE, COMPOSITION AND MECHANICAL BEHAVIOR AT HIGH TEMPERATURE OF THE OXYGEN-FREE HI-NICALON FIBER*

G. Chollon, R. Pailler, R. Naslain, Laboratoire des Composites Thermostructuraux (UMR-47, CNRS-SEP-UB1) Domaine universitaire, 3, allée de la Boëtie, 33600 Pessac, France.
P. Olry, Société Européenne de Propulsion BP 37, 33165 Saint-Médard-en-Jalles, France.

ABSTRACT

The oxygen free Hi-Nicalon Si-C-fiber* was studied on the chemical, structural and thermomechanical point of view as a function of annealing treatments (Tp; tp). The fiber exhibits a moderate structural evolution within the range Tp=1200-1400°C, namely SiC grain growth, organization of the free carbon phase and densification. A strong increase of the thermal dependence of the creep strain rate is observed for T≥1300°C, which might be partly related to the structural evolution of the fiber. Furthermore, heat treated fibers (Tp=1400-1600°C) exhibit a much better creep resistance, owing to their improved structural organization.

1- INTRODUCTION

The main drawback of commercial Si-C-O fibers, i.e., their limited thermal stability, is related to the presence of an amorphous silicon oxycarbide phase SiC_xO_y surrounding SiC crystals. It is well known that it decomposes with an evolution of silicon and carbon monoxides when heated above 1100-1200°C [1,2], but it is also responsible for the low creep resistance of the Si-C-O fibers with respect to equivalent SiC-based materials, owing to a viscous flow at temperatures as low as 1000-1200°C [3-4]. It thus appeared that oxygen and consequently, SiC_xO_y amorphous phase should be eliminated from the fibers composition in order to improve the thermal stability [5] and the creep resistance of the fibers [6]. Thermomechanical behavior of oxygen-free fibers has been described by several authors but rarely explained through structural analyses.

The aim of the present study is to correlate the mechanical behavior of as-received and heat-treated oxygen-free Hi-Nicalon fibers* at high temperatures with their chemical and stuctural properties.

2- EXPERIMENTAL

Annealing treatments were performed with a radio frequency heating furnace. Samples were heated (30°C.min^{-1}) and maintained at the annealing temperature Tp (1200≤Tp≤2000°C) during a time tp (1≤tp≤10h), under pure argon (100 kPa).

The elemental analysis of the fibers was performed by electron probe microanalysis (EPMA). Composition-depth profiles from the fiber surface were recorded by Auger electron spectroscopy (AES), the spectrometer being equipped with an Ar^+ sputtering gun.

The structure of the monofilaments was studied by transmission electron microscopy (TEM) [7] and X-ray diffraction (XRD, $\lambda Cu-K_\alpha$). The apparent mean grain size (L) of the β-SiC crystalline phase was calculated from the width at mid-height (D) of the diffraction peaks according to equation $L=K.\lambda/(D.\cos\theta)$, θ being the Bragg angle and K a constant (K=1). The density of the samples was measured according to helium-pycnometry on filament tows. The tensile strength in air of as-received and heat-treated fibers was measured at room (gauge length L_0=25mm) and high temperatures (T≤1450°C, L_0=200mm) according to a "cold-grips" tensile system. The fibers were submitted to bend stress relaxation tests (BSR) [3]. A bending stress is applied to the fiber (with a given curvature R_0), which is then submitted to a heat treatment (Tp, tp=1h). After cooling, relaxation results in a residual curvature R (with R≥R_0). Relaxation strength can be quantified by the parameter defined as m=1-(R_0/R). Creep tests under argon were carried-out on single fiber samples. The creep apparatus with cold grips, designed by Bodet et al., has been described elsewhere [4-6]. The testing temperatures were in the range 1000<T<1600°C and the applied load was σ=1GPa. Apparent activation energy for creep and thermal gradient corrected creep strain were determined by an equivalent isothermal gauge length calculation from the elongation data and the temperature profiles calibration.

3- RESULTS AND DISCUSSION

3-1- Chemical and structural analyses

The chemical composition of Hi-Nicalon fibers remains almost unchanged after heat-treatments up to Tp = 1800°C. It is close to: Si: 41.0 at.%; C: 58.1 at.% and O: 0.9 at.%. The fiber has a very low oxygen content and an excess of free carbon (≈17 at.%). A slight decomposition nevertheless takes place at high temperatures, resulting in a silicon depletion, first located at the surface (fig. 1) and gradually extending toward the core. This feature might be related to SiC decomposition and/or active oxidation, due to residual oxygen from the annealing atmosphere or the fiber.

In the as-received state, the Hi-Nicalon fiber consists of β-SiC crystals ranging from 2 to 15 nm and turbostratic carbon with layers of small size (La=2-3 nm) and a stack number N ranging from 5 to 8 [7]. The carbon domains form an intergranular network and frequently set flat upon SiC crystals. After a T_p=1400°C; tp=1h heat-treatment, SiC crystals reach a maximal size of 18-25nm. Free carbon is also better organized, than in the as-received fiber. The number of layers in the coherent stacks is nearly the same whereas the extent of the carbon layers is slightly larger (La=3-6nm). For T_p=1600°C, the maximum SiC grain size reaches ≈50nm and joined crystals are more often observed. The organization of free carbon is still improved, with N now reaching 7-12, as well as an extension of the carbon domains, with La=5-10nm.

The β-SiC mean grain size calculated from XRD pattern is about 5 nm in the as-received fiber. It grows to about 10 nm for T_p=1600°C (tp=1h) and 20 nm for T_p=1800°C (fig. 2). The density of the as-received fiber is ρ=2.77g.cm^{-3}. Annealing treatments at increasing temperatures result in a densification of the fibers (fig. 2) starting at about T_p=1400°C.

The Hi-Nicalon fiber has a high thermal stability owing to the extremely low amount of unstable SiO_xC_y phase. However, a slight structural evolution of both SiC and free carbon starts between Tp=1200 and 1400°C. It becomes more noticeable as tp and Tp are raised. The β-SiC grain growth might be governed by the reduction of the specific grain area through a diffusion mechanism of atomic silicon or traces of $SiO_{(g)}$ and $CO_{(g)}$.

3-2- Mechanical behavior

The tensile strength at ambient is almost unchanged for Tp=1400°C; tp=1h (fig. 3). It decreases as Tp and tp are raised, first slightly (1400<Tp<1600°C) and then dramatically. The superficial decomposition and/or the SiC coarsening may introduce new flaws and, as a consequence, a decrease of the tensile strength.

While the stress-elongation curves at T=25°C show a typical brittle elastic behavior for as-received fibers, they show a marked non-linear behavior as T is raised, as well as a strong decrease of stiffness (fig. 4). The former feature is confirmed by cycling tests at T=1450°C since each cycle induces an additional residual elongation in the unstressed fiber (fig. 5a). The visco-plastic behavior is reduced after annealing at Tp=1400°C (tp=1h) and has almost disappeared for Tp=1600°C (tp=1h), with an elastic behavior up to the rupture at T=1450°C (fig. 4). When carried out on a heat treated fiber (Tp=1600°C; tp=1h), a cycling test at T=1450°C shows that the fiber has lost its non-linear mechanical behavior (fig. 5b). No significant residual strain is induced by the stress-cycling, and the stress-elongation curves remain elastic during the whole test. A previous heat-treatment (Tp=1600°C; tp=1h) clearly improves the BSR thermal resistance of the fiber by about 200°C, with respect to the as-received fiber (fig. 6).

Creep elongations as a function of time are reported in fig. 7. An apparent steady state creep domain (stationary strain rate) is observed for T≤1400°C in these short term tests. For higher testing temperatures (T≥1500°C), the elongation rate continuously decreases up to the rupture. The steady-state creep strain rate $(d(\varepsilon_{0T})/dt)$, reported in an Arrhenius plot (fig. 8) shows two distinct linear domains, corresponding to two different apparent activation energies: Q=220±17kJ.mol^{-1} for 1000≤T≤1250°C, and Q'=700±30kJ.mol^{-1} for 1250≤T≤1400°C. This change remains for the moment unexplained, since the creep mechanism is still unknown. Tentatively, it could be partly related to the structural evolution of the fiber starting from Tp=1200-1400°C.

The creep behavior of heat treated fibers under argon (T_p=1400 and 1600°C) has also been characterized. The creep curves for as-received and heat treated fibers are shown in fig. 9, for a testing temperature T=1400°C and an applied stress σ=1GPa. All creep curves involve a steady state creep domain, whatever the heat-treatment temperature. It clearly appears that the steady state creep strain rate gradually decreases as T_p is raised. The strain recorded for heat treated fibers (Tp=1600°C; tp=1h) is much less than that for the as received fiber, for 1200≤T≤1500°C (fig. not shown). Furthermore, the heat-treated fiber exhibits a steady-state creep domain for T≤1500°C, while the creep rate of the as-received fiber continuously decreases up to the rupture.

As it has already been reported [8], the annealing treatment induces a marked increase of the creep strength at high temperature. The structural change described above results in a stiffening of the fiber and in a decrease of the creep strain rate, for a given stress and testing temperature. Furthermore, the heat-treatment seems to have also a structural stabilization effect during the creep test at high temperature. Unlike for the untreated fiber, the fibers which were heat-treated at Tp=1600°C are not expected to undergo any structural change during the test up to T=1500°C, and as a consequence, steady state creep behavior is still observed for T=1500°C.

4- CONCLUSION

The present work has confirmed the much higher thermal stability of the Hi-Nicalon fiber with respect to the Si-C-O Nicalon fiber. However, despite their high chemical stability (owing to their very low amount of oxygen), Hi-Nicalon fibers are subject to a slight densification and a structural evolution of both SiC and free carbon phases when annealed at high temperatures.

The complex thermal dependence of creep rate for as-received fibers is not yet fully understood but might be related to the structural evolution occurring above their maximum processing temperature.

Heat treatments of the fiber at increasing temperatures (Tp) result in a better creep resistance (owing to their gradually improving structural organization) and in a decrease of the room and high temperature strengths probably caused by new flaws.

Aknowledgements

This work has been supported by CNRS and SEP. The authors gratefully thank F. Laanani and M. Monthioux from LMM, CNRS from Pau for the TEM data, R. Bodet for his valuable advices for conducting creep tests and Nippon Carbon for providing the Hi-Nicalon fibers.

References

1. T. Mah, N. Lecht, D.E. Mc Cullum, J.R. Hoenigman, H.M. Kim, A.P. Katz and H.A. Lipsitt, J. Mater. Sci., 19 (1984) 1191-1201.

2. T.J. Clark, R.M. Arons, I.B. Stamatoff and J. Rabe, Eng. Adv. Sci. Proc., 7-8 (1985) 901-929.

3. J.A. Di Carlo, Comp. Sci. and Techn., 51 (1994) 213-222.

4. R. Bodet, J. Lamon, N. Jia and R. E. Tressler, J. Am. Ceram. Soc., in press.

5. K. Okamura, M. Sato, T. Seguchi and S. Kawanashi, in "Controlled Interphases in Composite Materials", H. Ishida, ed., Elsevier, (1990) 209-218.

6. R. Bodet, X. Bourrat, J. Lamon and R. Naslain, J. Mat. Sci., 30 (1995) 661-677.

7. F. Laanani, M. Monthioux, C. Guimon, G. Chollon and R. Pailler, to be published in J. Europ. Ceram. Soc..

8. H.M. Yun, J.C. Goldsby and J.A. DiCarlo, submit. to Ceram. Eng. Sci. Proc..

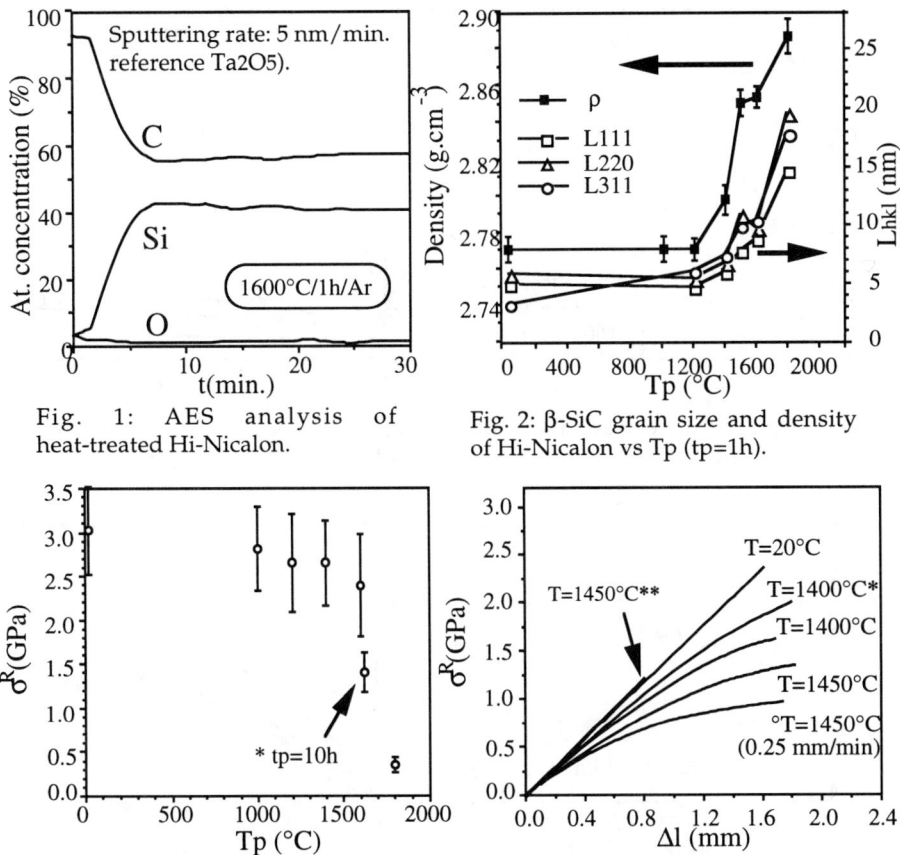

Fig. 1: AES analysis of heat-treated Hi-Nicalon.

Fig. 2: β-SiC grain size and density of Hi-Nicalon vs Tp (tp=1h).

Fig. 3: Tensile failure strength at ambient vs Tp (tp=1h except *).

Fig. 4: Stress-elongation curves vs T and Tp (tp=1h, *Tp=1400°C, **Tp=1600°C). (v=1mm/min except °)

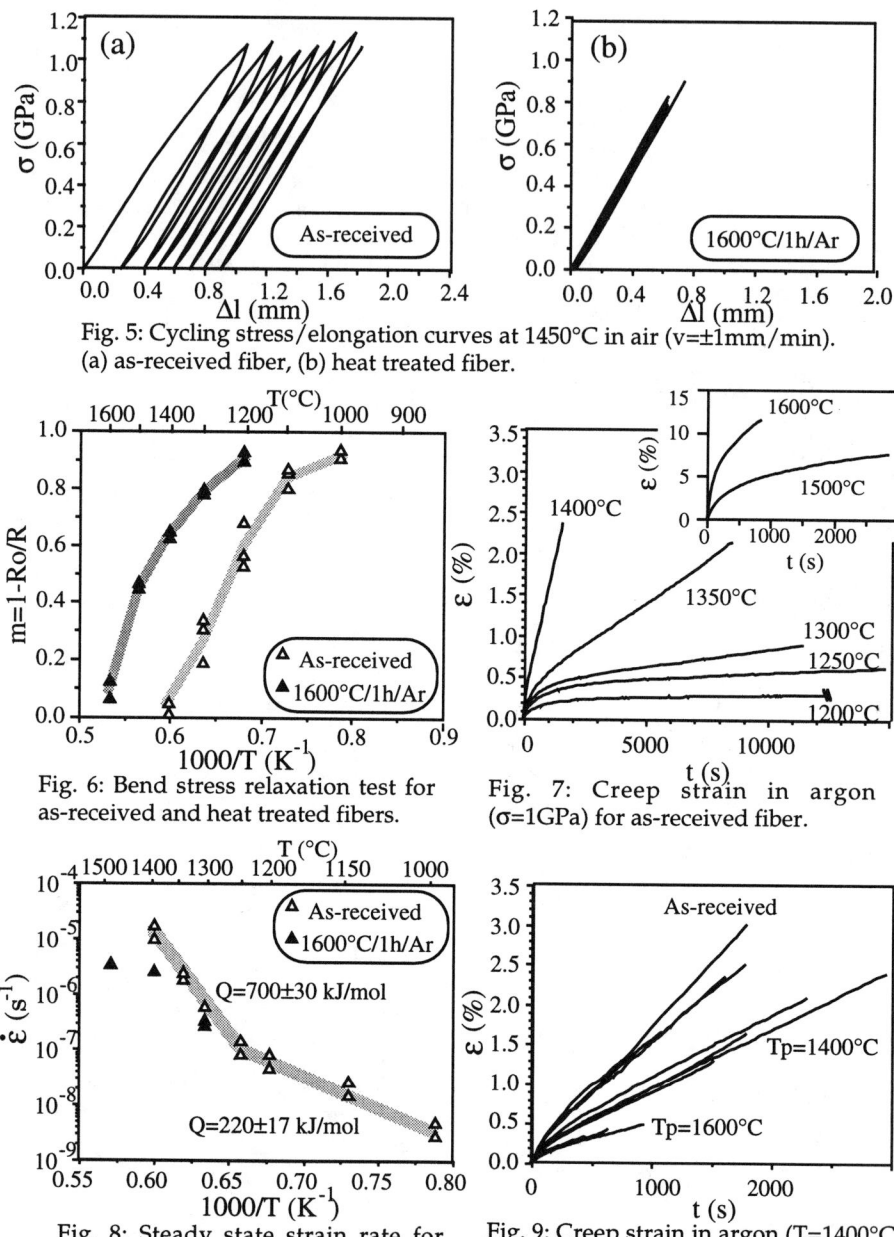

Fig. 5: Cycling stress/elongation curves at 1450°C in air (v=±1mm/min). (a) as-received fiber, (b) heat treated fiber.

Fig. 6: Bend stress relaxation test for as-received and heat treated fibers.

Fig. 7: Creep strain in argon (σ=1GPa) for as-received fiber.

Fig. 8: Steady state strain rate for as-received and heat treated fibers.

Fig. 9: Creep strain in argon (T=1400°C; σ=1GPa) for as-received and heat treated fiber (tp=1h).

STRUCTURE AND THERMAL EVOLUTION OF SiC-BASED FIBERS WITH LOW OXYGEN CONTENT

G. Chollon, R. Bodet, R. Pailler and X. Bourrat
Laboratoire des Composites Thermostructuraux, UMR47 CNRS-SEP-UB1
3 Allée la Boëtie, 33600 Pessac, France.

1. INTRODUCTION

Polycarbosilane-derived ceramic grade fibers have been commercialized for a decade under the trade name Nicalon (CGN). However, some problems associated with this fiber have limited its ultimate use as reinforcement of CMCs. Besides SiC (49 mol%), a large amount of free carbon (28 mol%) and an amorphous $SiO_{1.12}C_{0.44}$ phase (23 mol%),[1-3] are also present in the fiber. At high temperatures (typically above 1200°C) the silicon oxycarbide phase decomposes, forming gaseous species such as CO and SiO, whose diffusion through the fiber and reaction with the free carbon are believed to create pores and other defects in the fiber.

New routes for the fabrication of ceramic fibers have been investigated. The main break-through concerns the curing[4] of PCS filaments by γ-rays or electron beam irradiation (EB). In this process, the cross-linkage between polymeric chains of the precursor is mainly achieved by forming Si-C bonds between polymeric chains, instead of Si-O-Si bridges as in oxygen curing. Compared to fibers cured in air, the strength of fibers cured by electron irradiation remained almost unaffected after heat treatment to 1400°C. A slight increase in Young's modulus was even noticed at these temperatures.[4]

In this study, the thermal stability in argon of an home-made experimental SiC-based fiber (O : 3 at%), prepared by electron irradiation of PCS filaments[5], was investigated on the chemical and structural point of view and compared to the Hi Nicalon Si-C based fiber (O : 0.5 at%) known to be obtained by the same process.

2. EXPERIMENTAL PROCEDURE

The experimental fiber processing has been described elsewhere[5]. Hi-Nicalon fiber was studied, as-received and also after annealing at 1600°C for 1 and 10 hours in argon. A full paper on the structure/mechanical properties relationship of this fibre have been given elsewhere[6].

X-ray diffraction (XRD) was used to record the crystallisation of the fiber vs the heat treatment temperature (HTT) (Cu-Kα / Siemens D5000 diffractometer). Chemical analyses using Auger electron spectroscopy (AES) and electron-probe microanalysis (EPMA) were carried out on the as-received and heat treated fibers as well as the experimental fibre. Structures and morphologies were characterized by means of scanning electron microscopy (SEM : JEOL 840 A) and transmission electron microscopy (TEM : JEOL 2000 FX). For that purpose, each sample was glued with silver paint on a copper grid and ion milled with a Ar^+ sputtering gun (Gatan 600). TGA analyses were run in argon between 1200 and 1400°C to record the weight loss of fibers versus heat treatment time.

3 RESULTS
3.1 Experimental fiber

Elemental analysis were performed on the experimental fibers after EB curing as a function of heat treatment temperature (Tables I). With EPMA technique, hydrogen is not taken into account (which might be present in significant amount below $1400°C^7$). The low oxygen concentration during all the processing results from the constant anaerobic conditions; it appears as the condition for a stable chemical composition in the bulk up to 1600°C (one hour, under argon or nitrogen).

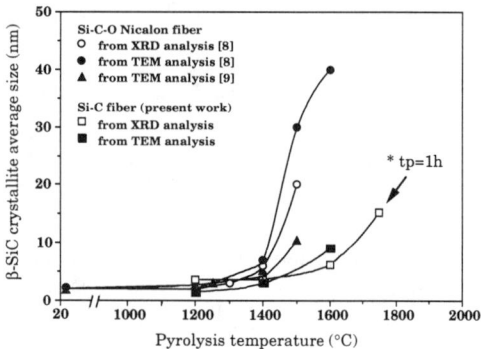

Fig. 1: β-SiC crystallite apparent size as a function of pyrolysis temperature Tp for Si-C-O (Nicalon) and EB-cured Si-C fibers (tp=15min., except *) (as assessed from XRD analysis).

Tp (°C)	900 / Ar	1000/Ar	1200/Ar	1400/Ar	1600/Ar	1600/N2
Si(at.%)	39.5	40	40	39	40	39
C (at.%)	57.5	57	57	58	57	58
O (at.%)	3	3	3	3	3	3

Table I: Elemental composition of EB-cured ex-PCS fibers pyrolysed at increasing temperature Tp as derived from EPMA (hydrogen is not taken into account).

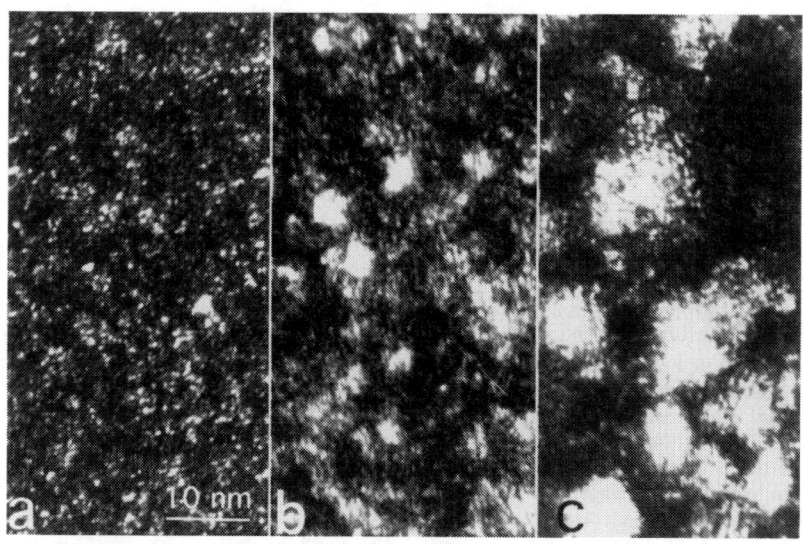

Fig.2 Experimental fiber (TEM SiC_{111} dark-field) : a) annealed 15 mn at Tp = 1200°C, b) same at Tp = 1400°C and c) same at Tp = 1600°C.

with (i) a mean SiC apparent grain size of about 4 nm (Fig. 2b), and (ii) an intergranular phase, consisting of turbostratic carbon layers. This carbon is rather well organized as turbostratic stacks of 3-5 carbon layers, partly surrounding SiC crystals. Both SiC and carbon crystallites do not exhibit any preferential orientation. The SiC grains keep on growing between 1400 and 1600°C. The crystallite size is of the order of 8-10 nm as determined locally (statistical average class) by TEM dark-field analysis (Fig. 2c). However, kinetics of growth of SiC are much lower than those reported for oxygen-cured fibers at 1600°C[8]. Additionally, some reorganization of the intergranular carbon is observed for Tp = 1600°C: (i) the extent of the turbostratic layers is larger (few tens of nm), (ii) a polygonal shell ordering is still visible with a larger size, and (iii) an additional poorly organized carbon is observed.

3.2 Hi Nicalon fiber

The average elemental composition of the as-received fiber, obtained using EPMA, as the oxygen weight percentage, i.e. 0.5 wt%, is in good agreement with the manufacturer data. This fibre also contains an excess of carbon relative to stoichiometric SiC. Neglecting oxygen and considering stoichiometric SiC, the molar composition of the Hi-Nicalon fiber can be estimated as follow : 35 mol% of free carbon and 65 mol% of SiC.

The structure of the as-received Hi-Nicalon fiber was investigated by TEM analysis. Figures 4a and b show the bright- and dark-field images of the fiber obtained near the central region. As shown on the SAD pattern in the inset of Figure 4b, three distinct rings corresponding to the 111, 220 and 311 reflections of ß-SiC are apparent. In the 111-centered dark field image, ß-SiC crystals can be easily recognized. A value of ≈5 nm has been obtained for the average ß-SiC grain size. However, grains as large as 10 nm were easily found from dark-field analysis of the fiber

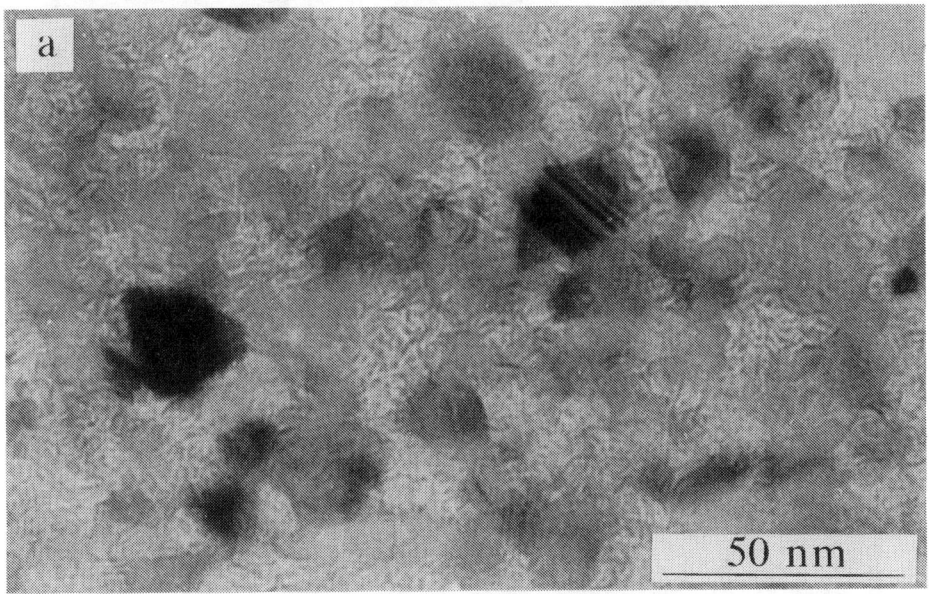

Fig.4 Hi Nicalon fiber : a) TEM bright field.

For Tp<1200°C, the fiber is observed to be amorphous by X-ray diffraction, since no diffraction peak is detected. For Tp≥1200°C, the three main lines, assigned to 111, 220 and 311 reflections of β-SiC are observed (corresponding respectively to 0.251, 0.154 and 0.131 nm d-spacings). From Tp=1200 to 1750°C, the material undergoes crystallization. A slight increase, in the grain size (calculated from the width of the (111) reflection) is observed from 1200 to 1600°C, with a mean values of about 4 nm at 1200°C and 6 nm at 1600°C (Fig. 1). Beyond this temperature, the grain size increases more rapidly (about 10 nm at 1750°C for tp = 1 h) but remains much more lower than for Si-C-O fibers[8,9].

According to TEM analyses, the organization of fibers with a low oxygen content is basicaly different as compared to the regular CGN fibre[9]. For Tp = 1200°C, the fiber consists of very small β-SiC crystallites : i.e. 2-3 nm, a result in agreement with XRD analysis (Fig. 2a). Lattice fringe imaging also shows free carbon in the bulk, as isolated aromatic layers or turbostratic stacks of layers (Fig. 3). When treated up to 1400°C, the fiber is polycrystalline

Fig.3 Experimental fiber, Tp = 1400°C 15 mn (TEM lattice fringes technique).

Significant quantities of aromatic carbon structures have been detected in the Hi-Nicalon fiber using high resolution TEM. Arrows in Figure 5 show this turbostratic carbon which appears in the form of wrinkled, continuous layers; either as a single layer or stacked in thin piles (up to 5) between SiC grains. Compared to the CGN fiber, the Hi-Nicalon fiber, as the experimental one, was found to contain better organized turbostratic carbon.

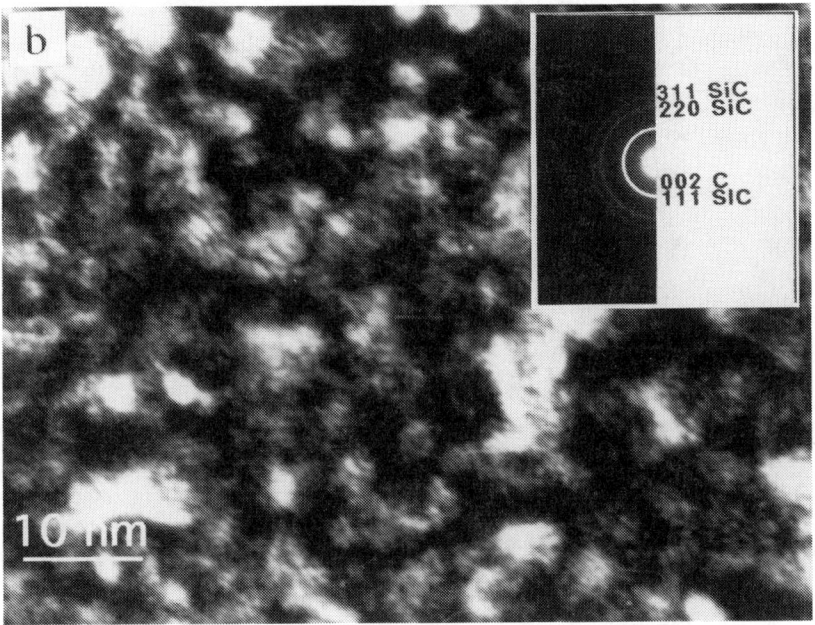

Fig.4 (continued) Hi Nicalon fiber : b) SiC_{111} dark-field and selected area diffraction.

TEM analyses show that the Hi-Nicalon fiber structure remained stable below ~1300°C. A progressive growth of the SiC crystallites from 5 nm to 20-40 nm occurred with increasing temperature during annealing (or creep[6] : the rate of crystallization was very rapid during creep at 1600°C, i.e. 0.1 hour under an applied stress of 0.7 GPa). TEM observations also revealed evidence that the carbon layer size increases during annealing in the Hi-Nicalon fiber : annealed unstressed fibers along with crept specimens exhibited similar trends[6].

4. DISCUSSION - CONCLUSION

Experimental fibers obtained from the pyrolysis of the EB-cured PCS were compared to the commercial Hi Nicalon SiC based fibers, known to be obtained by the same process. Both fibers have a low oxygen content. The former remains amorphous up to about 1000°C. The SiC mean crystal size, as measured by TEM dark field or XRD technique is equal to about 2-3 nm. Fibers also contain turbostratic carbon. It is apparently more organized in both experimental and Hi Nicalon fiber than in the CGN fiber. Even if negligable, the small amount of silicon oxycarbide still present (≈ 3 O at. % and 0.5 at% in the experimental and Hi Nicalon fibers, respectively) is suspected to be responsible for the grain growth in the range of 1200 to 1600°C (1400 to 1800°C for Hi Nicalon).

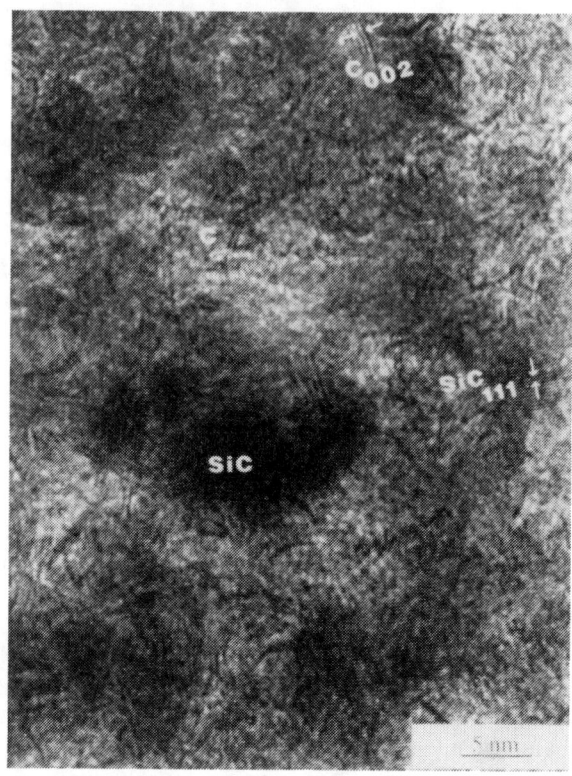

Fig.5 Hi Nicalon fiber (TEM lattice fringes technique).

ACKNOWLEDGEMENTS

The authors acknowledge the French Ministry of Research and Société Européenne de Propulsion for their financial support to this project. The authors would also like to thank M. Lahaye for the AES analyses.

REFERENCES

1- L. Porte, A. Sartre, J. Mat. Sci., 24, 271-75, 1989.
2- C. Laffon, A.M. Flank, P. Lagarde, M. Laridjani, R. Hagege, P. Olry, J. Cotteret, J. Dixmier,L.Miquel, H. Hommel, A.P. Legrand, J. Mat. Sci., 24, 1503-12, 1989.
3- R. Bodet, J. Lamon, N. Jia, R.E. Tressler,"Microstructural Stability and Creep Behavior of Si-C-O (Nicalon) Fibers in Carbon Monoxide and Argon Environments", submitted to J. Am. Cer. Soc.
4- K. Okamura, M. Sato, T. Segushi and S. Kawanish in "Controlled Interphases in Composite Materials" (H. Ishida, ed.) (1990) 209-218, Elsevier.
5- G. Chollon, M. Czerniack, R. Paillet, X. Bourrat, R. Naslain, J. P. Pillot and R. Canet, submitted to MRS.
6- R. Bodet, X. Bourrat, J. Lamon and R. Naslain, J. Mat. Science 30 (1995) 661- 677.
7- C. Gerardin, "Caractérisation par résonance magnétique nucléaire de matériaux céramiques à base de carbure ou carbonitrure de silicium obtenus par voie polymérique", thesis, University of Paris VI, 1991.
8- Y. Sazaki, Y. Nashina M. Sato and K. Okamura, J. Mater. Sci., 22 (1987) 443-448.
9-. P. Le Coustumer, M. Monthioux and A. Oberlin, J. Eur. Ceram. Soc., 11 (1993) 95-103

SiC -BASED FIBERS WITH LOW FREE CARBON CONTENT

A. Tazi Hémida, R. Pailler and R. Naslain
Laboratoire des Composites Thermostructuraux, UMR 47 (CNRS-SEP-UB1)
Domaine Universitaire, 3 Allée de la Boétie, 33600 Pessac, France
J.P. Pillot, M. Birot and J. Dunoguès
Laboratoire de Chimie Organique et organométallique, URA 35 (CNRS-UB1)
Université Bordeaux 1, 351 Cours de la Libération, 33405 Talence Cédex, France

Abstract
2,4-dichloro-2,4 disilapentane is used as a starting material in the synthesis of novel polycarbosilane precursors of silicon carbide-based ceramics containing a low excess of free carbon. These polymers can be melt spun, cured by electron beam irradiation in the solid state, and pyrolyzed at 1400°C to provide near stoichiometric ceramic fibers (typically C/Si ≈ 1.07) and almost free of oxygen (O% at ~1) which exhibit a high Young's modulus (320 GPa).

I-Introduction
Since it has been recognized that an excess of free carbon might be responsible for a decrease in the thermomechanical stability of SiC-based ceramic fibers [1-3] (e.g. creep behavior), there is a current interest in the preparation of stoichiometric silicon carbide from the pyrolysis of organosilicon polymers [4,5]. There are very few precursors leading to SiC-based ceramics with an atomic ratio C/Si ~ 1 and most of them exhibit serious drawbacks [6]. For example, the polycarbosilane (PCS) derived from disilacyclobutane [7,8] requires an expensive multistep synthesis whilst the direct pyrolysis of poly(methylsilane) leads to ceramics containing free silicon [9]. In addition, the preparation of fibers is often tedious or even not possible because of poor rheological properties of the corresponding preceramic polymers. In a recent approach in this field, the Yajima's PCS [10] has been used for the preparation of nearly stoichiometric silicon carbide fibers, according to a modified "Hi-Nicalon"* process [11].
The aim of this paper is to depict :
(i)the synthesis of original precursors with a C/Si ratio < 1.5 and their conversion by pyrolysis into bulk Si-C(O) ceramic samples,
(ii)The spinning , curing and pyrolysis of these precursors,
(iii)The study of the mechanical properties of these fibers at room temperature.
(*) Hi-Nicalon is an experimental fiber from Nippon Carbon

To the extent authorized under the laws of the United States of America, all copyright interests in this publication are the property of The American Ceramic Society. Any duplication, reproduction, or republication of this publication or any part thereof, without the express written consent of The American Ceramic Society or fee paid to the Copyright Clearance Center, is prohibited.

II-Synthesis of 2,4-dichloro-2,4-disilapentane

The starting material in the present approach is the 2,4-dichloro-2,4-disilapentane. The synthesis of this monomer has previously been reported from dibromomethane [12]. This compound was prepared according to a new and cheaper route using dichloromethane, in the presence of magnesium powder and zinc (30% mass) and a large excess of methyldichlorosilane [13]. The pure product was recovered in 35% yield upon distillation (66°C/42 mmHg) :

$$CH_2Cl_2 + 2\,CH_3SiHCl_2 + 2Mg \xrightarrow[THF]{Zn} Cl\text{-}\underset{H}{\overset{CH_3}{\underset{|}{Si}}}\text{-}CH_2\text{-}\underset{H}{\overset{CH_3}{\underset{|}{Si}}}\text{-}Cl + 2\,MgCl_2$$

III-Synthesis of precursors

Two routes were used to produce polymers with low oxygen content and $1.2 < C/Si(\text{theoretical}) < 1.5$:

III-1 Copolymer route

The first step of the procedure consists in the copolymerization of 2,4-dichloro-2,4-disilapentane and methylphenyldichlorosilane by condensation in the presence of sodium, so the copolymer I is obtained. The substitution of phenyl groups by hydrogen atoms then proceeds in two steps : the first is the cleavage of the Si-Ph bonds at 70°C with HCl/AlCl$_3$ which leads to the formation of the chlorinated polymer II. Then the copolymer III is obtained upon reduction of Si-Cl bonds by LiAlH$_4$:

$$0.6\ Cl\text{-}\underset{Ph}{\overset{CH_3}{\underset{|}{Si}}}\text{-}Cl + 0.4\ Cl\text{-}\underset{H}{\overset{CH_3}{\underset{|}{Si}}}\text{-}CH_2\text{-}\underset{H}{\overset{CH_3}{\underset{|}{Si}}}\text{-}Cl \xrightarrow[toluene]{2\,Na} \underset{Ph}{\overset{CH_3}{-(Si)}}_{0.6}\underset{H}{\overset{CH_3}{-(Si\text{-}CH_2\text{-}\underset{H}{\overset{CH_3}{Si}})}}_{0.4}\ +\ 2\,NaCl$$

(I) $\xrightarrow[\text{(AlCl}_3)\text{, }70°C]{HCl}$ II $-(\underset{Cl}{\overset{CH_3}{Si}})_{0.6}-(\underset{H}{\overset{CH_3}{Si}}-CH_2-\underset{H}{\overset{CH_3}{Si}})_{0.4}$ + 0.6 Ph-H

II $\xrightarrow[Et_2O]{LiAlH_4}$ $-(\underset{H}{\overset{CH_3}{Si}})_{0.6}-(\underset{H}{\overset{CH_3}{Si}}-CH_2-\underset{H}{\overset{CH_3}{Si}})_{0.4}$ (III)

Heating III to 300-350°C yields a solid material (precursor 1) that can be melted without decomposition. This transformation proceeds smoothly under atmospheric pressure in the absence of a catalyst.

GPC results : Mw = 2550, I_p = 1,8. The ^{29}Si NMR spectrum (INEPT) shows three signals : -70 ppm ($SiCSi_2H$), -30 ppm ($SiSiC_2H$) and -14 ppm (SiC_3H).
The infrared spectrum of the precursor 1 is shown in fig.1 and the observed absorption bands are listed in table 1 :

$1/\lambda$ (cm^{-1})	Intensity	A$^+$	$1/\lambda$ (cm^{-1})	Intensity	A$^+$
2950	s	vaCH$_3$	952	m	δSi-H
2923	w	vaCH$_2$	932	w	δSiH$_2$
2888	s	vsCH$_3$	885	vw	ρCH$_3$
2850	w	vsCH$_2$	869	s	rCH$_3$
2077	s	vSi-H	815	vw	ωSiH$_2$*
1406	m	δaCH$_3$	772	s	rCH$_2$
1350	w	δCH$_2$	683	m	vSi-CH$_3$
1243	s	δsCH$_3$	640	vw	tSiH$_2$*
1035	s	ωCH$_2$	487	w	rSiH$_2$*

s : strong ; m : medium; w : weak; vw : very weak ;* : end groups ; A$^+$:Assignment

Table 1 : Main absorption bands observed in the IR spectrum of precursor 1

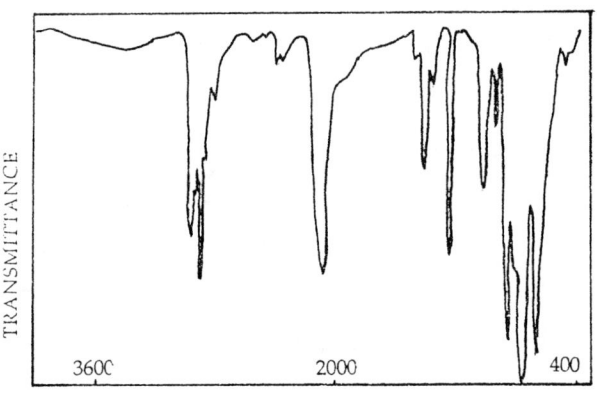

WAVENUMBER (cm^{-1})

Fig. 1: Infrared spectrum of the precursor 1.

III-2 Homopolymer route

The autocondensation of 2,4-dichloro-2,4-disilapentane in the presence of sodium yields poly(2,4-disilapentane-2,4 diyl) IV, then IV is treated by LiAlH$_4$ to remove remaining Si-Cl bonds giving the homopolymer V. So, a treatment at temperatures in the range 300-350°C yields the precursor 2 :

$$n\ Cl-\underset{H}{\underset{|}{Si}}-CH_2-\underset{H}{\underset{|}{Si}}-Cl \xrightarrow[\text{toluene}]{2n\ Na} -(\underset{H}{\underset{|}{Si}}-CH_2-\underset{H}{\underset{|}{Si}})_n + 2n\ NaCl$$

$$IV \xrightarrow[Et_2O]{Li\ AlH_4} V$$

IV-Bulk pyrolysis of the precursors

Precursor	Ceramic yield(%)	C (% at)	Si (%at)	O (% at)	C/Si ratio
1	60	52.5	46	1.5	1.14
2	79	51.3	47.5	1.1	1.08

Table 2 : Ceramic yields and elemental composition of ceramics obtained after pyrolysis at 1000°C.

The ceramic yield of precursor 2 is relatively high (fig.2) owing to the presence and reactivity of a number of Si-H bonds. The oxygen content is rather low and the C/Si ratio close to 1. The ceramics remains amorphous when the pyrolysis temperature (T_p) reaches 1000°C, but β-SiC crystallites appearing at 1100°C become larger and larger when T_p increases (2.3 nm at 1100°C and 14 nm at 1600°C). The elemental compositions are stable in this temperature range (table 3).

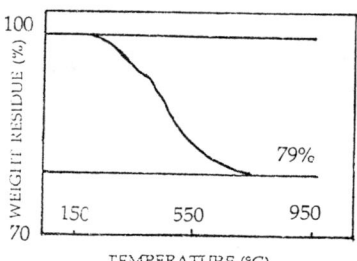

Fig.2 : TGA curve of precursor 2 in an argon flow

Pyrolysis temperature (°C)	Si (% at)	C (%at)	O (% at)	C/Si (at)
1000°C	47.5	51.3	1.1	1.08
1100°C	47.6	51.3	1.2	1.08
1200°C	47.8	51.0	1.2	1.07
1600°C	48.2	51.0	0.9	1.06

Table 3 : Elemental composition of ceramics prepared from precursor 2 at different temperatures.

V-Preparation of ceramic monofilaments

The precursor was prepared according to the copolymer route (as depicted in section III-1). It was spun in a glove box in the molten state at T = 230°C, with a laboratory-scale apparatus equipped with a single spinneret, in order to prepare a continuous monofilament of 22 μm in diameter (after pyrolysis). The green filaments were cured by electron irradiation under dry nitrogen according to the following procedure : (i) They were first introduced, under nitrogen atmosphere , in aluminum containers subsequently tightly sealed, then (ii) the containers were submitted to an irradiation dose of 1000 Mrad and finally (iii) the irradiated
containers were opened in the glove box in order to avoid any contact with oxygen and moisture.

The cured filaments were set in a silica boat, transferred to a furnace, the silica tube of which was directly connected to the glove box, annealed at 400°C for 12 h under argon (P = 100 KPa) and pyrolysed up to T_p = 1000°C (heating rate = 60°C/h; plateau at 1000°C : 1 h). After such a treatment, the fibers were no longer sensitive to oxygen and moisture and could be handled in the open air.

For T_p> 1000°C, the fibers were then set in a graphite crucible, heated rapidly (30°C/mn) at T_p with a radio frequency coil inductor under a high purity argon flow at a pressure of 100 KPa and maintained at this temperature for t_p=1h.

VI-Mechanical characteristics

The variations of the tensile strength and Young's modulus (at room temperature) of filaments as a fonction of the pyrolysis temperature are shown in table 4 :

Pyrolysis temperature (°C)	E (GPa)	σ (MPa)
1000°C	250	1300
1200°C	300	1850
1400°C	320	1600

Table 4 : Mechanical properties at room temperature of filaments pyrolysed under argon atmosphere.

The tensile strength (gauge length : 10 mm) undergoes a maximum (1800 MPa) for $T_p = 1200°C$, then it slightly decreases as T_p is raised (no attempt was presently made to optimize the tensile failure strength by improving processing conditions). The Young's modulus increases from 250 to 320 GPa when T_p is raised from 1000 to 1400°C.

The results of the elemental analysis performed by EPMA at 1000°C are :
Si = 44.5 % (at)   C = 51.5 % at   O = 4 % at

VII-Conclusions

From the experimental data reported in sections II to VI, the following conclusions can be drawn :
(i) Si-C(O) model filaments with a low free carbon content have been prepared for the first time from an original tailored precursor, by melt spinning, electron beam irradiation curing and pyrolysis under argon at a temperature of 1400°C,
(ii) these fibers are characterized by a high Young's modulus (up 320 GPa) as it could be expected from a near stoichiometric SiC fiber. It is thought that the failure stress could be further increased by reducing the filament diameter and the processing conditions. A more complete characterization of the fiber is in progress.

Acknowledgements

This work has been partly supported by SEP. The authors are indebted to P. Olry from SEP for valuable discussion.

References

1 G. Simon and A. R. Bunsell, J. Mater. Sci; **19**, 3649, 1984.
2 M. Takeda, Y. Imai, H. Ichikawa, T. Ishikawa, N. Kasai, T. Seguchi and K. Okamura, Ceram . Eng. Sci. Proc; **13**, 209, 1992.
3 W. Toreki, C. D. Batich, M. D. Sach, M. Saleem and G. Choi, Mat. Res. Soc. Symp. Proc; **271**, 761, 1992.
4 D. Seyferth, T. G. Wood, H. J. Tracy and J. L. Robison, J. Am. Ceram. Soc; **75**, 1300,1992.
5 Z. F. Zhang, F. Babonneau, R. M. Laine, Y. Mu, J. F. Harrod and J. A. Rahn, J. Am. Ceram. Soc; **74**, 670, 1991.
6 R. M. Laine and F. Babonneau, Chem. Mater; **5**, 260, 1993.
7 T. L. Smith, US Pat. 4, 631, 179, 1986.
8 H. J. Wu and L. V. Interrante, Macromolecules, **66**, 1840, 1992.
9 D. Seyferth, G. E. Koppetsch, T. G. Wood, H. J. Tracy, J. L. Robinson, P. Czubarow, M. Tasi and G. Woo, Polym. Prepr; **34,** 223, 1993.
10 S. Yajima, K. Okamura, J. Hayashi, and M. Omori, Chem. Lett; 1209, 1975.
11 M. Takeda, J. Sakamoto, Y. Imai, H. Ichikawa, and T. Ishikawa, 18th Annual Conference on Composites and Advanced Ceramic Materials, 9th-14th January, 1994, Cocoa Beach, FL.
12 D. J. Cooke, N. C. Lloyd and W. J. Owen, J. Organomet. Chem; **22**, 55,1970.
13 J. P. Pillot, C. Biran, E. Bacqué, P. Lapouyade and J. Dunoguès, Brevet Fr 86-07-814, 30-05-1986.

NOVEL SYNTHESIS AND CHARACTERIZATION OF SILICON NITRIDE FIBERS

Ulrich Vogt, Karl Berroth and Georg Engeli
Swiss Federal Laboratories for Materials Testing and Research (EMPA)
Überlandstr. 129, CH-8600 Dübendorf, Switzerland

ABSTRACT

A high temperature CVD-process has been developed to produce Si_3N_4-fibers for high temperature applications. The reaction $3SiO_2 + 4NH_3 \rightarrow Si_3N_4 + 6H_2O$ operates at temperatures above 1400°C by evaporation of SiO_2 to SiO and dissociation of NH_3 into active nitrogen and hydrogen atoms. The active nitrogen atoms react with unbounded Si atoms thus forming Si_3N_4 fibers. The growth process is mainly influenced by the reaction temperature, reaction time and the gas flow rate in the reaction chamber. Depending on the reaction parameters temperature and time, amorphous or crystalline fibers can be grown. The length of the amorphouse fibers is up to 15 cm and the diameter in the range of 1-15 µm, depending on the reaction parameters. The oxygen content of the fibers decreases with an increase of crystallinity. 29-Si NMR analysis shows that the oxygen in the fibers is not present as silica, but in the composition of non-stoichiometric silicon oxynitride. The beginning of oxidation in air could be established at 1230°C by DTA/TG analysis. Annealing experiments in argon at 1500°C for 16 hours show no crystallization or recrystallization of the fibers. Tensile strength measurements of flexible amorphous fibers with 15 µm in diameter yield values up to 5 GPa.

1. INTRODUCTION

The development of new materials for high temperature applications requires a new generation of high performance ceramic fibers. Commercially available SiC fibers based on polycarbosilane precursor or vapor deposited SiC have a long service time in air only at temperatures below 1000°C. Above 1200°C structural changes are due to crystallization or recrystallization [1]. Therefore, a European research group (CEFIR) is calling for the development of a new fiber generation for applications up to 1400°C with a room temperature strength of 2.5-3 GPa. The Si_3N_4 fiber described in this paper should serve for applications up to 1200°C in air and at least 1400°C in non-oxidizing atmosphere.

The formation of silicon nitride and silicon oxynitride powder from silica in a stream of ammonia has been described in several publications [2-6], the new developed gas-phase process from EMPA is described in detail in [2]. The decomposition of ammonia at high temperatures is an important reaction for this process. According to Bartnitskoya [3] at 1000°C-1500°C the gas mixture consists of atomic and molecular nitrogen, hydrogen, as well as NH- and NH_2-radicals, produced by incomplete dissociation of ammonia. Atomic nitrogen and hydrogen which are formed by dissociation of ammonia have a much higher activity than the molecular species. These active hydrogen atoms cleave the SiO-bonds in SiO_2 at high temperatures and the active nitrogen atoms react with free silicon atoms to silicon nitride. The dissociated oxygen is eliminated from the surface as a volatile species. It should be noted that the flowing system is in a dynamic equilibrium, since some of the products are continuously being removed. The influence of substrate materials on the formation of silicon nitride has been described by Gray and Hendry [4]. They have grown Si_3N_4 whiskers on substrates of carbon and recrystallized alumina to determine their influence on product phase and morphology. Gribkov supposed that whiskers which grow from a vapor-gas phase may involve two deposition schemes: vapor-solid (VS) and vapor-liquid-solid (VLS) [5]. It is generally assumed that in the first case, condensation from the vapor phase directly into the solid phase (VS), whiskers grow by an axial-screw-dislocation mechanism corresponding to the Sears model [6]. In the second model (VLS) no screw dislocation is required.

Amorphous Si_3N_4 fibres grown from a vapor phase by a LCVD process is described by Wallenberg [7]. Hüttinger et al. grow high strength high modulus monocrystalline filaments by a CVD method, starting with silicon powder [8].

2. EXPERIMENTAL RESULTS

2.1 Fiber Synthesis

For experimental work a tube furnace with an inner tube diameter of 43 mm and a tube length of 1 m was used. The influence of the reaction parameters such as temperature, time, gas flow rate, type of silica and metallic additives were tested.

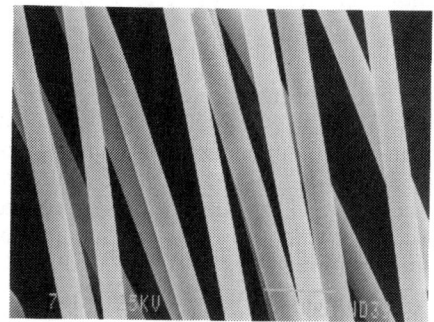

Figure 1
flexible amorphous fibers
—— 10 μm

Figure 2
crystalline fibers
—— 10 μm

As substrate materials SiC crucibles were used, as metallic additives Ti, Fe, V and Cr which were mixed with the silica. As precursor three different amorphous silica powders were tested.

At temperatures below 1400°C the reaction rate is very slow, making 1450°C the optimum temperature for the synthesis of amorphous fibers as shown in figure 1. At temperatures above 1500°C crystalline α-Si_3N_4 fibers can be grown (figure 2). The type of silica precursor as well as the type of metallic additive has no significant influence on the reaction. The main effects with respect to a low oxygen content of the fibers can be achieved by an increase in reaction temperature, reaction time and gas flow rate. Low reaction temperatures and short reaction times lead to amorphous fibers with a high oxygen content. With an increase in reaction temperature and time the oxygen content decreases (figure 3). Lower oxygen contents can also be achieved by a higher gas flow rates. At high gas flow rates, an oxygen content of 4 wt% can be achieved after a reaction time of 4 h, whereas a low gas flow rate requires nearly 18 h at a reaction temperature of 1480°C (figure 4). An increase of the α-Si_3N_4 phase can be accomplished mainly by a rise of the reaction temperature and time and this again coincides with a decrease in the oxygen content of the fibers (figure 9).

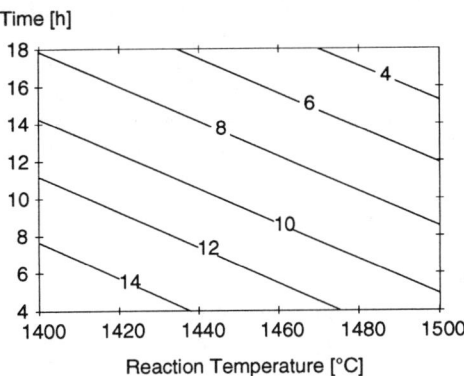
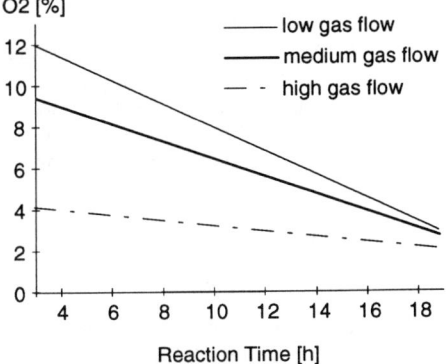

Figure 3:
Iso-lines of the oxygen content [wt%] of Si_3N_4 fibers at different reaction times and temperatures at low gas flow rates

Figure 4:
Oxygen content [wt%] of Si_3N_4 fibers in relation to the reaction time and gas flow rate at 1480°C reaction temp.

2.2 Fiber Characterization

The fibers were analyzed by SEM, XRD, 29-Si MAS NMR and they were also characterized with respect to their nitrogen- and oxygen content with a LECO TC 463 analyzer. The oxygen content of the fibers varies from 0.8 wt% up to 18 wt%, depending on the reaction parameters temperature, time and gas flow rate. Amorphous fibers have oxygen contents in the range of 5-15 wt%, while in crystalline or partially crystalline fibers oxygen contents between 0.8 wt% and 12 wt% were observed (figure 9).

2.2.1 Solid State NMR-Spectroscopy

29-Si Solid State NMR spectra were acquired with Magic Angle Spinning technique. The instrument was a Bruker ASX 400 spectrometer (29-Si resonance at 79.49 MHz). The pulse sequence was a single pulse of 10°, followed by acquisition and 100 seconds relaxation-recycle delay. The samples were of 100-400 mg, the MAS rate 4000 Hz. Typically 180 pulses were sampled. NMR allows the detection and identification of the crystalline as well as the amorphous phase content of the fibers. But quantification of the phases would require extremely long measuring times because silicon nitride has a relaxation time of 43 minutes. Therefore we did not measure with full relaxation, but used a reference sample of known phase content. So it was possible to do an approximate calculation of the phase content of the Si_3N_4 fibers. The reference sample was made of three different crystalline phases: $\alpha-Si_3N_4$ (UBE SN E-10, 36wt%), Si_2N_2O (research sample, 31wt%) and SiO_2 (quartz, 33wt%) (figure 5 D). Three different fiber samples were measured, with oxygen contents of 12.6 wt%, 6.5 wt% and 0.8 wt% (figure 5, A-C).

The fibers consist of $\alpha-Si_3N_4$ with a chemical shift of -46.8 and -48.9 ppm and amorphous Si_3N_4 with a chemical shift of -46.4 ppm. The amount of amorphous Si_3N_4 increases with the oxygen content of the fibers. SiO_2 with a chemical shift of -110 ppm was not detected, even in fibers with an oxygen content of 12.6 wt%. Therefore we presume, that the oxygen is present as a silicon oxynitride phase with a composition of $Si_3N_{4-x}O_{1.5x}$. This phase has been detected at a chemical shift of -56.5 ppm (figure 5 A) and has been described as 'non-stoichiometric silicon oxynitride' by Carduner et al. [9]. Amorphous fibers contain more than 50 wt% of this phase and crystalline fibers only traces.

2.2.2 XRD Analysis

X-ray powder diffractograms were taken of ground fibers on a Siemens D500 diffractometer with $Cu_{K\alpha}$ radiation. XRD confirmed the results from NMR concerning the cristallinity of the fibers. The only crystalline phase that could be detected for fibres which had been grown below 1500°C was $\alpha-Si_3N_4$ (figure 6). Fibers which had been synthesized at temperatures above 1500°C also contain traces of $\beta-Si_3N_4$. The higher the crystallinity of the fibers, the lower the oxygen content.

The amorphous non-stoichiometric phase which had been identified by NMR was not visible in XRD. No other phases such as Si_2N_2O, SiO_2 or $Si_3N_{4-x}O_{1.5x}$ were visible by XRD, thus confirming the NMR results. The peak intensity distribution of crystalline $\alpha-Si_3N_4$ fibers was different from UBE $\alpha-Si_3N_4$ powder (figure 6, C-D). This is due to the morphology of the fibers. From SEM images it was observed that the fibers are of rectangular cross-section and cannot be ground to homogenous particles during XRD sample preparation. Because of the preparation method the fibers are oriented in the direction of the sample holder. Therefore all hkl-lines with $l \neq 0$ are of reduced intensity.

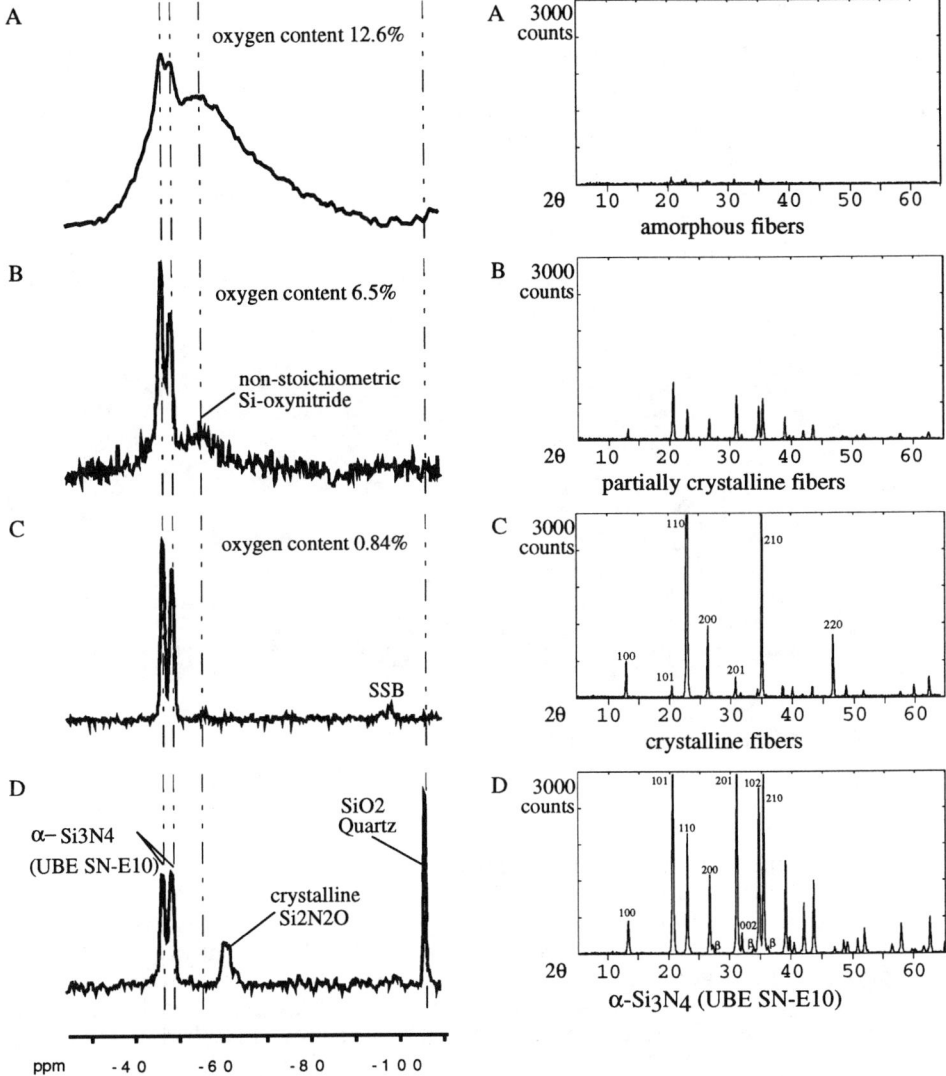

Figure 5: NMR spectroscopy

A: amorphous fibers
B: partially crystalline fibers
C: crystalline fibers
D: reference sample consisting of Si_3N_4 (UBE SN E-10), Si_2N_2O and quartz powder

Figure 6: Powder X-ray diffraction

A: amorphous fibers
B: partially crystalline fibers
C: crystalline fibers
D: reference sample Si_3N_4 powder (UBE SN E-10),

2.2.3 Thermal Annealing

Amorphous Si_3N_4 fibers were heated in air and argon for investigations concerning their high temperature behavior. To determine the start of oxidation, DTA/TG experiments were carried out with amorphous fibers in air up to 1500°C in a Netzsch STA 409. At temperatures up to 1200°C there is only a small increase in weight, passive oxidation starts at 1230°C. After annealing in air for 16 hours at 1000°C and 1200°C the fibers show no visible change of the surface structure (figure 7) and XRD proves them to be amorphous. After annealing in argon at 1500°C for 16 hours there is also no visible change of the smooth fiber surface and the fibers remain amorphous (figure 8). No crystallization could be observed by XRD and REM.

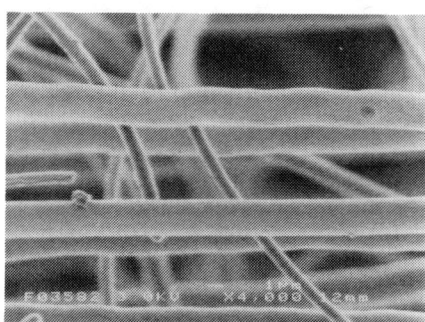
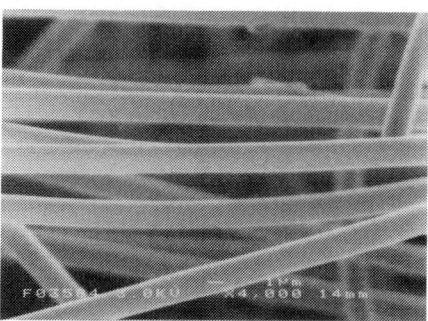

Figure 7:
Fiber surface after annealing in air at 1200°C for 16 h

Figure 8:
Fiber surface after annealing in argon at 1500°C for 16 h

2.3 Mechanical Strength Measurements

For room temperature tensile tests, we used a screw-driven universal tensile testing machine with a special measure cell unit. The tests were carried out in accordance with Standard ENV 1007-4, JIS R 7601 and ASTM D 3379. The fibers were glued onto paper holders with a commercially available rubber cement with a gauge length of 10 mm. For the tensile tests the paper holders were fixed in the test apparatus where the force at failure was measured. The fiber fracture was investigated by SEM and the diameter measured in order to calculate the tensile strength.

The diagram in figure 10 shows a large spread in strength of the measured fibers. It is an average of fibers from different production lots which are different in surface morphology as well as fiber diameter. Nevertheless, the trend of the tensile strength data for the measured Si_3N_4 fibers can be seen clearly. Fibers with diameters of 60 µm show values of 1 GPa, while fibers with diameters of 15 µm can reach the high tensile strength of 5 GPa. Only amorphous fibers without surface damage as shown in figure 1 reach the high "upper strength limit". Fibers with visible surface defects like bends, kinks or even intergrowth only reach lower values.

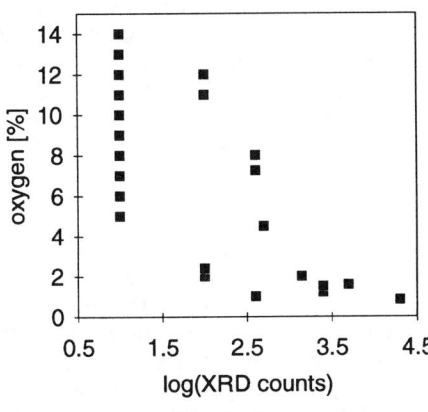

Figure 9
Correlation between oxygen content and crystallinity of Si_3N_4 fibers (background noise 10 counts)

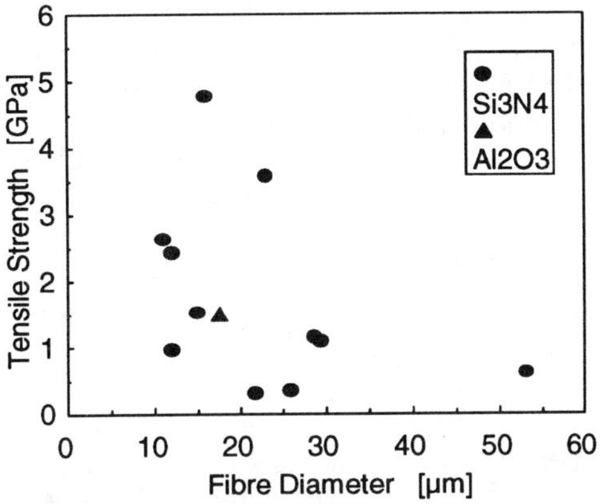

Figure 10
Tensile strength vs. Si_3N_4 fiber diameter at room temperature.
For comparison a Sumitomo Al_2O_3 fiber

3. DISCUSSION

The formation of Si_3N_4 fibers can be controlled by the reaction parameters temperature, time and ammonia gas flow rate. Below 1400°C the reaction is very slow. Up to a temperature of 1500°C the synthesized fibers contain crystalline Si_3N_4, amorphous Si_3N_4 and non-stoichiometric silicon oxynitride of the composition $Si_3N_{4-x}O_{1.5x}$ with x<<1. Under the current standard reaction conditions (1450°C, 16 h) the amorphouse fibres are 10-15 cm in length with diameters of 1-5 μm.
The amount of crystalline Si_3N_4 fibers increases with the reaction-temperature and -time, above 1500°C the fibers consist mainly of crystalline α-Si_3N_4. The non-stoichiometric oxynitride phase is completely amorphous and could therefore be identified only by NMR and not by XRD. Increasing ammonia gas flow rate coincides with an increase in nitridation rate and a lower oxygen content of the fibres because of the higher reduction potential due to the hydrogen formed. It should be noted that the flowing system is in dynamic equilibrium since some of the products are continuously being removed. Metallic additives such as Fe, Cr, V or Ti, which are common catalysts for the ammonia synthesis, also have a positive influence on the fiber growth. It is believed that the catalytic effect of metals on the decomposition of NH_3 also enhances the fiber growth rate. Also SiC substrates promote the fiber growth. According to Gray and Hendry [4], a nucleation process on a suitable substrate is assumed, with a heterogeneous nucleation from the gas phase on the substrate. In air the fibers are stable up to 1200°C with no surface defects or crystallization, but a slow increase in oxygen content. Passive oxidation in air starts at 1230°C. In argon the fibers are stable up to 1500°C without crystallization during 16 hours annealing time. Investigations concerning the room temperature strength show the potential of the fibers. A representative average of fibers were tested. Fibers with diameters of 30 μm achieve the strength goal for the new fiber

generation of 2.5 to 3 GPa, fibers with diameters of 15 μm achieve strength values up to 5 GPa. The production of amorphous weavable fibers with diameters of 2-15 μm and room temperature strength of 5-6 GPa will be the goal of the ensuing research activities. Investigations with focus on the growth mechanism and the mechanical behavior at room- and elevated temperatures will be carried out in the future as well as fundamental work concerning the exact fiber composition.

4. CONCLUSION

Si_3N_4 fibers can be produced from silica and ammonia via a gas phase reaction at elevated temperatures. Important parameters are reaction temperature, reaction time, ammonia gas flow, substrate materials and metallic additives. Depending on the reaction parameters the fibers grow amorphous or crystalline with diameters between 1-15 μm and up to 15 cm in length. Amorphous fibers have oxygen contents between 5-15 wt%. They consist of amorphous Si_3N_4 and a non-stoichiometric silicon oxynitride phase. Crystalline fibers can contain less than 1 wt% oxygen and consist of α-Si_3N_4 with traces of β–Si_3N_4. Amorphous fibers with 15 μm in diameter and no visible defects can reach strength values up to 5 GPa.

REFERENCES

1) G. SIMO0N AND A.R. BUNSEL, Creep behaviour and structural characterization at high temperatures of Nicalon SiC fibers, J. Mater. Sci., **19** (1984) 3658-3670)
2) U. VOGT, H. HOFMANN AND V. KRÄMER, Synthesis of Si_3N_4 Fibers by a Gas-Phase Process; Key Engineering Materials Vol. 89-91 (1994) 29-34
3) T. S. BARTNITSKOYA, P. P. PIKUZA, E. S. LUGOVSKAYA and T. YA. KOSOLAPOVA, Formation of Silicon Nitride Fibers from Silicon Oxide in a Stream of Ammonia; Translated from Izvestiya Akademii Nauk, SSSR, Neorganicheskie, Materialy, **18**, No. 10 (1982) 1733-35.
4) D. GRAY and A. HENDRY, The Influence of Substrate Material on the Formation of Silicon Nitride; Journal of the European Ceramic Society, **10** (1992) 75-82.
5) V. N. GRIBKOV, V. A. SILAEV, B. V. SHCHETANOV, E. L. UMANTSEV and A. S. ISAIKIN, Growth Mechanism of Silicon Nitride Whiskers; Soviet Physics-Crystallography, **16**, No.5 (1972) 852-854
6) G. W. SEARS, A Growth Mechanism for Mercury Whiskers; Acta Metallurgica, **3** (1955)
7) F.T. WALLENBERG, P.C. NORDINE, Amorphous silicon nitride fibers grown from the vapor phase, J. Mater. Res. Vol. 9, No. 3, 1994
8) K. J. HÜTTINGER, T.W. PIESCHNIK, Monocrystaline Si_3N_4 Filaments, Adv. Mater., Vol. 6, Nr. 1, 1994
9) K. R. CARDUNER, R. O. CARTER, M. E. MILBERG, G. M. CROSBIE, Determination of Phase Composition of Silicon Nitride Powders by 29-Si MAS-NMR, Anal. Chem., **59**, (1987), 2794-2797

CERAMIC FIBERS FOR FUNCTIONAL COMPOSITE MATERIALS

Ulyanova T.M., Titova L.V.
Institute of General and Inorganic Chemistry Academy of Sciences of Belarus. Minsk, 220072, Surganov St., 9,
Belarus Republic

The formation processes of calcium and zirconium titanate fibers from salt impregnated hydrated cellulose filaments by thermal treatment have been investigated, and their structure has been studied. Ceramic materials based on these fibers were obtained and their characteristics were determined.

1. Introduction

The important factors controlling the characteristics of ceramic materials during their processing are the purity and dispersity of initial reagents. For composite materials reinforcements composition structure and adhesion properties have an essential role. It was thought of interest to prepare calcium and zirconium titanates fibers, to study their structure, properties and the possibility of their use as functional ceramic materials.

II. Experimental procedure

Ceramic fibers were obtained by impregnation of hydrated cellulose filaments with a mixture of solutions of titanium chloride and calcium salt with a variable molar component ratios, in recount on metal oxides, from 1:1 to 1:1.45 and zirconium and titanium chloride with the ratio 1:1. Dried salt-containing fibers were heated in oxidizing atmosphere according to a specific technology discribed elsewhere (1). The processes occurring during the heating of the impregnated filaments were studied by thermal analysis. Heat treatment of hydrated cellulose fibers was carried out in air up to 1000°C with a heating rate of 5 degree/min on a sample mass of 0.5g. Alumina was used as a standard substance. Materials structure was investigated with an IR Perkin-Elmer spectrometer in 250-40000cm^{-1} frequency range, cesium iodide being used as immersion medium. X-ray phase analysis was carried out on Dron-3 unit (Cu-K$_\alpha$-radiation). Fiber surface morphology was studied by using scanning electron microscope SEM-100 U with a magnification of 500-5000 (preliminary, silver was sprayed on the samples). Nitrogen and benzene adsorption method (BET) was used for determination of specific surface area.

III. Results and discussion

Thermal treatment investigation of cellulose fibers, containing titanium and calcium salts has shown that dehydration of polymeric matrix began at 190-200°C. With temperature increase the fiber oxidation occurred with heat evolution, but within the 380-410° C temperature region a deep endoeffect was observed which was assigned to $CaTiO_3$ formation. That was confirmed by X-ray phase analysis. At 900°C the orthorhombic calcium titanate (a=0.544, b=0.764, c=0.538nm) was the predominant phase in fibers. Generally speaking, titanium and calcium chlorides introduced in polymer fiber, accelerated the initial stage of dehydration and thermal polymer destruction. When formed, titanium oxide catalyzed the cellulose oxidation process. Calcium chloride and then calcium titanate covering the fibers, decelerate carbon burnout and shift the cellulose oxidation by 150-200°C to higher temperatures.

Experiment showed that formation of $CaTiO_3$ fibers proceeded at all studied ratios of titanium and calcium salts, in the 400-550° C temperature range. A more stechiometric product was obtained when a solution with 1:1.3 and 1:1.45 a molar ratios of metal was used. At other component ratios, TiO_2 excess was observed. The initial cellulose fiber had a ribbed form, therefore ceramic fibers repeated the precursor form. With increase of annealing temperature, change in size of fiber diameter and porosity occurred. Specific surface area of ceramic fibers changed from 75-80 m^2/g at 450-550° C to 1 m^2/g at 1100-1200°C. In the 1200-1300°C temperature range a growth of oxides grains in fiber was observed with a size increasing from 0.1-0.2μm to 1 μm. At 1500°C the grains were sintered into conglomerates, then into nonporous fiber. The introduction of titanium and zirconium salts led to cellulose dehydration, already at 125-130°C, and at 300-350° C its oxidation took place. On the exoeffect background caused by polymer oxidation two endoeffects were observed at 350-380°C and 400-450°C associated with dissociation of titanium and zirconium chlorides, respectively, and the formation of metal oxides nuclei. With temperature increase ,and of phase transitions of titanium oxide (from anatase to rutile), zirconia (from tetragonal phase into monoclinic one) were observed. Cellulose oxidation and ceramic fibers formation were achived at 900° C. For quantitative estimation of change in fiber phase composition, the integrated reflex intensities of ZrO_2 M($1\bar{1}1$), ZrO_2 T(111), TiO_2 A(101), TiO_2 R(110), $ZrTiO_4$ (111) and their relations were determined (Fig. 1). As indicated in (1), zirconia formation at 400°C occurred first in the tetragonal structure which was then converted into monoclinic when temperature rising up to 600-650°C. The formation of zirconium titanate took place in the temperature range of 1150-1200° C. A growth of metal oxides grains in fibers was observed simultaneously with temperature rising. Their size increased from 0.3 µm at 900°C to 0.7-1.5 µm at 1300° C. At 1400° C, a longitudinal, splitting of fibers took place. The fiber fragments at the expense of volume and surface diffusion processes were sintered with each other forming a network structure. $ZrTiO_4$ fibers being formed were nonporous with ellipse-shaped section (Fig. 2). At 1500-1600° C, the processes of zirconium titanate recrystallization and crystal structure stabilization occurred. By changing the temperature conditions of heat treatment ceramic fibers of definite composition with a given properties were obtained. It should be noted that the grains of fibers annealed at temperatures not above 900° C had a size value of about 0,1-0,2 µm and were characterized by monodispersity which allowed us to obtain superdispersive active powders when these fibers were ground. Fibers with 1500°C annealing temperature could be used as reinforcement in composite materials. They had a low porosity and specific surface area, but exhibited high strength and hardness. For preparation of dielectric and ferroelectric ceramics, samples were prepared as beams with 5x5x70 mm dimensions, disks and cylinders of 10 mm in diameter and of 3.5 and 25 mm height. Pressing was carried out in steel die at pressure from 130 to 500 MPa. Samples were annealed at 1100-1500°C. It is worthy of note that fibrous dispersions annealed at 500-900°C were pressed at a lower pressure than powders from fibers obtained within the 1100-1300°C temperature range. High porosity and activity were associated with fibers annealed at low temperatures a feature which permitted to press the material without temporary binder. Density, porosity, strength and electrophysical properties of the samples were determined after their annealing. The results show that main influence on ceramics density was related to annealing temperature. Pressure value applied to half-product also affected ceramics properties, in TiO_2-CaO system at 1:1, 1:1.2 ratios, its influence was higher than for compositions at 1:1.3 and 1:1.45 ratios. Dependence of ceramics characteristics of 1:1.45 composition with pressure is shown in fig.3. Electrophysical characteristics of $CaTiO_3$ samples were determined. When material density increased, its specific volume resistance increased, too, but values of dielectric permittivity and dielectric loss decreased. High density materials were obtained from fibers annealed at 300, 700 and 1200°C in ZrO_2-TiO_2 system. In these temperature ranges the formation of amorphous oxides, followed by phase transition and compound formation was observed. This corresponded to disordered state of materials. In these cases, a diffusional solid phase reaction proceeded actively that allowed us to obtain a high density material using traditional ceramic technology. $CaTiO_3$ and $ZrTiO_4$ samples heat treated at 1500°C were found to have a density of $3.96 \cdot 10^3$ and $4.85 \cdot 10^3 kg/m^3$, respectively; their strength was found to be of 90-115MPa, and their specific resistance $(5\text{-}7) \cdot 10^{10}$ ohm.m. Dielectric permittivity of these materials were of 150 and 80, respectively, but dielectric loss changed from 0.0004 to 0.0005.

Conclusion

Due to particular properties, fibers prepared according to the preset work may be used as fillers in compositions and as dispersions for functional ceramics.

Acknowledgment

The investigation was supported by the Belarussian Foundation of Fundamental Research.

References.

1. I.N.Yermolenko, T.M.Ulyanova, P.A.Vityaz, I.L. Fyodorova, p.70, 85 Fibrous high temperature ceramic materials. Navuka & Technika. Minsk, 1991.

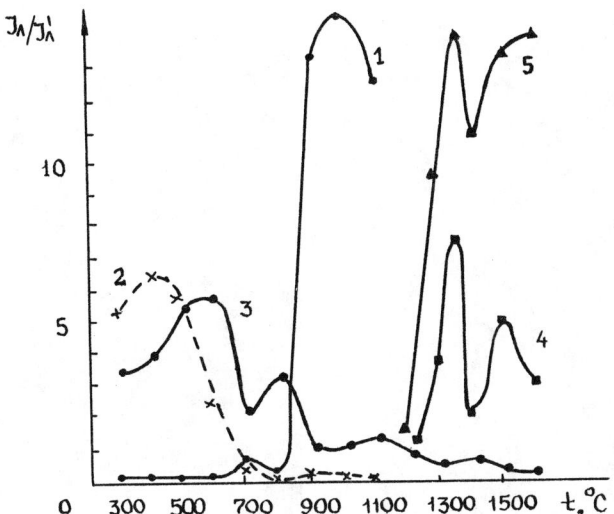

Fig.1. Temperature variation of intensity phase ratio: TiO_2R/TiO_2A-1, ZrO_2T/ZrO_2M-2, ZrO_2M/TiO_2R-3, $ZrTiO_4/TiO_2$-4, $ZrTiO_4/ZrO_2M$-5.

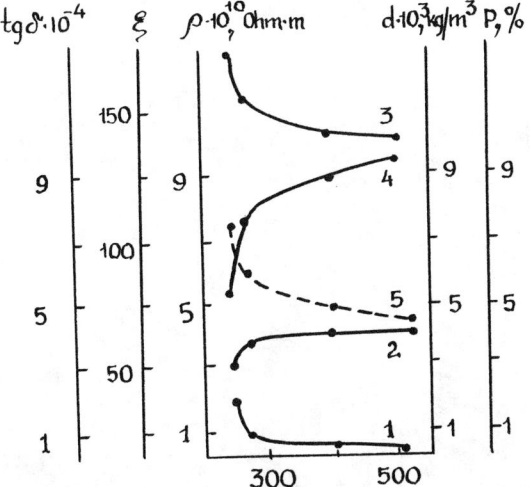

Fig.3. Axial pressure variation of $CaTiO_3$ ceramic of porosity-1, density-2, dielectric permettivity -3, specific resistance-4, dielectric loss-5.

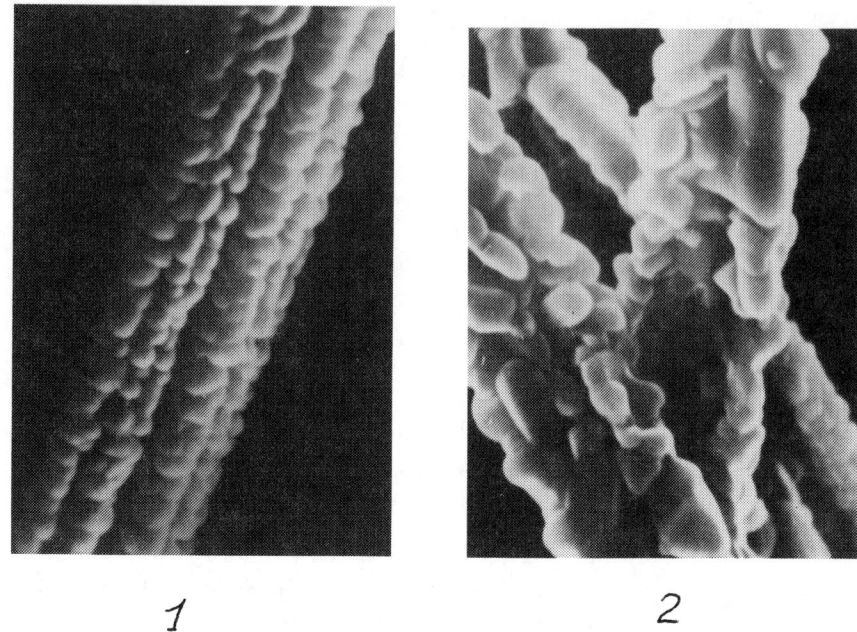
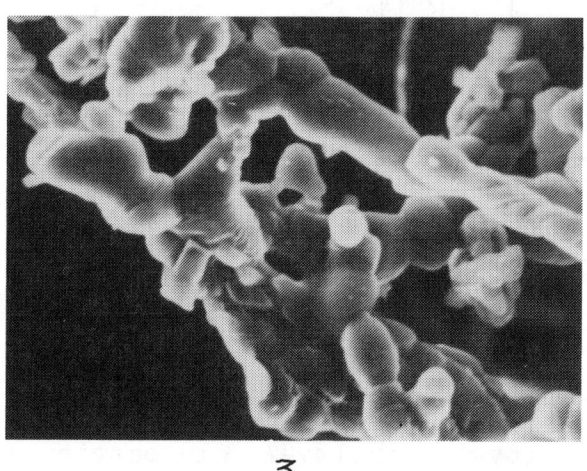

Fig.2. Microstructure of ZrO_2-TiO_2 fibers, annealed at temperatures: 1100 - 1, 1200 - 2, 1400°C - 3.

ENVIRONMENTAL EFFECTS ON CREEP AND STRESS-RUPTURE PROPERTIES OF ADVANCED SiC FIBERS

H.M. Yun[1], J.C. Goldsby[2], and J.A DiCarlo[2]
[1]Cleveland State University, [2]NASA Lewis Research Center,
Cleveland, Ohio, USA

INTRODUCTION

Small diameter polycrystalline SiC fibers are of high technical interest for reinforcement of ceramic matrix composites. These fibers have high room temperature stiffness and strength, and the potential for retaining these properties in oxidizing environments up to 1400 °C [1,2]. However, with current SiC fibers, it has been observed that the influence of environment on the high temperature creep and rupture behavior of individual fibers can be very complex. For example, Bodet et al. [3] reported better creep resistance of ceramic grade Nicalon fiber in a controlled CO environment than in argon. Yun et al. [4] reported better rupture resistance of Nicalon and Hi-Nicalon in air than in vacuum. Rugg [5] reported better rupture resistance of the Carborundum fiber in nitrogen than in air. In general, in comparison to inert conditions, testing in oxygen can increase or decrease fiber creep resistance, while having similar or opposite effects on rupture resistance. These effects can vary significantly with stress, temperature, and the SiC fiber grain boundary structure, second phase and impurity compositions, stoichiometry, as well as open porosity. As an initial step to better understand these factors, the objectives of this study were 1) to measure the effects of air and argon environments on the creep and rupture behavior of various advanced SiC fibers, 2) to determine whether these effects can be correlated among the fiber types, and 3) to suggest probable sources for these effects based on microstructural mechanisms.

EXPERIMENTAL PROCEDURE

The SiC fibers examined in this study were the Dow Corning, Carborundum, and Hi-Nicalon fibers, and an annealed version of the Hi-Nicalon fiber. The Dow Corning and Carborundum fibers were near-stoichiometric with nominal diameters of 10 and 33 μm, respectively; while the Hi-Nicalon fiber (from Nippon Carbon Co.) was carbon-rich (about 36 wt. % excess carbon plus 0.5 wt. % oxygen) with a 14 μm nominal diameter. Because the process temperature of the Hi-Nicalon fiber was below the maximum test temperature of this study (1400 °C), Hi-Nicalon fibers were also tested after pre-annealing at 1600 °C for 1 hour in argon. The annealed Hi-Nicalon had a larger grain size than the as-received Hi-Nicalon (about 80 versus 4 nm [6]) with effectively the same diameter as the Hi-Nicalon. The grain sizes of the Dow Corning fiber and Carborundum fibers, 0.1 μm and 2 μm, respectively, were larger than both versions of Hi-Nicalon. Other key properties of these fibers are reported elsewhere [1,6,7,8].

Measurement of fiber tensile creep and rupture were made both in ambient air and in 1.4

atmosphere argon at 1200 and 1400 °C for times from 0.05 up to 300 hours. The load was applied by dead weight at temperature, and the creep deformation was measured by an LVDT. Hot zone lengths were 1 and 4 inches for the $MoSi_2$-heated air furnace and 4.5 inches for the graphite-heated argon furnace. Effects of hot zone length could not be detected within the typical scatter of the air creep and rupture results. More details of the fiber testing rigs are reported elsewhere [4,9].

RESULTS

The measured creep deformation is shown in Fig. 1 for the four different SiC fibers in (a) argon and (b) air at 1400 °C and 270 MPa. Unless indicated by an arrow, the fibers ruptured at the end of the creep curves. As shown, all fibers displayed various degrees of transient behavior (primary stage creep) and some appeared to reach steady-state behavior. The Carborundum fibers were the most creep resistant, typically showing higher creep in air than in argon at 1400 °C. However, in contrast to Carborundum, creep of the Dow Corning fiber was higher in argon than in air. The as-received Hi-Nicalon fiber showed more creep than the Dow Corning and Carborundum fibers, typically failing at strain levels above 1% in the primary region. Creep was similar in air and argon. Finally the annealed Hi-Nicalon fiber tested in argon displayed much lower creep than the as-received fiber. However, in air the annealed Hi-Nicalon crept nearly the same as the as-received fiber.

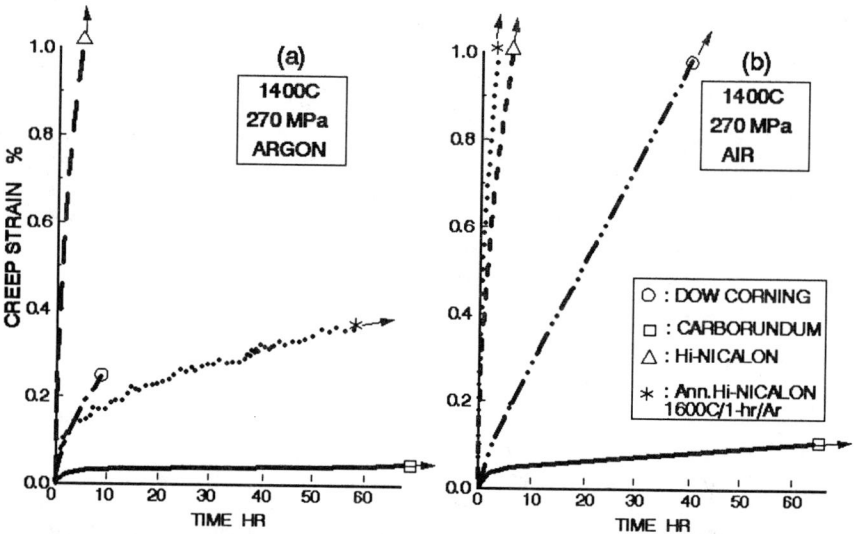

Fig.1 REPRESENTATIVE CREEP CURVES AT 1400 °C FOR POLYCRYSTALLINE SiC FIBERS IN (a) ARGON AND (b) AIR

From the creep curves, it was also possible to determine the creep strength of the various SiC fibers, that is, the maximum time-dependent stress level needed to reach a certain creep strain at a given temperature. For comparison of the various fibers, the 0.1% creep strengths in air and argon are shown in Fig. 2-a. For the log-log plots, the data were best fit to straight-lines for two temperatures: 1200 and 1400 °C, and two environments: argon and air. At a given time and

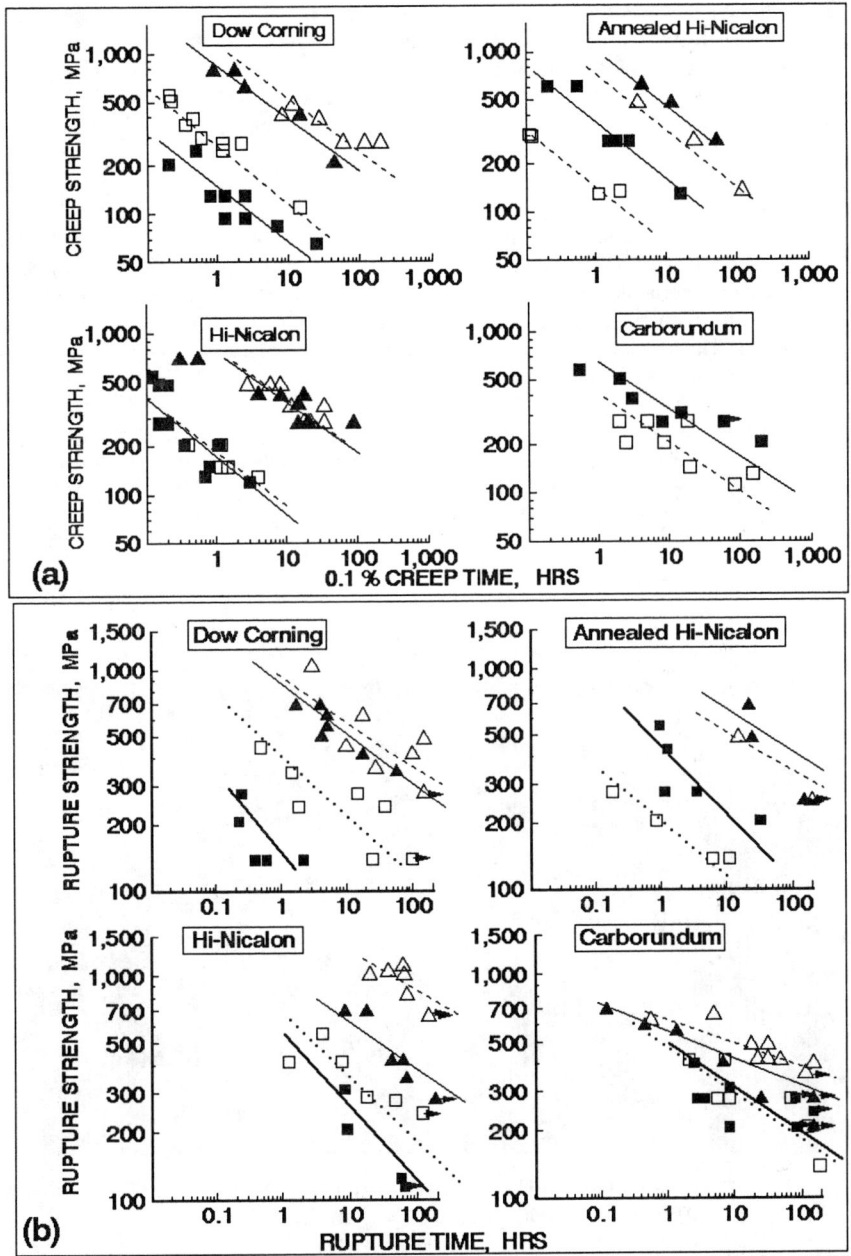

Fig.2 (a) 0.1 % CREEP STRENGTH AND (b) RUPTURE STRENGTH OF SiC FIBERS IN AIR (OPEN SYMBOLS) AND ARGON (CLOSED SYMBOLS) AT 1200 C (△) AND 1400 C (□).

temperature, the higher the creep strength, the more creep resistant was the fiber. In general, the strengths for the Dow Corning and Hi-Nicalon fibers were lower than those for the Carborundum fibers. At 1200 °C, the difference in creep strength in air and argon was small for all fiber types. However at 1400 °C, the Carborundum and annealed Hi-Nicalon fibers showed a lower creep strength in air than in argon; while the Dow Corning fiber displayed a lower strength in argon than in air.

The rupture time versus applied stress is shown in Fig. 2-b at 1200 and 1400 °C in air and argon for the four SiC fibers. At 1200 °C, the rupture times were environmentally independent for the Dow Corning and annealed Hi-Nicalon fibers, but longer in air than argon for the as-received Hi-Nicalon and Carborundum fibers. At 1400 °C, rupture times were environmentally independent for the Carborundum fiber, but higher in air for the Dow Corning and as-received Hi-Nicalon fibers. The annealed Hi-Nicalon showed an opposite effect at 1400 °C, that is, longer rupture times in argon than in air.

Typical fracture and surface photomicrographs are shown in Fig. 3 for the Carborundum and Hi-Nicalon fibers. As expected, a relatively uniform surface oxide layer was formed on the smooth Hi-Nicalon fiber; while the oxide layer of the Carborundum fibers was very rough resembling the initial surface of the fibers.

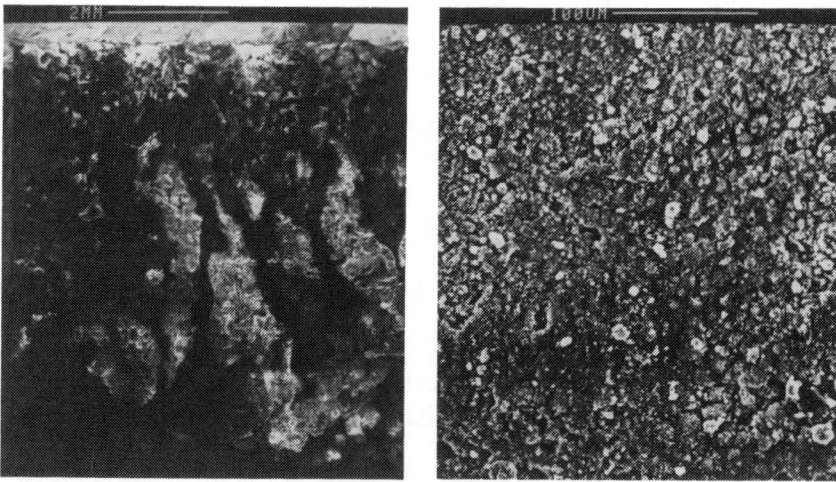

Fig.3 SEM PHOTOMICROGRAPHS OF CREEP AND RUPTURE TESTED SiC FIBERS;
A) CARBORUNDUM, 1400C / 75HOURS / ARGON
B) CARBORUNDUM, 1400C / 85HOURS / AIR
C) Hi-NICALON, 1400C / 30HOURS / ARGON
D) Hi-NICALON, 1400C / 35HOURS / AIR

DISCUSSION

Currently, without any detailed microscopic analysis of the tested fibers, one can only suggest underlying sources based on physical and chemical mechanisms typically invoked to explain similar environmental effects in bulk SiC monoliths. In the discussion that follows, it is assumed that possible air effects related to reduced fiber diameter and increased effective fiber stress are not significant. This is believed to be the case because experimental observations and theoretical calculations indicate that even for the most severe test conditions (100 hrs at 1400 °C), the thickness of the oxide layer, ~1 µm, could not account for the magnitude of the observed effects.

Regarding fiber creep, since this property is bulk controlled and typically related to grain boundary sliding, it follows that any observed environmental effects should be related to changes in the viscosity of grain boundary phases located throughout the fiber. Thus for those fibers which creep more in air than in argon, e.g. Carborundum and annealed Hi-Nicalon, one possible mechanism might involve oxygen attack of oxidizable phases along surface-exposed grain boundaries. For example, for carbon-rich boundaries, volatile carbon monoxide can be formed, allowing easy ingress of oxygen and the formation of low-viscosity silica-containing phases in the boundaries [10]. However, the likelihood of this effect occurring to any great degree is negated by the as-received Hi-Nicalon results which show for a fiber with high carbon content, little difference in creep behavior between air and argon testing. More likely, it would appear that oxygen ingress into the bulk of the Carborundum and annealed Hi-Nicalon fibers was provided by some degree of porosity open to the fiber surface. Indeed, such porosity can be expected in the annealed Hi-Nicalon fiber which during annealing should undergo some weight loss due to the thermal decomposition of process generated Si-C-O phases [6,9]. Likewise, optical observations of the Carborundum fiber show observable porosity both on the surface and in the structure. For the Dow Corning fiber which near 1400 °C creeps more in argon than in air, it is difficult to understand how grain boundary phases are altered during inert testing. One possible mechanism is that under argon there may occur a gaseous decomposition of some high viscosity phase in the boundaries which pins their sliding motion. During air testing, this decomposition effect may not readily occur due to the formation of a protective surface oxide which could serve as a hermetic seal on the fiber.

Regarding environmental effects on fiber fracture, it is well known that if the strength of a SiC monolith is controlled by process-generated surface flaws, the growth of an oxide layer on the monolith surface under zero stress conditions can blunt the surface flaws and increase the monolith strength. However, if stress is applied during oxide growth, oxygen can possibly enhance the slow crack growth of the surface flaws by weakening bonds at the crack tips of the surface flaws. This oxygen-related effect, which typically occurs at stresses greater than 50 % of the monolith fast fracture strength, has been observed to reduce rupture time and degrade rupture strength of SiC monoliths[11]. At lower stress levels, generally rupture is no longer controlled by slow crack growth of process-generated flaws, but by creep-induced internal flaw growth (cavitation). Thus any environmental effect which enhances creep at low stress levels should also reduce fiber rupture time and rupture strength. When these general observations are applied to the complex rupture results of this study, they suggest that for those environments which enhance creep at low stress levels (e.g. Dow Corning in argon and annealed Hi-Nicalon in air), the degraded rupture time in comparison to the opposite environment is probably related to enhanced flaw growth by creep cavitation. For the higher stress levels, the observation of improved rupture time in air (e.g., as seen for the Carborundum and as-received Hi-Nicalon fibers) suggests that the beneficial effect of oxide blunting of surface flaws can be more important than the degrading effects of oxygen-enhanced flaw growth either by bond breaking or by porosity-related creep.

CONCLUDING REMARKS

The results of this study support previous findings that the creep and rupture strength of SiC fibers are influenced in a complex manner by test environment and by fiber surface and internal microstructures. Generally it would appear that the silica layer which forms on the dense Dow Corning and Hi-Nicalon fibers during air testing can protect the fibers from internal oxygen attack and from thermal degradation, while also blunting surface flaws, thereby improving and/or maintaining fiber creep and rupture strength. On the other hand, the open porosity of the Carborundum and annealed Hi-Nicalon fibers allows oxygen ingress during air testing, generally resulting in lower creep and rupture strengths. However, at stress levels near the fast fracture strength, the rupture strength of the coarse grained Carborundum fiber appears to be improved by silica formation.

REFERENCES

1. J.A. DiCarlo and S. Dutta: Continuous Ceramic Fibers for Ceramic Composites. HANDBOOK ON CONTINUOUS FIBER REINFORCED CERAMIC MATRIX COMPOSITES, eds. R.Lehman, S.El-Rahaiby, and J.Wachtman, Jr., Ceramic Information Analysis Center, Purdue University, West Lafayette, Indiana, 1995
2. J.A. DiCarlo and H.M. Yun: Issues for Creep and Rupture Evaluation of Ceramic Fibers. PLASTIC DEFORMATION OF CERAMICS, eds. R.Bradt, C.Brooks, and J.Routbort, Plenum Publishing, New York, 1995
3. R.L. Bodet and R.E. Tressler: Effects of Chemical Environments on the Creep Behavior of Si-C-O Fibers. HIGH TEMPERATURE CERAMIC MATRIX COMPOSITES, HT-CMC 1, eds. R.Naslain, J.Lamon, D.Doumeingts, Woodhead Publishing Limited, France, Sept. 20-24, 1993, pp.75-83
4. H.M. Yun, J.C. Goldsby, and J.A. DiCarlo: Tensile Creep and Stress-Rupture Behavior of Polymer Derived SiC Fibers, ADVANCES IN CERAMIC-MATRIX COMPOSITES II, eds. J.P.Singh and N.P.Bansal, The American Ceramic Society, 1994, pp.17-22.
5. K.L. Rugg: Subcritical Crack Growth in Carborundum Alpha Silicon Carbide Fibers at Elevated Temperatures, Master Thesis, The Pennsylvania State University, 1994.
6. M. Takeda, J. Sakamoto, Y. Imai, H. Ichkawa, and T. Ishkawa: Properties of Stoichiometric Silicon Carbide Fiber Derived from Polycarbosilane, Ceramic Eng. Sci. Proc., 15, 1994, pp.133-141
7. J. Lipowitz, J.A. Rabe, and G.A. Zank: Polycrystalline SiC Fibers from Organosilicon Polymers, Ceramic. Eng. Sci. Proc., 12, 1991, pp.1819-31
8. G.V. Srinivasan and V. Venkateswaren: Tensile Strength Evaluation of Polycrystalline SiC Fibers, Ceramic Eng. Sci. Proc., 14, 1993, pp.563-78.
9. H.M. Yun, J.C. Goldsby, and J.A. DiCarlo: Effects of Thermal Treatment on Tensile Creep and Stress-Rupture Behavior of Hi-Nicalon SiC Fibers, Ceramic Eng. Sci. Proc., 16, The American Ceramic Society, 1995.
10. G. Gratwohl and F.Thuemmler: Creep of Reaction Bonded Silicon Nitrides. CERAMICS FOR HIGH PERFORMANCE APPLICATIONS - II, eds. J.J.Burke, E.N.Lenoe, and R.N.Katz, The Metals and Ceramics Information Center, Columbus, Ohio, 1977, pp.573-91.
11. E.J. Minford, D.M. Kupp, and R.E. Tressler: Static Fatigue Limit for Sintered Silicon Carbide at Elevated Temperatures, J. Am. Ceram. Soc., 66 [11], 1983, pp.769-73.

SiC FIBERS FROM MODIFIED POLYORGANOSILANES: PRODUCTION, PROPERTIES AND APPLICATION

V. V. Ivanov, T. V. Tikchonovich, S. A. Bashkirova, L. G. Polyak, N. M. Balagurova, E. A. Chernyshev

The State Scientific Research Institute of Chemistry and Technology of Organoelement Compounds (GNIIChTEOS), 38, sh. Entuziastov, 111123 Moscow (Russia)

ABSTRACT

New modified polyorganosilanes with functional groups (Si-H and C≡C) in polymer chain have been prepared by Wurtz reaction by 1,4-bis(chloromethylsilyl)benzene and bis(chlorodimethylsilyl)acetylene interaction with sodium. These modified polyorganosilanes undergo controlled cross-linking under UV-irradiation, which results in their molecular weight increase up to desired value for SiC fibers production. The preparation of SiC fibers with improved thermal stability without surface oxidation is possible.

INTRODUCTION

Search and development of highly efficient sources of continious oxide-free SiC fibers is an urgent target today.
Only organosilicon polymers with alternating Si and C atoms in main chain can serve as such sources. Polycarbosilane (PCS), produced by means of two-staged technique is a source for commercial SiC fibers ("Nicalon") [1].
Molecular weight limit of polycarbosilanes obtained by thermal molecular rearrangement is about 2,000 (M_n) and

herefore inorganic residue yield is also low. Besides, to provide for shape retention, the commercial production of "Nicalon" SiC fibers requires surface oxidation and oxidative cross-linking prior to pyrolysis [2,3]. Thus, this method does not permit to produce SiC fibers with minimum content of oxygen and thermal stability higher then 1000^{0}C.

Recently, investigators have explored alternative cross-linking methods without surface oxidation to produce fibers with improved mechanical and high-temperature properties [4-6]. Thus, electron beam irradiative cross-linking is the technique that does not involve PCS oxidation [4], however because of difficulties in experiments this method is hard to reach for commercial production.

As a novel approach to process realization chemical modification of commercial PCS through ethynyl functional groups introduction has been suggested that allows to perform cross-linking during UV-irradiation [5].

Recently SiC fibers have been prepared from PCS fibers cured by only heat-treatment with unsaturated hydrocarbon vapors [6].

Nevertheless PCS produced by thermal molecular rearrangement of polydimethylsilane (PDMS) through two stage is used in all foregoing processes.

Although the polymers involving $(-SiC \equiv CSi-)_n$ and $(-SiC_6H_4Si-)_n$ are known [7-10], the polymers containing Si-H functional groups was not previously investigated. The authors have developed a novel method of thermally stable SiC fibers production directly from modified polyorganosilanes, containing functional groups in polymer chain, such as Si-H and $C \equiv C$. These functional groups provide a possibility to obtain SiC fiber precursors of required molecular weight with high yields (80%) and to perform thermal cross-linking or by oxygen-free ultraviolet irradiation. Previously we proposed to use polysilane (PS) with allyl chain ends for these purposes [11].

1. EXPEREMENTAL

^{1}H- and ^{13}C-NMR spectra were recorded on a Bruker AM-360 spectrometer. IR and TGA data were obtained respectively on a Bruker IFS 113 V and Du Pont 990 Analyzer.

The number average molecular weight (M_n) of PS with polystyrene standard was measured with a liquid chroma-

tograph and GPC column, by detection at a wavelength of 254 nm.
The polymeric samples were exposed to ultraviolet radiation (200-400 nm) inside a glove box under argon atmosphere. Pyrolysis was conducted in a graphite furnace under flowing argon atmosphere.
Mechanical propeties of the fibers were identified in a testing machine "INSTRON 1122" at room temperature with 10 mm gauge length for 30 filaments.
The structure of the fibre was examined by electron microscope BS-301.

2. RESULTS AND DISCUSSION

New modified polyorganosilanes with functional groups (Si-H and C≡C) in polymer chain have been prepared by Wurtz reaction by 1,4-bis(chloromethylsilyl)benzene and bis(chlorodimethylsilyl)acetylene interaction with sodium in accordance with the following technique [12]:

$$\begin{array}{cc} H \quad H \\ ClSiC_6H_4SiCl \\ CH_3 \quad CH_3 \end{array} + \begin{array}{cc} CH_3 \quad CH_3 \\ ClSiC{\equiv}CSiCl \\ CH_3 \quad CH_3 \end{array} \xrightarrow[-NaCl]{Na} \begin{array}{cc} H \quad H \\ (SiC_6H_4Si)_n \\ CH_3 \quad CH_3 \end{array} \begin{array}{cc} CH_3 \quad CH_3 \\ (Si{\equiv}CSi)_m \\ CH_3 \quad CH_3 \end{array}$$

where: n=10-40
       m=10-40

PS structure is proved by NMR and IR-spectra. ^{1}H NMR (CDCl$_3$, 20°C) δ = 0.23 (12H, s, CH$_3$Si), 0.64 (6H, s, CH$_3$Si), 3.70-4.81 (br s, SiH), 7.28 (4H, s, ring protons); ^{13}C NMR (CDCl$_3$, 20°C) δ = 128.9, 127.7, 115.8, -2.5, -4.4, -5.0; IR (CHCl$_3$, cm^{-1}) 3047, 2955, 2924, 2899, 2849, 2081, 1427, 1379, 1128, 1105, 843, 824.
The main advantage of modified PS is functional groups (Si-H and C≡C) ability to polymer chains cross-linking under UV irradiation. This allows to increase polymer precursors molecular weight to desired value by means of controlled cross-linking on the one point, and carry out polymer fibers surface cross-linking without oxidation on the other point. Thus, total yield of fiber-forming polymer (PS) with M_n = 10.000 can reach 80%.
Fibers have been spun from the melt of polymer samples. The cross-linking of the fibers surface under UV-irradiation occured rapidly (10 min) and result in a thermally rigid coating that allows shape retention during pyrolysis in inert atmosphere. TGA-analysis of modified PS fibers cross-linked by UV-irradiation is shown in Fig.1 (ceramic yield 80%). Comparative TGA data for PCS

prepared by the technique [1] is shown in Fig.1 (ceramic yield 50%).
The cross-linking of fibers from modified PS can be conducted by partial surface oxidation in inert gas flow which results in unmelted fibers with oxygen content up to 5%. But ceramic yield is lower (70%,Fig.1).
The fibers cross-linked by UV-irradiation (A) and partial surface oxidation (B) then heat treated in a argon atmosphere from room temperature to $1400^0 C$ at a heating rate of $100^0 Ch^{-1}$ and held at this temperature for 2 h. The tensile strength, X-ray diffraction and elemental analysis of the SiC fiber were then measured. The fibers (A and B) acquire the following characteristics: (A)- d=12 μm, tensile strength 3.0 GPa, elemental composition $SiC_{2.02}H_{0.05}$, the highest usable temperature in air up to $1500^0 C$; (B)- d=0.12 μm, tensile strength 2.8 GPa, elemental composition $SiC_{2.14}H_{0.09}O_{0.11}$, thermal stability in air up to $1400^0 C$.
The grain size of the fibers was 3 nm (A) and 7 nm (B). SiC/SiC fibrous composites thermally stable at maximum $1500^0 C$ have been obtained on the basis of sinthesized SiC fibers.

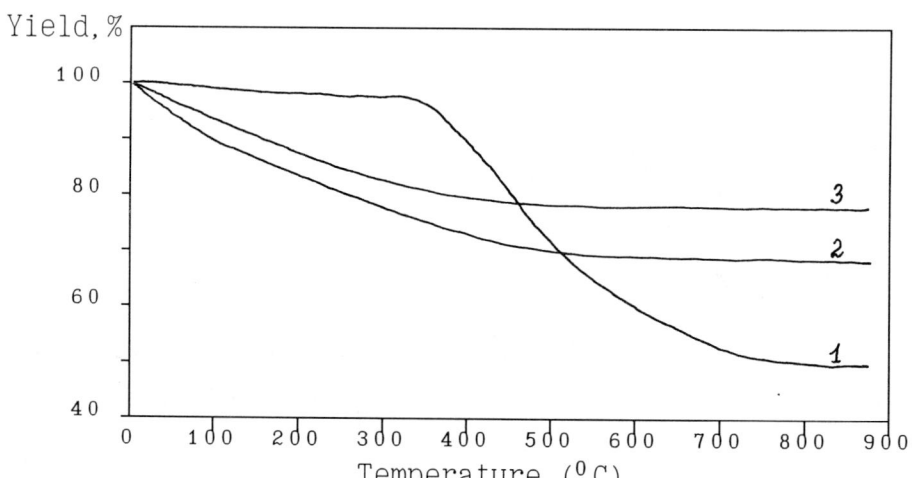

Fig.1 Thermogravimetric analysis: 1) PCS precursor from PDMS, 2) modified precursor (PS) after partial oxidation, 3) modified precursor (PS) after UV-irradiation

CONCLUSION

It was demonstrated that modified polysilanes with functional groups (Si-H, C≡C) in polymer chain were possible to produce by Wurtz reaction with high yield. It was found that modified polysilanes functional groups ability for polymer chains cross-linking under UV-irradiation allows to increase their molecular weight to desired value for SiC fibers production as well as to carry out their surface cross-linking without oxidation.

SiC fibers with the following characteristics were obtained: d = 12 μm, tensile strength 3.0 GPa, highest usable temperature in air - 1500°C.

ACKNOWLEDGEMENTS

The authors are extremely grateful to doctor I. M. Sytova for her help in our work.

REFERENCES

1. S. Yajima, J. Hayashi, M. Omori, "Continuous silicon carbide fiber of high tensile strength", Chem. Letters, 931-934 (1975)
2. S. Yajima, K. Okamura, J. Hayashi, M. Omori, "Synthesis of continuous SiC fibers with high tensile stregth", J. Am. Ceram. Soc., **59**, No. 7-8, 324-327 (1976)
3. G. Simon, A. R. Bunsell, "The creep of silicon carbide fibres", J. Mater. Sci. Lett., **2**, No. 2, 80-82 (1983)
4. M. Takeda, Y. Imai, H. Ichikawa, T. Ishikawa, T. Seguchi, K. Okamura, "Properties of the low-oxygen-content silicon carbide fiber after high temperature heat treatment", Ceram. Eng. Sci. Proc., **12**, 1007-1018 (1991)
5. K. J. Thorne, S. E. Johnson, H. Zheng, J. D. Mackenzie and M. F. Hawthorne, "Chemically designed, UV curable polycarbosilane polymer for the production of silicon carbide", Chem. Mater., **6**, 110-115 (1994)
6. Y. Hasegawa, "Synthesis of thermally stable Si-C fibre", pp 59-65, Proc. 6-th Eur. Conf. on composite mater., HT-CMC-1, (R. Naslain et al. ed.), Woodhead Publ. Limited, (1993)
7. V. V. Korshak, A. M. Sladkov, L. K. Luneva, Bull. Russ. Acad. Sci. Div. Chem. Sci., p 728 (1962)
8. I. I. Wun Shin, W. M. Risen, J. Organometal. Chem., **260**, 171 (1984)

9. S. Ijadi-Maghsood, Yi. Pang, Th. J. Barton, "Efficient, "ona-pot" synthesis of silylene-acetylene and disilylene-acetylene preceramic polymers from trichloroethylene", J. Polym. Sci., Polym. Chem., **28**, 955-965 (1990)
10. T. J. Barton, "Polyacetylene-polysilane and polyacetylene-polysilylene ceramic precursors", Pat. 4 940 767, US (1990)
11. S. Bashkirova, T. Tikhonovitch, L. Ulyanova, E. Chernyshev, "Polyorganosilanes: SiC pyrolysis and ceramic application", pp 193-197, "High temperature ceramic matrix composites HT-CMC-1" (R. Naslain et al. ed.), Woodhead Publ. Limited, (1993)
12. I. Tverdokhlebova, V. Ivanov, S. Bashkirova, E. Chernyshev, V. Menshov, O. Sutkevich, J. Vysokomolek. Soedineniya, ser. A., **36**, 1424-1428 (1994)

MODELS FOR THE THERMOSTRUCTURAL PROPERTIES OF SiC FIBERS

J. A. DiCarlo[1], H. M. Yun[2], G. N. Morscher[3], and J. C. Goldsby[1]
[1] NASA Lewis Research Center,
[2] Cleveland State University,
[3] Case Western Reserve University, Cleveland, Ohio

INTRODUCTION

To understand, predict, and optimize the thermostructural performance of ceramic matrix composites (CMC), descriptive mechanism-based models for the mechanical and physical properties of the reinforcing fiber are required. Currently the achievement of this goal is complicated by the fact that at temperatures as low as 800°C, continuous-length polycrystalline ceramic fibers display complex time-dependent deformation and fracture behavior [1,2]. The objective of this paper is to show that despite this complexity, simple generic analytical models can be developed which under certain conditions closely describe key fiber thermostructural properties as a function of time, temperature, and applied stress. The basic assumption employed here is that the properties are controlled by thermally-activated mechanisms. Because of current technical interest, focus is placed on the performance of advanced SiC fibers at creep strains below 1%. The models are applied to the creep-related properties of creep strain, creep strength, and stress relaxation, and to the fracture properties of rupture strength and rupture time.

MODEL THEORY

For this initial modeling study, only those time-dependent properties will be considered which are measured on individual fibers under conditions of constant temperature, applied stress (or strain), environment, and gauge length. In this case, it is assumed that if a fiber property $P(t,T)$ is controlled by thermally activated mechanisms, then the time t and absolute temperature T variables are not independent, but are interrelated by the relation [3]:

$$\Theta = t \exp(-Q/RT) \qquad (1).$$

Here R is the universal gas constant (8.314 J/mol·K), and mechanistically the Θ and Q variables can be considered as the characteristic relaxation time and activation energy, respectively, for each thermally-activated microstructural entity controlling P. In general, the values for Θ and Q can be broadly or narrowly distributed depending on the complexity or simplicity of the primary mechanism [4]. A further complication is that these distributions can also change with applied stress (or strain).

In order to experimentally determine the underlying Θ and Q distributions for a given thermostructural property P, one would need first to hold applied stress (or strain) constant at a level of practical interest and then measure the property as a function of time at various temperatures of practical interest. Then by Arrhenius plots of **ln t** versus **1/T** at constant **P** values, one should obtain a series of straight lines for each **P** value. Assuming each particular **P** line is described by Eq. 1, the slope of the line would yield the applicable **Q** value and the intercept at $1/T = 0$ the applicable Θ value [cf. Ref. 5]. With this procedure, one can then determine and plot **P** versus **Q** and **P** versus Θ to obtain the desired distributions. As a final step, one would change the applied stress level and repeat the above procedure in order to determine the stress dependencies of the distributions.

The advantage of the above approach is that, under test conditions of practical interest, it has been generally observed that the Θ_c and Q_c distributions for the **creep-related** properties of polycrystalline ceramic fibers are not stress dependent and that Q_c for each fiber type is effectively single-valued [6]. From a basic viewpoint, this implies that ceramic fiber time-dependent deformation is produced by the motion of many microstructural entities (e.g. grains) whose internal movements are controlled by the same creep activation energy, but occur at different times due to a wide or continuous distribution in relaxation time. From a modeling viewpoint, these observations and Eq. 1 imply that the time and temperature variables for each fiber type can be combined into a single temperature-compensated time or creep variable Θ_c which is stress independent [3]. Thus, in general, a fiber creep-related property can be described by independent functions of applied stress σ and the Θ_c variable. Likewise, as will be discussed, when thermal activation theory is applied to the time-dependent **rupture-related** properties of a variety of fibers, similar modeling simplicities also arise in that Θ_r now is effectively single-valued and a stress-dependent Q_r becomes the prime time-temperature variable.

MODEL APPLICATION TO SiC FIBER PROPERTIES

To illustrate these modeling results for some creep-related properties of advanced SiC fibers, Fig. 1 shows typical tensile creep curves for four as-produced fiber types measured at 140 MPa at 1400°C in air [7]. These fibers varied significantly in composition, microstructure, and processing approach [1]. Nevertheless, Fig. 1 shows that all fiber types displayed a large transient or primary creep stage, particularly in the technically important strain range below 1%. As discussed elsewhere [7], it is recognized that not all of this transient behavior was stress-

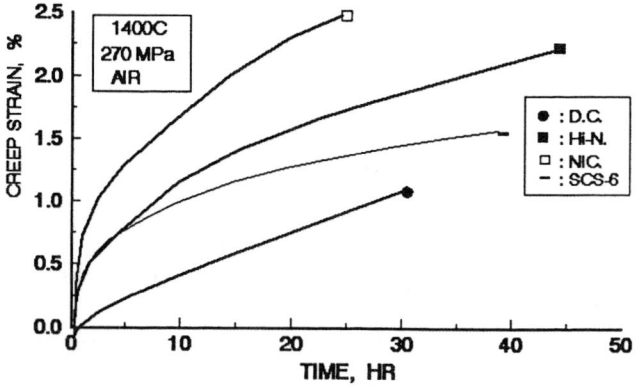

Fig. 1. Typical creep curves for advanced and first-generation as-produced SiC fibers.

induced because, due to low processing temperatures, some of these fibers probably experienced microstructural changes during testing. Also, there can be measurable effects of the test environment on fiber creep and rupture behavior [8]. Nevertheless, for those cases where these effects were eliminated or not expected, a relatively large transient stage was still observed in all SiC fibers.

When the modeling procedures were applied to fiber tensile creep data below 1% strain [7], it was observed that the stress, time, and temperature dependencies for fiber creep strain ε_c could be accurately described by simple power functions of σ and Θ_c; that is,

$$\varepsilon_c = A\,\sigma^n\,\Theta_c^p = A\,\sigma^n\,[t\,\exp(-Q_c/RT)]^p \qquad (2).$$

Table I lists the empirical creep constants A, n, p, and Q_c which were found to best fit Eq. 2 to the air creep data for the four SiC fibers in their as-produced condition. Clearly, the values for these constants will be subject to change, for example, as fiber creep data bases increase, as fiber production approaches improve, or, as discussed above, when the fibers are pre-treated and tested under other conditions which may better simulate the microstructural and environmental conditions experienced during CMC service. Notwithstanding these factors, Table I does show some interesting trends such as n values near unity, p values near 0.4 (transient effect), n/p values equal to 3, and creep resistances generally increasing with Q_c. The basic and technical implications of these trends will be discussed in future papers.

Table I. CREEP-RUPTURE PARAMETERS FOR AS-PRODUCED SiC FIBERS

FIBER TYPE	CREEP PARAMETERS [a]				RUPTURE PARAMETERS [a]		
	A	n	p	Q_c kJ/mol	E	$B \times 10^4$	C
Hi-NICALON	1×10^4	1.8	0.58	600	7.8	1.4	42
DOW CORNING	7×10^1	2.1	0.75	455	6.3	1.0	42
NICALON	1×10^2	1.2	0.40	500	6.0	1.0	42
SCS-6	2×10^2	1.1	0.37	580	4.6	0.5	42

a. Best fit to Eqs. 2 and 6 with strain = %, stress = MPa, time = sec, temperature = kelvin

Besides creep strain, another important fiber deformation property is the fiber creep strength, σ^*, which is defined as the maximum allowable tensile stress for a fiber to reach an application-limited creep strain, ε^*, for a given service time and temperature. From Eq. 2, it follows that

$$\ln \sigma^* = (1/n)\cdot(\ln \varepsilon^*/A) - (p/n)\cdot(\ln \Theta_c) \qquad (3).$$

An example of how accurately Eq. 3 and the empirical constants model describe creep is shown in Figs. 2a and 2b which plot, respectively, the 1% creep strength of Hi-Nicalon and the 0.1% creep strength of Dow Corning fibers as measured in air at various times and temperatures. Such plots can be practically useful for comparing fibers and estimating their behavior for time-temperature conditions outside of the original envelope of test conditions [7].

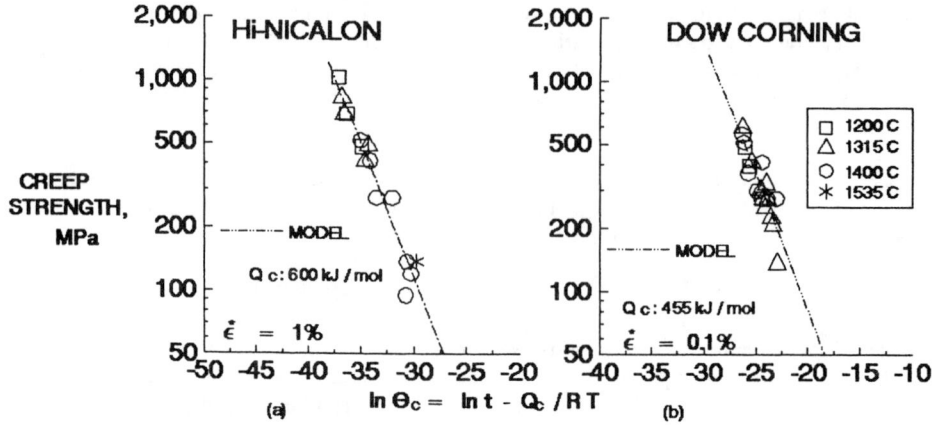

Fig. 2. Model fit to creep strength data for (a) Hi-Nicalon and (b) Dow Corning fibers in air.

For further understanding CMC behavior, it is also important to model the creep-related fiber property of stress relaxation both under tension [9] and bend [10] conditions. It is typically measured by subjecting a material to a constant strain at a constant temperature and measuring the time-dependent degradation of elastic stress in the material due to creep. Because it is very difficult to perform this test on individual fibers in tension, a simple bend test was developed to measure the fiber bend stress relaxation ratio m_b which is the ratio of the local time/temperature-dependent elastic stress in the fiber to the initial elastic stress [11]. For modeling this property, it has been postulated that since the early stages of SiC fiber creep are nearly linear in stress (n = 1) and predominately transient over a long time scale, creep can be primarily attributed to a broad distribution of anelastic (i.e. recoverable) mechanisms [6]. This would then imply that stress relaxation should be independent of applied strain so that for fibers with uniform microstructures,

$$m_b \approx m_t \approx [1 + \varepsilon_c/\varepsilon_o]^{-1} \qquad (4).$$

Here m_t is the tensile stress relaxation ratio that would be measured in a tension test with an initial applied strain of ε_o, and ε_c is the tensile creep strain that would be measured in a tension test at an applied stress of $\sigma = M\varepsilon_o$ where M is the fiber tensile elastic modulus.

Thus the Eq. 4 model implies that for certain fiber microstructures and test conditions (constant applied strain, temperature, environment), it should be possible to use the tensile creep model of Eq. 2 and the empirical constants of Table I to predict the stress relaxation of SiC fibers under pure tension and bend conditions. Recent studies have examined the validity of this approach by applying Eqs. 2 and 4 to m_b data to generate the four empirical creep constants for a variety of SiC fibers [6,12,13]. Excellent agreement was found for the time/temperature constants of Table I (that is, p and Q_c); but as expected from the Eq. 4 assumptions, close agreement for the A and n constants was found only for fibers with uniform microstructures and n values near unity.

Regarding the time-dependent fracture properties of SiC fibers, one can assume that the flaw-growth mechanisms controlling fiber rupture are also thermally activated with each flaw-growth entity characterized by a certain rupture relaxation time Θ_r and rupture energy Q_r. Using rupture strength S_r as the thermally activated property, recent studies of SiC fibers have examined Arrhenius plots of the natural log of rupture time t_r versus $1/T$ at constant S_r values [7]. In contrast to the Arrhenius plots for the creep properties, the rupture plots displayed straight lines whose slopes decreased significantly with increasing S_r values and whose intercepts at $1/T = 0$ were approximately the same. From a modeling viewpoint, these observations suggest that unlike the creep energy Q_c, the rupture energy Q_r is stress dependent and broadly distributed; while the rupture relaxation time Θ_r is effectively single valued. From a basic viewpoint, this supports a mechanistic model in which Q_r decreases with increasing stress intensity at the crack tip of the multi-sized strength-limiting flaws within a fiber.

From the SiC fiber rupture results, it follows that the primary variable controlling time-dependent fracture is Q_r. Thus by Eq. 1, the rupture time and temperature variables can now be interrelated by the equation

$$Q_r = RT \cdot (\ln t_r + C) \qquad (5)$$

where $C\ (= -\ln \Theta_r)$ is effectively a constant. Best fitting the Arrhenius plot results, it was determined that rupture strength versus rupture energy can be closely related by the following equation:

$$\ln S_r = 2.3E - (\beta/R) \cdot Q_r \qquad (6).$$

Here E and β as well as C in Eq. 5 are rupture constants that are determined empirically. Effectively Eq. 6 is a Larson-Miller representation of fiber rupture strength with the variable Q_r equivalent to the Larson-Miller variable [14]. Eq. 6 was empirically fit to SiC fiber data for average rupture strength versus rupture time [7]. These data were measured in air at two or more temperatures with a fiber gauge length of 25 mm. Best fit rupture constants are listed in Table I. Since the fitting procedure was found to be insensitive to the C value, the same average value was chosen for all fiber types ($C = 42$ for time = sec). With these constants and rupture model Eqs. 5 and 6, one can closely estimate at a given temperature either SiC fiber average rupture strength for a given rupture time, or the fiber average rupture time for a given applied tensile stress. Currently the model and constants apply only for as-produced individual fibers tested in air under conditions of constant stress, temperature, environment (air), and gauge length (25 mm). The degree of accuracy of the rupture model is illustrated in Fig. 3.

CONCLUDING REMARKS

Although only in its initial stage, this study has demonstrated some useful results regarding the development of fiber thermostructural property models. First, the use of thermal activation theory allows simplification of the time-temperature variables and sheds light on underlying mechanisms. Second, for SiC fibers, the assumption of anelastic theory allows simple correlation of creep and relaxation models and may also serve as a base for fiber property modeling under time-varying test conditions. Finally, the use of thermal activation theory for modeling fiber rupture not only appears successful, but provides a bridge to Larson-Miller models. Clearly, increased efforts are required to develop more general SiC fiber models which can account for all potential CMC service conditions, including time-varying stresses, temperatures, environments, and gauge lengths.

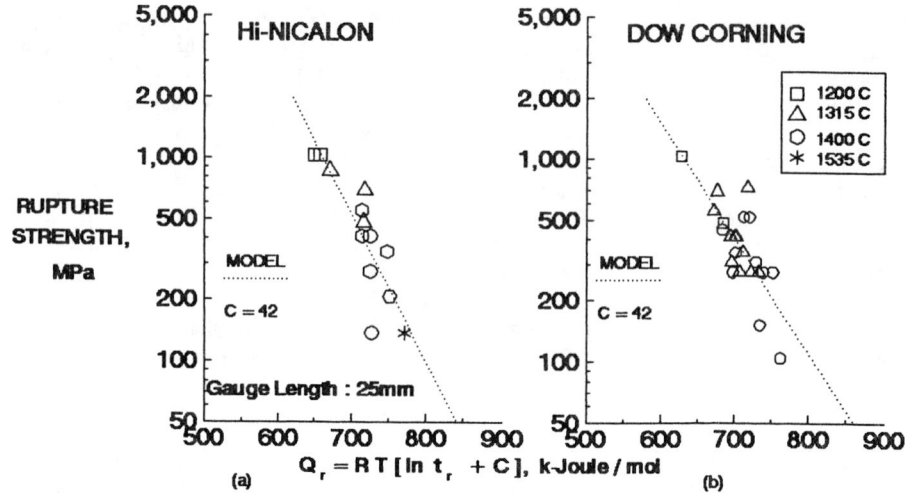

Fig. 3. Model fit to rupture strength data for (a) Hi-Nicalon and (b) Dow Corning Fibers in air.

REFERENCES

1. J.A.DiCarlo and S.Dutta: Continuous Ceramic Fibers for Ceramic Composites. HANDBOOK ON CONTINUOUS FIBER REINFORCED CERAMIC MATRIX COMPOSITES, eds. R.Lehman, S.El-Rahaiby, and J.Wachtman, Jr., CIAC, Purdue University, West Lafayette, Indiana, 1995.
2. R.E.Tressler and J.A.DiCarlo: Creep and Rupture of Advanced Ceramic Fiber Reinforcements. Proceedings of HT-CMC-2, Santa Barbara, 1995.
3. H.I-Lieh Huang, O.D.Sherby, and J.E.Dorn: Activation Energy for High Temperature Creep of High Purity Aluminum. Transactions AIME, J. of Metals, Oct., 1385-1388 (1956).
4. A.S.Nowick and B.S.Berry: ANELASTIC RELAXATION IN CRYSTALLINE SOLIDS. Academic Press, New York, 1982.
5. J.A.DiCarlo: Creep of Chemically Vapour Deposited SiC Fibers. J. Mater. Sci., 2, 217-224 (1986).
6. J.A.DiCarlo and G.N.Morscher: Creep and Stress Relaxation Modeling of Polycrystalline Ceramic Fibers. NASA TM 105394 (1994).
7. J.A.DiCarlo, H.M.Yun, and J.C.Goldsby: Creep and Rupture Behavior of Advanced SiC Fibers. Proceedings of ICCM-10, Vancouver, 1995.
8. H.M.Yun, J.C.Goldsby, and J.A.DiCarlo: Environmental Effects on Creep and Stress-Rupture Properties of Advanced SiC Fibers. Proceedings of HT-CMC-2, Santa Barbara, 1995.
9. J.A.DiCarlo: Creep Limitations of Current Polycrystalline Ceramic Fibers. Composites Sci. and Tech., 51, 213-222 (1994).
10. G.N.Morscher and H.Sayir: Bend Properties of Sapphire Fibers at Elevated Temperatures; I: Bend Survivability. Mat. Sci. and Eng. A190, 267-274 (1995).
11. G.N.Morscher and J.A.DiCarlo: A Simple Test for Thermomechanical Evaluation of Ceramic Fibers. J.Am. Ceram. Soc. 75[1], 136-140 (1992).
12. G.N.Morscher, C.A.Lewinsohn, C.E.Bakis, and R.E.Tressler: A Comparison of Bend Stress Relaxation and Tensile Creep of CVD SiC Fibers. To be published in J. Amer. Cer. Soc.
13. G.N.Morscher, H.M.Yun, and J.C.Goldsby: Bend Stress Relaxation and Primary Creep of an Alpha-SiC Fiber. PLASTIC DEFORMATION OF CERAMICS, eds. R.Bradt, C.Brooks, and J.Routbort, Plenum Publishing, New York, 1995.
14. J.B.Conway, NUMERICAL METHODS FOR CREEP AND RUPTURE ANALYSIS. Gordon and Breach, New York, 1967.

THE MACHINING OF CMC MATERIALS USING YAG LASER.

I P Tuersley, A P Hoult and I R Pashby.
Warwick Manufacturing Group, University of Warwick, Coventry CV4 7AL UK.

ABSTRACT.

Three different Ceramic Matrix Composites (CMC) have been subjected to a systematic series of processing trials using a pulsed Nd-YAG laser, allowing an analysis to be made of some of the most influential factors. By careful comparison of the results obtained from the *specific* cases, trends are identified that enable conclusions to be made about the processing of CMC systems in general.

INTRODUCTION.

The well-documented problems associated with the use of monolithic ceramic materials, in terms of low fracture toughness and catastrophic failure mode, have been addressed in recent years by the development of a number of Ceramic Matrix Composite (CMC) systems. Their advantages over the monolithic compositions usually depend on the specific application, but the problems of cutting and shaping ceramics to produce component forms has, if anything been exacerbated by the combination of fibre and matrix phases that may possess different physical and chemical properties. Much of the need to develop a novel means of machining ceramic materials lies in the fact that despite significant advances in the material microstructural systems, the production of component forms is still largely limited to diamond grinding. This is a costly, time-consuming and inevitably damage-inducing process, and can therefore only be considered to present a partial solution to the problem. There is an obvious benefit in the use of non-contacting processing methods, for example Electro-Discharge Machining (EDM) and lasers. The high hardness and modulus of ceramic materials generate extremely high normal forces in any conventional cutting tool, leading to either unacceptably high tool wear rates or worse, component failure. EDM as a technique is restricted to materials with an electrical resistivity of less than 100Ωcm, which precludes many engineering ceramics. The use of lasers does not impose such general limitations on materials; the principal benefits including the wholly non-contacting nature of the process (no tool wear, no mechanical stresses imposed on workpiece), and the fact that it is capable of sophisticated geometrical manipulation using CAM techniques. The main problem stems from the extreme thermal input of the process and the behaviour of materials that are generically refractory in nature.

There is therefore a great deal of interest in the factors that may contribute to the successful processing of CMC materials using lasers. The approach that has been adopted in this work is to specify a single *type* of laser (to limit the incident radiation to a single wavelength and thus the material/laser coupling considerations to a single case) and to systematically investigate the

---

To the extent authorized under the laws of the United States of America, all copyright interests in this publication are the property of The American Ceramic Society. Any duplication, reproduction, or republication of this publication or any part thereof, without the express written consent of The American Ceramic Society or fee paid to the Copyright Clearance Center, is prohibited.

processing of three different material compositions that between them may be considered representative of the range of CMC systems. The YAG laser may not be the most appropriate type of laser for each material case but the intention is to permit comparison across the range of CMCs such that the factors contributing to any determined results may be applied to other compositional variations.

MATERIALS STUDIED AND LASER CONFIGURATION.

The three examples studied all use the same fibre reinforcement, Nippon Carbon's Nicalon™ 201-grade silicon carbide, reflecting the predominance of SiC fibres in composites that are primarily intended for high temperature applications. The three matrix phases are a borosilicate glass (very similar to commercially available Pyrex™, producing a *Glass*-Matrix Composite, GMC), a Magnesium-Alumino-Silicate (MAS, Glass-Ceramic Matrix Composite or GCMC) and Chemical Vapour Infiltrated (CVI) SiC (resulting in a true Ceramic Matrix Composite, CMC). The first two materials were prepared by filament winding into unidirectional sheets. Twelve such sheets, when laid-up in alternate 0-90° layers and hot pressed form a plaque 3,5-5mm thick, and thicker or thinner plates may be fabricated by adjusting the number of plies accordingly. In the case of the MAS, the matrix is produced by recrystalizing a glass composition, requiring careful control of the process thermal cycle [1]. The SiC-SiC material was produced by 'filling' a cross-weave mat of the fibre by CVI, as outlined in [2]. Once again, twelve sheets of the cross-weave form a flat plaque approximately 3mm thick.

All of the trial sets have been performed with the Lumonics JK701 Nd-YAG laser. It has been used in the (Low Divergence) LD2 Resonator arrangement, giving high beam quality at increased average power settings. An 80mm focal length cemented achromat was used for the final stage lens. The JK701 has an internal power meter, but this takes its measurement from a point just after the resonator and before the final optic train. An external power meter - a water-cooled Coherent 'Fieldmaster' was used to verify the internal meter's reading and evaluate the power losses occurring in the final optical stages. Positioning of the workpiece was achieved by the use of a Unidex 400 CNC controller connected to a 3-axis workstation.

Oxygen, nitrogen and argon were investigated as choices of assist gas. These were manifolded to allow a maximum supply pressure of approximately 8bar, measured from a point close to the nozzle orifice.

RESULTS OF TEST PROGRAMME.

Whilst the specific results for both the material removal and the surface quality vary significantly among the three separate composites, there are some clear trends that emerge when a full comparison is made.

Initial trials performed a survey of the effects of *peak* and *average* power on the laser's penetrative abilities with the different materials. The various parameters that contribute to the laser pulse profile were varied individually within limits constrained by the tuned design of the resonator. In many cases the effects of the pulse repetition rate were isolated by using single pulses to produce blind holes. These were subsequently sectioned, and examined using optical and scanning electron microscopy. Having determined that the geometry in various instances is consistent, the *depth* of the holes produced was recorded as a measure of the penetration achieved. One set of results, plotted as Peak Power versus Depth of Hole produced, for three different levels of pulse energy, are shown in Figure 1. There is clearly a substantial difference in the response of the three separate materials to the laser, typically a 40-50% increase in penetration obtained with the MAS matrix material when compared with the glass matrix, and an even greater improvement with the SiC/SiC material - in some instances as high as 90%. As all of the materials contain the same fibre reinforcement phase ie SiC fibre, this raises a question concerning the relative 'machinability' of the three matrix phases. The laser emits a coherent beam of light of 1.06μm wavelength; the extent to which various materials absorb and react to such a discrete form of energy differs, and in this context is referred to as 'coupling'. Trials were performed on monolithic samples of the three matrix materials and confirmed that there is a high discrepancy in the coupling characteristics in these cases. The Pyrex material is in fact almost completely transparent to the YAG laser, to the extent that it may not be processed at all as a monolithic material. The laser may couple with impurities in the glass, but as the surrounding material is unaffected this simply results in brittle cracking as the bulk of the glass is unable to accommodate the thermal expansion. The MAS material does couple successfully with the laser, in fact every bit as well as the SiC monolithic. It would seem, therefore that the processing of the Pyrex matrix material is achieved by the conduction of heat from the absorbing SiC fibre phase to the matrix glass. The difference in the penetration of the MAS and the SiC matrix materials may be as a result of better matching of thermal properties in the case of the SiC/SiC composite.

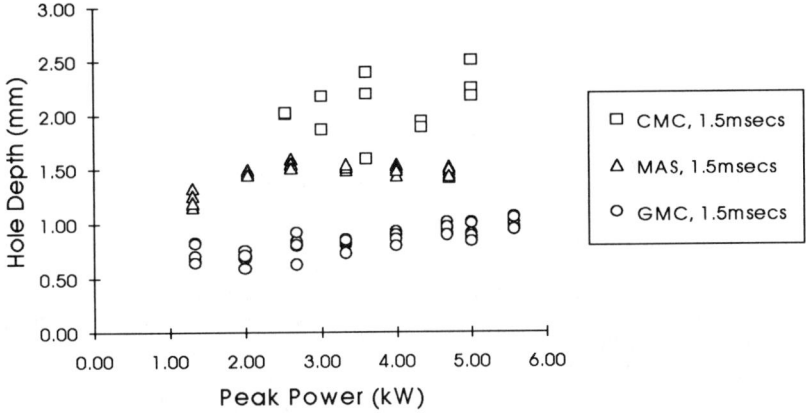

Figure 1. Peak Power vs Hole Depth. Three materials, three pulse energies.

The degree of scatter of the data is significant, typically 10% of the measured penetration. The explanation for this is in the homogeneity of the composite plaques. Measurements made on the first two material compositions after processing give a value of approximately 30-35vol.% for the fibre content, but due to the method of production, local variations are estimated to lie within the range 20-65vol.%. The SiC/SiC material does not have such a high variation as the fibre is supplied as a pre-woven mat, but this material shows a significant amount of porosity (the material is not fabricated under pressure) and once again, this disrupts the homogeneity. As the local fibre content is the determining factor in the processing of the glass matrix material and the laser does not couple with voids in the SiC/SiC composite, these processing variations could well account for the scatter in data.

The cut speed trials show (Figure 2) that as the material thickness increases, the maximum speed of cut for full penetration is reduced, but in a nominally hyperbolic nature rather than linear. This may be due to the heat flow within the plaques during continuous cutting; in the thinner plates, this will have a significant 'pre-heating' effect that may aid the cutting process [3]. The greater degree of coupling of the SiC/SiC material is reflected by an overall increase in the maximum cut speed. It should also be noted that the maximum speed recorded for the thinnest specimens was limited by the repetition rate of the laser; the value taken is the speed at which the cut ceased to be continuous and became a perforated line due to the cut speed exceeding the overlap of the individual pulses. A related series of tests identified very little effect on the cut speed of using different assist gases for any of the materials.

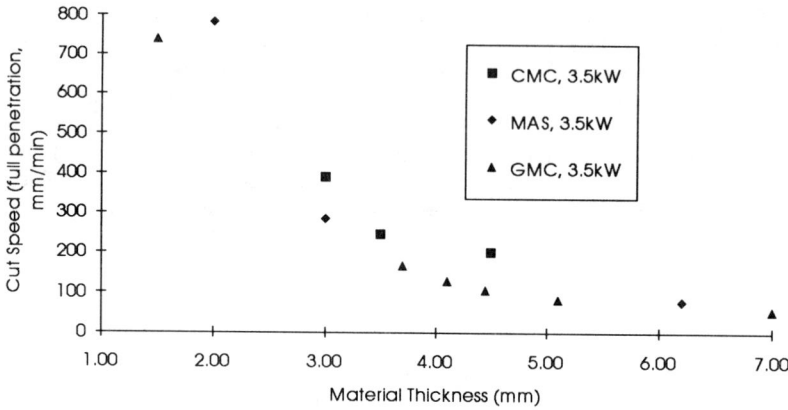

Figure 2. Maximum cut speed for various thicknesses of the three materials.

Four different assist gases were tried, nitrogen, argon, oxygen and air, over a range of pressures, 2-8bar, resulting in substantial differences in the kerf quality. The glass matrix material is very susceptible to oxidative degradation and in all cases a silicate glass 'dross' was produced, particularly evident on the underside of the cut. When oxygen was used (Figure 3), the build-up of this material was sufficient to partially re-seal the cut. The use of nitrogen and argon did reduce this substantially, the resulting cut surface being quite generally acceptable. Compressed air was tried as a cost effective alternative as it had been speculated that the oxygen content of the air may get absorbed by the most reactive molten/vaporised material and subsequently be ejected from the cut along with the process debris. This would then leave a predominantly nitrogen-rich atmosphere to 'flush' the surface of the bulk material. This was found to be the case when the tests were examined. The air-assisted cut was very similar to the inert gas examples, again producing an acceptable cut that may, depending upon requirements benefit from a 'deburring' operation to remove the remnants of the ejected dross.

The cut surfaces of the MAS composite were similar to those of the Pyrex, but to a lesser degree, reflecting the greater stability of the crystallised matrix to an oxidising environment. None of the cuts re-sealed, but there were still significant deposits of alumino-silicate glass on the processed surface when oxygen was used. Once again, the use of air proved to be an economic alternative to nitrogen or argon, with only a slight drop in cut quality.

The results obtained from the SiC/SiC material were a significant improvement. The nitrogen (Figure 4), argon and air produced particularly clean cuts (with, as before, a slight degradation when using air), but even the oxygen cut surface did not show excessive silicate deposits. It should be mentioned that in the case of this material, there may be an advantage to generating a thin layer of glass. The material depends upon the bonding of the matrix to the fibres for many of its physical and mechanical properties, and this is enhanced by microstructurally engineering a deposited layer of pure carbon on the surface of the fibres [4,5]. This sub-micron layer is *particularly* susceptible to oxidation, and so the material is usually sealed by a layer of glass as a final manufacturing step to prevent the ingress of oxygen to the interfacial layer. Cutting the material exposes the carbon layer, and so the generation of a silicate glass (if it could be shown to inhibit diffusion of atmospheric oxygen at high temperatures) could be beneficial.

Figure 3. Heavily oxidised kerf surface, GMC material, 3bar O_2 gas assist

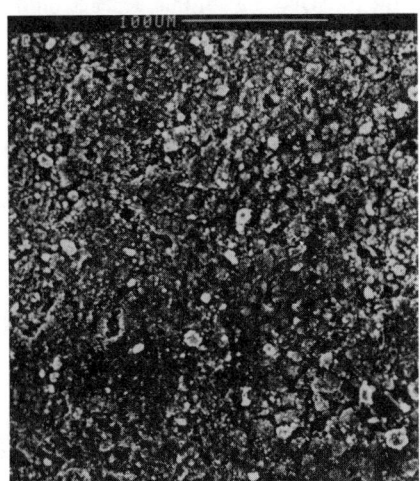

Figure 4. Processed surface of SiC/SiC material using 3bar N_2 gas assist.

CONCLUSIONS

The use of a pulsed, Nd-YAG laser to process three related but significantly different glass and ceramic matrix composites has been demonstrated. Consideration has been given to both the laser's abilities to penetrate the material with different pulse parameters and the resulting processed surface quality. The material removal rate for a given composite is primarily dependant on the average power which may be limited by the tuneable condition of the laser. To produce a continuous, through cut using a pulsed laser it is necessary for the individually pulsed holes to (at least partially) overlap; failing this a perforated line is produced. With factors such as the beam diameter and intensity having discrete maxima, there is a limit to the speed at which the material may be cut, determined by the maximum pulse repetition rate permissible at any tuned condition.

The material itself is a significant factor to the rate of material removal. Certain structures are simply transparent to the laser energy's wavelength of $1.06\mu m$. This is the case with certain borosilicate (Pyrex) compositions. The effect of this has repercussions for the quality of the processed surface as well, as the removal of these phases depends upon the conduction of heat from a phase that *does* absorb the laser energy. As this is necessarily achieved with a limited efficiency, the cut surface shows more evidence of redeposited and molten material. Fortunately, silicon carbide (which is most commonly used for fibres in these composites) absorbs this wavelength particularly well. The coupling efficiency of the constituent phases is therefore an influential factor.

Although the type and pressure of the gas assist do not seem to significantly influence the rate of material removal, it does have an overriding effect on the quality of the processed surface. Most of the materials are, to varying degrees, subject to oxidation at high temperatures. In the case of silicate compositions, the use of oxygen as an assist gas is to be avoided. Large volumes of silicate glass are generated, which may partially or completely re-seal the cut. More acceptable results are obtained from the use of nitrogen or argon (ie gases with which the material does not readily react at high temperatures), but in the cases tried, air may be recommended as an economic alternative. Once again, this depends on the specific requirements of the process, as there is a slight degradation in the cut quality. Of the materials tested, the most promising results were obtained from the SiC/SiC composite material. Despite the porosity shown by this composite, it couples with the laser energy more completely than either the Pyrex or MAS matrices, and it has significantly less susceptibility to oxidation at high temperatures, resulting in a particularly clean cut surface.

ACKNOWLEDGEMENT.

Financial support for this work, part of an investigation into the machining of advanced composites using traditional and laser techniques has been gratefully received from collaborating industrial partners and the UK DTI under the MTTMP programme

REFERENCES.

1. A Chamberlain, M W Pharaoh & M H Lewis; *"Novel Silicate Matrices for Fibre Reinforced Ceramics."* in Proc. 17th Annual Conf. on Composites and Advanced Ceramic Materials (Cocoa Beach, Fl. USA) Jan. 1993 Vol.14 Nos.9-10, pp939-946.
2. D P Stinton, T M Besmann & R A Lowden; *"Advanced Ceramics by Chemical Vapour Deposition Techniques"*, Cera. Bull., Vol.67 No.2, 1988 (A.Cer.Soc.) pp350-355.
3. Y Arata; *"Some Fundamental Properties of High Power Laser Beam as a Heat Source (Report 3) - Metal Heating by Laser Beam."* in 'Plasma, Electron & Laser Beam Technology', ASM (Carnes Publication Inc.) USA 1986 pp245-262.
4. T-I Mah, M G Mendiratta, A P Katz & K S Mazdiyasni; *"Recent Developments in Fibre-Reinforced High Temperature Ceramic Composites."* Cera. Bull., Vol.66 No.2 1987 (A. Cer.Soc.) pp304-308.
5. K M Prewo; *"Fibre-Reinforced Ceramics: New Opportunities for Composite Materials."* Cera. Bull., Vol.68 No.2 1989 (A.Cer.Soc.) pp395-400.

TEM AND EDX INVESTIGATIONS OF EXPERIMENTAL SiC_f - YMAS COMPOSITES

F. DOREAU, J. VICENS and J.L. CHERMANT
LERMAT, URA CNRS n°1317, ISMRA,
6 Boulevard du Maréchal Juin, 14050 CAEN Cedex, France

ABSTRACT

This paper deals with microstructural characterization which is of prime importance to understand and thus control mechanical behavior of composites. It appears that SiC_f-YMAS composites exhibit a totally crystallized matrix formed by two major phases (cordierite and yttrium silicate). The fiber/matrix interface is made of a carbon sublayer and a transition layer at the fiber side.

INTRODUCTION

Ceramics are potential materials for high-temperature applications. Unfortunately they are too brittle and not tough enough. So ceramic matrix composites (CMCs) have been developed. CMCs with a glass-ceramic matrix are investigated for applications up to 1273K. Matrices of such materials are generally complex [1]. An extensive study by transmission electron microscopy has to be performed to determine the different phases of the matrices. Furthermore, an accurate knowledge of the fiber/matrix interface is of prime importance to understand the mechanical behavior and to establish relationships between microstructure and properties for these composites.
In this paper the results of recent investigations in transmission electron microscopy (TEM) on a SiC_f-YMAS glass-ceramic composite developed by Céramiques & Composites (Bazet) and ONERA (Palaiseau) in France are described. After a short presentation of the material and of its fabrication, results concerning the matrix phases are presented and particular attention will be given on the study of the interfacial zone between fiber and matrix. Several "interphases" are observed and the results of their analyses are discussed.

MATERIALS AND EXPERIMENTAL PROCEDURE

The SiC_f-YMAS composites were fabricated from SiC Nicalon NLM 202 fibers (Nippon Carbon, Tokyo) and an aluminosilicate of magnesium and yttrium for the matrix prepared by oxide melting (SiO_2: 36%, Al_2O_3: 34.4%, MgO: 9.6%, Y_2O_3: 20%, in weight). After desizing fibers were impregnated by a slurry containing the matrix and organic binders. They were wound on a hexagonal mandrel which gives 6 prepreg sheets after cutting. Eight plies were stacked and hot-pressed at ~ 1573K. After pressing the dimensions of the composite plates were 300 x 300 mm^2, with a thickness close to 3.5 mm. Porosity was less than 1%. The fiber content was measured by automatic image analysis and found to be 43% with a standard deviation of 6%. The average value of the fiber diameter was 14.4 μm with a relatively narrow distribution for this NLM 202 fiber batch. Finally the fiber distribution inside the plate was seen fairly homogeneous.

For transmission electron microscopy studies, disc-shaped samples were mechanically ground, dimpled up to 10 μm thickness and thinned by ion-milling (Ar^+, 6KV). A Jeol 200 CX (200 KV) and a Topcon EM 002B (200 KV, Cs = 0.4 mm) electron microscopes equipped with a Kevex microanalyser have been used.

TEM AND EDX ANALYSIS

Matrix microstructure

An image in bright field of the matrix is shown in figure 1a. As confirmed by electron diffraction studies, the matrix appears totally crystallized into different phases forming a complex arrangement. TEM observations reveal the existence of two main phases (dark and white). The dark crystals have an elongated and dendritic shape. This may indicate that this phase first crystallizes from the glass.

The chemical composition of these two phases has been determined by EDX analyses. The white and dark crystals were identified respectively as a magnesium aluminosilicate $Mg_2Al_4Si_5O_{18}$ (cordierite) and yttrium disilicate $Y_2Si_2O_7$. Spectra of these two main phases (taken in the areas arrowed in figure 1a) are shown in figures 1c and 1d.

Secondary phases have been also observed (fig. 1f and 1g) and identified by EDX as spinel $MgAl_2O_4$ (fig. 1b) and mullite $Al_6Si_2O_{13}$ (fig 1e). Several crystals of ZrO_2 and TiO_2 are often observed inside the cordierite crystals (ZrO_2) or at the fiber/matrix interface (TiO_2). These phases are added as nucleating agents during the fabrication of the composite.

The crystallographic structure of the two main phases has been determined by electron and X-ray diffractions and identified respectively as the hexagonal polymorph, a = 0.977 nm, c = 0.9352 nm, space group P6/mcc, for the cordierite phase (so-called indialite) and the monoclinic polymorph, a = 0.6875 nm, b = 0.8970 nm, c = 0.4721 nm, β = 101.74°, space group C2/m, for the yttrium silicate phase (so-called keiviite). The observation of the high temperature form of cordierite and yttrium silicate is in good agreement with the process temperature during the hot pressing cycle (~1573K).

Fiber/matrix interface

A bright field image of an interface between a Nicalon NLM 202 fiber and the matrix is presented in figure 2. Two distinct nano-scale sublayers are often observed at the fiber/matrix interface. In figure 2, both sublayers are continuous and the sublayer on the matrix side has a bright contrast while the one on the fiber side is dark. The average thickness is respectively about 80 nm for the dark and about 100 nm for the bright sublayer.

The chemistry of the sublayers has been determined from EDX analyses by stepping the probe (8.8 nm in size) across the fiber/matrix interface for the two cases. In the case of yttrium silicate/ fiber interface, spectra taken in the fiber (Fig. 3-1), in the dark sublayer (Fig. 3-2), in the bright sublayer (Fig. 3-3) and in the yttrium silicate crystal (Fig. 3-4) are displayed. The bright sublayer close to the matrix is carbon rich. Silicon and oxygen are also detected in the carbon layer with a small amount of magnesium, aluminum and yttrium which have diffuse from the matrix. The exact chemical nature of the oxide compounds has not yet been established (i.e. SiO_2 or SiC_xO_y).

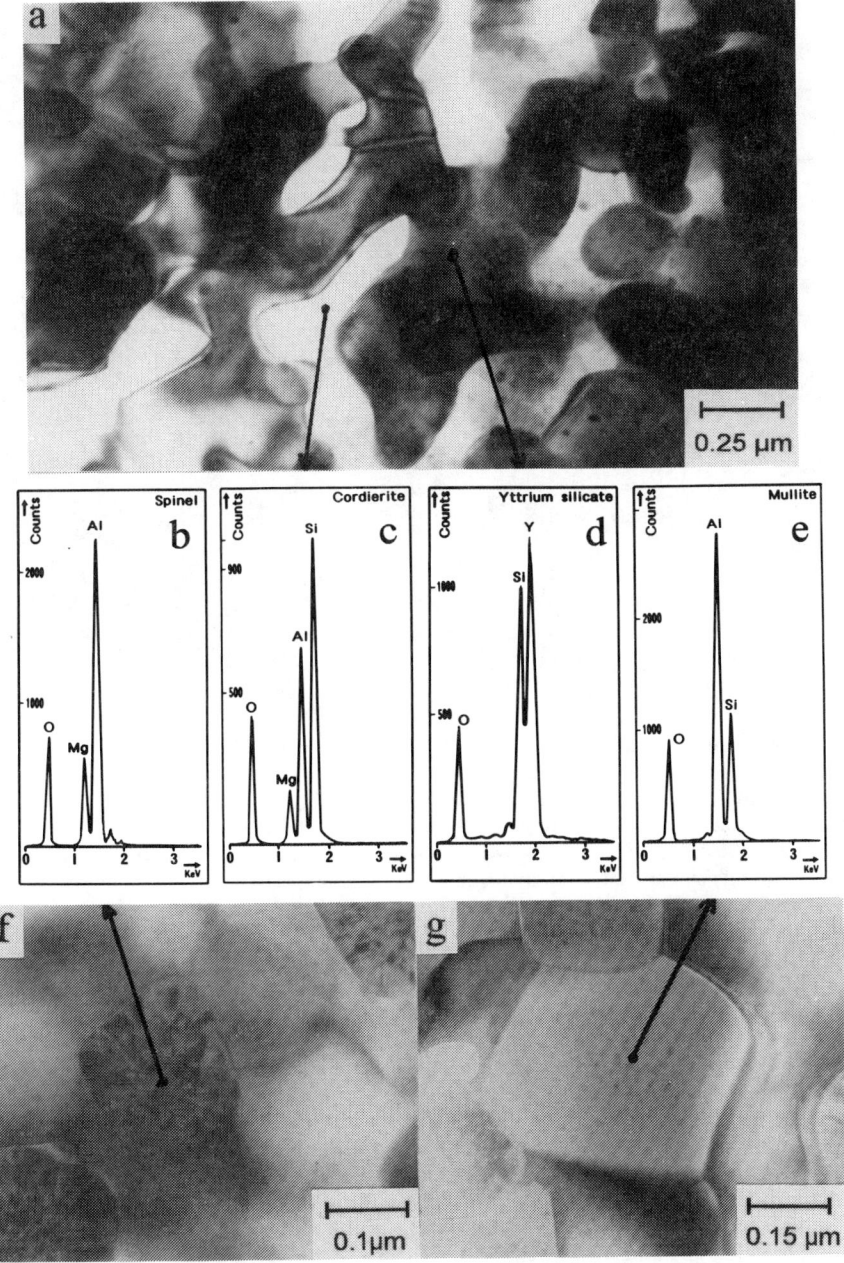

Figure 1: Matrix microstructure and results of the microanalyses.

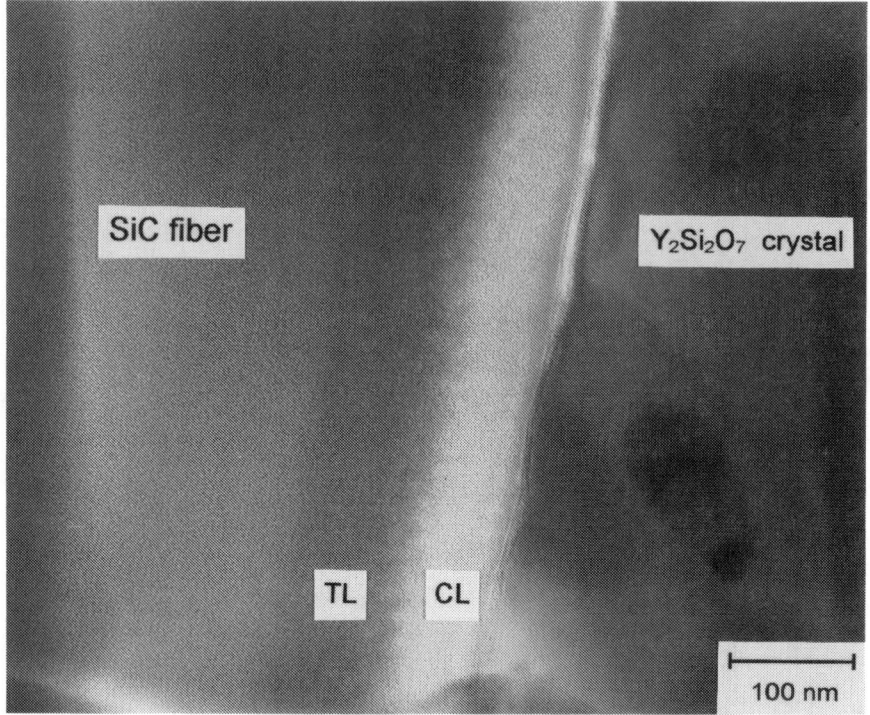

Figure 2 : Fiber/matrix interface (CL: carbon layer, TL: transition layer).

Compared to the spectra of the Nicalon fiber $((O/Si)_{at\%} = 0.43)$, the dark sublayer is characterized by a large amount of oxygen and significant quantities of magnesium and aluminum with O/Si, Mg/Si, Al/Si respectively equal to 0.75, 0.013, 0.072. The yttrium atoms do not diffuse into this sublayer. The enrichment in oxygen, magnesium and aluminum is even higher in the case of a cordierite/fiber interface.

Finally it appears that crystals of yttrium silicate and cordierite in contact with the carbon layer exhibit an enrichment in silica. Y/Si and Al/Si ratios are respectively equal to 0.6 and 0.2 at the interface. These ratios are equal respectively to 1 and 0.8 in the yttrium silicate and in the cordierite. In the case of a carbon/cordierite interface, the Al/Si ratio slowly increases from the interface to the core of the cordierite crystal.

In the case of the cordierite/fiber interface, only spectra taken in the cordierite crystals (Fig. 3-5 and 3-6) are displayed.

A transition layer (100 nm thick) between the carbon layer and the fiber has already been observed by Ponthieu et al. [2] and Lancin et al. [3] in SiC_f-LAS and SiC_f-LAS with Nb_2O_5 used as a diffusion barrier. This sublayer has been observed by HREM and has been characterized by the presence of SiC nanocrystals with a lower density than in the fiber core. EDX analyses showed also that an oxygen peak is observed close to the interface between the carbon layer and the transition layer.

The presence of a carbon sublayer on the matrix side is also confirmed by HREM observations performed on the same SiC_f-YMAS material. A thin carbon turbostratic layer (50 nm) has been

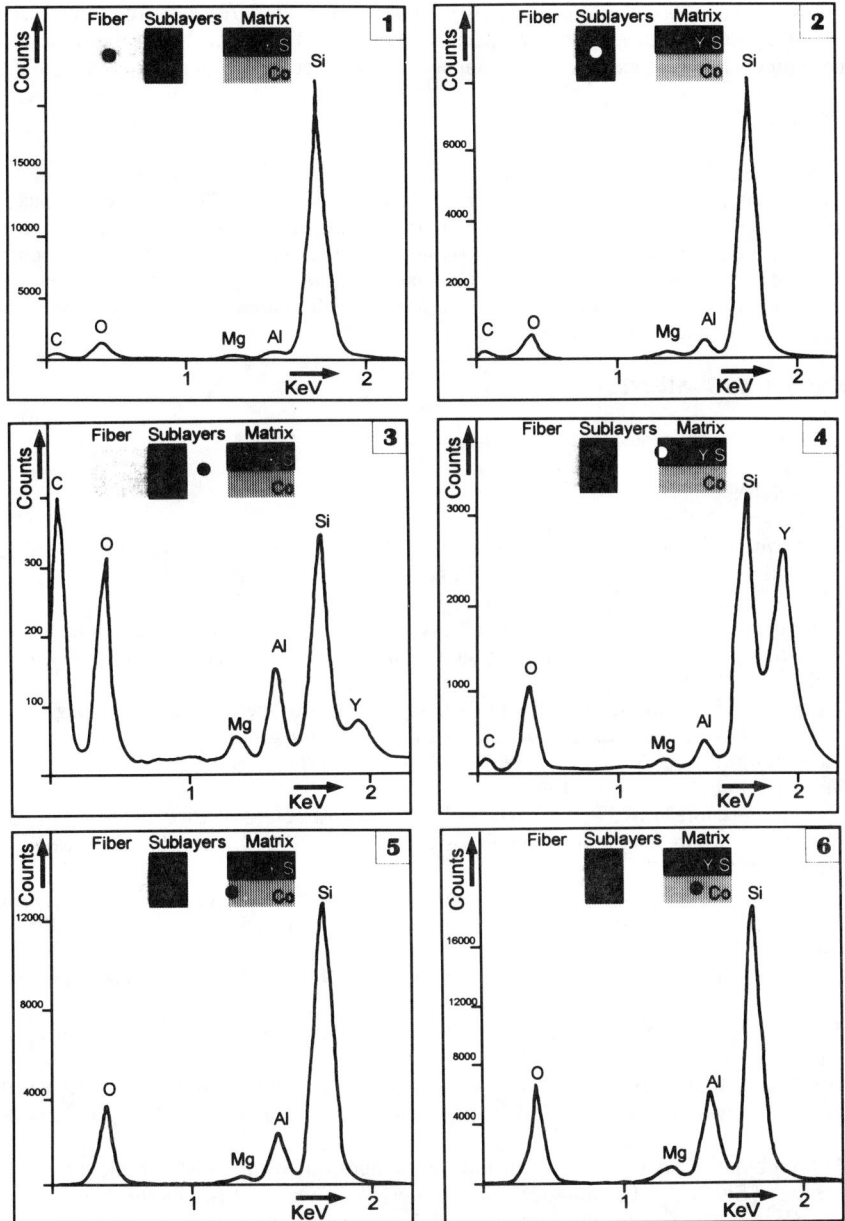

Figure 3 : Chemical composition of the different layers of the fiber/matrix interface. The point displayed on the drawing of the interface indicates the location of the probe, YS corresponds to yttrium silicate and Co to cordierite.

imaged with some (0002) basal planes (~10) parallel to the matrix/ carbon interface[4, 5]. The carbon layer arise from the reaction between the matrix and the fiber during the process, as it has been previously described for other composites reinforced with Nicalon fibers [2, 6-10].

CONCLUSION

The SiC_f -YMAS composite exhibits a totally crystallized matrix, formed by two major refractory phases: the cordierite and the yttrium disilicate. At the fiber/matrix interface, two sublayers are observed. One of them is rich in carbon and results on a chemical reaction between the oxygen rich species of the matrix and the fiber. The other one is a transition layer between the carbon sublayer and the fiber. Compared with the fiber, it corresponds to an oxygen rich zone with noticeable amounts of magnesium and aluminum. Finally, an enrichment in silica is observed at the carbon/matrix interface.

ACKNOWLEDGEMENT

This research was sponsored by DRET, under contract n° 92/1221 A.

REFERENCES

1. F. Doreau, H. Maupas, D. Kervadec, P. Ruterana, J. Vicens and J.L. Chermant, "The complexity of the microstructure in SiC fibers reinforced glass-ceramic matrix composites", *submitted to J. Europ. Ceram. Soc. 1995.*
2. C. Ponthieu, M. Lancin and J. Thibault-Desseaux, "Comparative study of the microstructure of the interphases in two SiC fibre/lithium alumino-silicate matrix composites", *Phil. Mag.* **A 62**, 605-615, 1990.
3. M. Lancin and C. Marhic, "Analysis by EDX and EELS of the interphases in glass or glass ceramic matrix composites", *Proceedings of the 13th International Congress on Electron Microscopy (Paris, July 17-22, 1994), ed. by B. Jouffrey and C. Colliex - Editions de Physique* **vol. 2B**, 841-842, 1994.
4. J. Vicens, F. Doreau and J.L. Chermant, "The microstructure of experimental SiC fibers-reinforced yttrium magnesium alumino-silicate (SiC_f -YMAS) materials", *J. Microscopy* **177**, 242-250, 1995.
5. F. Doreau, F. Gilbert, J. Vicens and J.L. Chermant, "On the creep behavior of the SiC_f - YMAS", *Proceedings of HT-CMC 2, Santa Barbara, Cf, USA, Aug. 21-24, 1995.* This volume.
6. R. Chaim and A.H. Heuer, "The interface between Nicalon SiC fibers and glass-ceramic matrix", *Adv. Ceram. Mater.* **2**, 154-158, 1987.
7. J.J. Brennan, "Interfacial characteristics of glass ceramic matrix/SiC fiber composites", *J. Phys. Colloque* **49 C5**, 790-809, 1988.
8. J. Homeny, J.R. Vanvalzah and M.A. Kelly, "Interfacial characterization of silicon carbide fiber / lithium-alumina-silica glass matrix composite", *J. Am. Ceram. Soc.* **73**, 2054-2059, 1990.
9. M. Monthioux, "Nano et microstructure de composites SiC/LAS-M - Nano and microstructure of SiC/LAS-M composites", *Rev. Comp. Mat. Avancés* **3 HS**, 69-90, 1993.
10. R.F. Cooper and K. Chyung, "Structural and chemistry of fibre/matrix interfaces in silicon carbide fibre reinforced glass ceramic composites: an electron microscopy study", *J. Mat. Sci.* **22**, 3148-3160, 1987.

Ceramic-matrix composites by a preceramic polymer route

Takeshi Isoda* and Takemi Yamamura**
*TONEN Corporation, Corporate Research & Development Laboratory, Nishitsurugaoka 1-3-1,Ohi-machi, Iruma-gun, Saitama-ken 356, Japan
**Ube Industries, Ltd. & Research Institute of Advanced Material Gas-Generator, Koishikawa IS Building, 4-2-6 Kohinata, Bunkyo-Ku, Tokyo 112, Japan

Abstract

High strength ceramic matrix composites(CMC) were obtained by two groups independently by a preceramic polymer-impregnation & pyrolysis (PCPI) method. TONEN improved a thermosetting infiltrate, Methylhydorocooligosilazane to result in three-point flexural strengths as high as 1250, 627 and 503 MPa for the UD, 2D and multi-directional Woven composites, respectively. Ube consolidated polytitanocarbosilane(PTC) as matrix precursor for CMC using a 3-D woven fabric of Si-Ti-C-O fiber. Tensile strength and ultimate strain to failure of PTC-CMC at room temperature reach 410 MPa and 0.85%, respectively. Tensile test at 1200 °C shows a possibility of maintaining the room temperature strength. The PCPI method should provide high performance CMC components, if the oxidation stability would be further improved.

Introduction

The excellent heat-resistant characteristics of non-oxide ceramic materials allow combustion engines to be driven at higher temperatures, leading to substantial improvement in thermal efficiency. Conventional monolithic ceramics have excellent mechanical strengths but also have disadvantages such as insufficient fracture toughness and sensitivity to thermal shock. By an addition of continuous fiber into a matrix as minimizing the damage of fiber, fiber-matrix interface and fiber geometric alignment during processing, the ceramic material acquires higher fracture energy, yielding damage tolerance and good thermal shock resistance.

Preceramic polymer-impregnation & pyrolysis method (PCPI) has been studied by many researchers and engineers because of the application of fabrication technologies for organic polymer matrix composite and lower processing temperatures during polymer-to-ceramic conversion than those during sintering of ceramic powder. However little successes have been reported to fabricate ceramic matrix composite (CMC) with strengths as high as those of CVI SiC CMC and glass matrix composites.

TONEN and UBE have demonstrated independently that PCPI method is potential to minimize fabrication cost and time and maximize the reinforcement by continuous fibers. Research activity at TONEN during these years resulted in the development of a new polysilazane, Methylhydorocooligosilazane (SNC)(1), for PCPI. UBE developed a 3-Dimensional (3-D) woven fabric of Si-Ti-C-O fiber suitable for high performance CMC. The consequence of these developments is that it enables the exploitation of very attractive CMC for high temperature applications.

The objective of the present paper is to describe a fabrication method for producing CMC through PCPI and the excellent properties of the products. The PCPI method and the raw materials should provide high performance CMC which may be applicable to combustion engines, if the oxidation resistance of the interface between fibers and matrix will be improved.

Experimental Procedure

1. Materials
1.1 Preceramic Oligomer and Polymer

SNC, a thermosetting oligomer, was synthesized by coammonolysis of the mixture between dichlorosilane (SiH_2Cl_2) and dichloromethylhydrosilane ($SiHCH_3Cl_2$) in pyridine. Typical data of SNC are shown in Table-1 and Figure-1 and 2.

Polytitanocarbosilane (PTC), the thermofusible precursor of Si-Ti-C-O fiber, was also used for PCPI. PTC was prepared in the same manner as described elsewhere (2-4) and its unit structures are shown in Fig. 3. Phenolic resin (PR) was also used for PCPI method in order to modify the interface between fibers and matrix.

1.2. Reinforcement materials
1.2.1 Fiber and Fabrics

Table-2 shows the properties and compositions of fibers used by TONEN for fabrication of SNC composites(5). Tensile strengths of the fibers were measured by the method JIS R 7601. The surfaces of fibers were coated with a mixture of pyrolytic carbon and silicon carbide by chemical vapor deposition (CVD) in the $SiCl_4$-CH_4 system. The thickness of the coated layer was about 0.1μm. Plain and multidirection weaves were also coated with carbon and silicon carbide multilayers by CVD by Mitsui Engineering and Shipbuilding Co.

A reinforcing fabric used for PTC-CMC is made of a new Si-Ti-C-O fiber (Tyranno Lox M, Table-3) which shows a higher strength around 1300°C in air due to its lower oxygen content of 13 wt.% than that found in the ordinary S-type Tyranno Fiber (the oxygen content; 18wt%). The reinforcement geometry of the orthogonal 3-D woven fabric is featured as shown in Fig. 4. The filament number ratio in the x, y, z direction is 1: 1: 0.13. The xy plane coincides with the plate plane and so the z axis does with the translaminar direction. A volume fraction of fibers, Vf ($Vf = vf/(vf + vm)$)) to the occupied volume of the fabric is about 40 %. The nominal size of the woven preform in mm is 240x 150x 6.

2. Fabrication of Ceramic Composite by Preceramic Polymer Impregnation and Pyrolysis Method (PCPI)

The unidirectional(UD) fiber-aligned SNC-CMC was fabricated in the same procedure as presented elsewhere in this volume. The UD prepregs each with a thickness of 0.2mm were cut and stacked into a 2-D structure of [0/90/90/0]2. Plain weave cloth tape was infiltrated with the precursor, pressed, cured, cut and formed in laminates. The stacked 2D (cross ply or plain weave laminate) preform was then treated in the same manner as UD preform to prepare a heat-treated preform. To obtain dense composites, the heat-treated preforms were then impregnated in a repeated manner with an undiluted pure SNC at a pressure ~0.1MPa, cured at ~250°C for 1-3 hours in N2 in a pressurized vessel, then pyrolyzed in a N2 atmosphere to 1350°C. The densification of the materials in the repeating processes was monitored by measuring bulk densities. A typical densification curve is shown in Fig- 5. Time length required for one cycle from the start of impregnation to the end of pyrolysis was 24 hours or less.

Impregnation of 3-D woven Si-Ti-C-O fabric has been done with a dilute organic solvent solution of PR or PTC By Ube. Then pyrolysis was carried out at 1300°C in argon. The PCPI process was repeated 8 times (2 times for PR and then 6 times for PTC) to obtain CMC with a density of 1.76 g/cm3 containing 20 vol% of voids. CVI method(6) using silane gas was also applied for matrix consolidation according to CERASEP method in order to compare with PCPI.

3. Characterization and Test methods

The products SNC-CMC were characterized by SEM and XRD, and three point flexural tests such as a room-temperature-test according to JIS R 1601, an inter-laminar shear strength test and an elevated temperature-test according to JIS R 1604. Open porosity was estimated by water absorption. The SNC-CMC specimens for flexural and tensile test were cut from a plate of a composite material consisting of approximately 50% of fibers uniaxially aligned in a matrix. All specimens were oriented with the fibers parallel to prospective applied tension. Fig-6 shows the dimensions of the specimens.

The fabricated PTC-CMC plates were sliced into strips with 25mm width and they were milled into the dumbbell shape indicated in Fig-7. Prepared specimens were subjected to tensile tests at ambient temperature and high temperature, 1200°C, in argon, according to methods reported previously (7).

Result and Discussion

Baseline results show that the SNC-CMC exhibits non-brittle and fiber pull-out fracture behavior with fairly good flexural strength at room temperature as shown in Fig-8, Fig-9 and Table-4. The UD composite of SNF that are pyrolyzed more than 8 times at 1350°C in N2 have a good flexural strength of higher than 1000MPa at room temperature. Excellent flexural strengths ($\approx$620MPa) are also obtained in the samples of SNF 2D composite, SNBF 2D composite and low oxygen Si-C-O F plain weave laminate composite. The apparent effectiveness of fiber strength in the composites was also extremely high. This could be explained as follows, i) a high volume fraction of fiber which is easily obtained, ii) the PCPI process which does not affect very much the reinforcement during matrix densification.

The stress (load)-strain relationship for a PTC-CMC specimen obtained by PCPI method is shown in Fig.-10. The strength and ultimate strain at failure reach 410 MPa and 0.85 %, respectively, which are high values in comparison with those ordinary CMC materials using 2-D woven fabric. Fig.-11 indicates the stress (load) - strain relation for CVI materials. The curve of this material shows a high strength of 320 MPa. Initial elastic modulus of the material, 134 GPa, is higher than that of PCPI materials.

In high temperature flexural test, the strengths at room temperature are maintained to 1250°C. There are no obvious differences in strength between SNF composite and SNBF composite, because both fibers would not degrade in present manufacturing temperature and pressure. After heat treatment at 1250°C for 200 hours in N2, the flexural strength of the SNF UD composite is still at 818 MPa which is equivalent to 78% of that of non-heat treated composites.

Two stress-strain relationships at 1200°C are shown in Fig.-12 for PCPI and CVI materials. In the results of PCPI material, the final failure mode is not tensile break in a parallel portion but a share-out at the specimen shoulder, and a partial load is still maintained after the peak value. Tensile strength of PCPI material at 1200°C is, therefore, expected to be over 340 MPa, an average of share-out strength, and presumably close to the ambient value. The strain reaches almost 1.5 % and modulus decreases from 70 to 29 GPa probably due to the combination of the fiber and matrix behavior.

In high temperature oxidation tests at 1250°C in air for 200 hours, the tensile strengths of ceramic fiber-RCC at room temperature were decreased to 100 MPa. There is no fiber-pull out observed in the SEM image of the fracture surfaces due to the oxidation of the interface layer between matrix and fiber. It is thus necessary to protect the interface layer from oxidation or to develop non-oxidizable interface with good matching to both matrix and fiber.

Mechanical strength of PCPI PTC-CMC material decreased gradually during heat-treating under oxidative atmosphere at above 1000°C for more than 100 hr (i.e. at 1200 and 1350 °C, respectively). Oxidative heat resistance and mechanical properties of these types PCPI and CVI materials have been recently improved by Ube Industries, Ishikawajima-Harima Heavy Industries Co., Ltd., and Research Institute of Advanced Material Gas-Generator by using Tyranno Lox M and furthermore Tyranno Lox E (Oxygen content: 5 wt%) fibers with the double layers-gradient structure near the surface and modifying PC's structure and the fabrication process. The detailed results concerning with these improvements are presented elsewhere in this volume.

Conclusion

SNF, SNBF, Si-C-O F and Si-Ti-O-F reinforced silicon nitride composites were obtained by PCPI method. Three-point flexural strength of the UD, 2D and multi-directional woven composites were as high as 1250, 627 and 503 MPa, respectively. This resulted from an improved infiltrate silazane, SNC, which has high fluidity, self-curability, high char yield and best-match to various fibers and interface layers. The CMC, which was composed of 3-D woven fabric of Si-Ti-C-O fiber and a pyrolyzed material of its precursor, polytitanosilane, was successfully fabricated. The tensile strength and ultimate failure strain of the material at room temperature and at a high temperature (1200°C) reach value higher than those of CMC materials obtained by CVI using 2-D woven fabrics.

This PCPI method does not require high mechanical pressure in firing and long processing time to obtain high density materials without size change. It has a great ability for a near-net-shape process. Conclusively, the PCPI method by using SNC, PTC and ceramic fibers as raw materials should provide high performance CMC components that may be applicable to automotive CGT etc., if the oxidation stability can be further improved.

Acknowledgement

The authors are grateful to Dr. Takashi Ishikawa of National Aerospace Laboratory for his helpful guidance and Mr. Masaki Shibuya of Ube Industries Ltd., Messieurs Osamu Funayama, Hiroki Morozumi and Kiyoshi Sato of TONEN and Mr. Teturo Hirokawa of Shikibo Ltd. for their coworks. A part of this work was conducted by the Petroleum Energy Center with a financial support from the Ministry of International Trade and Industry.

Reference

(1) Masaaki Ichiyama, Yuji Tashiro, Tomohiro Kato, Yoshio Kawashima, Koji Okuda, Hayato Nishii, Takeshi Isoda, "Random copolysilazane and producing the same", Japan Kokai Pat. 1991-31326, Feb. 12, 1991
(2) S. Yajima, T. Iwai, T. Yamamura, K. Okamura and Y. Hasegawa, "Synthesis of polytitanocarbosilane and its conversion into inorganic compounds", J. Mater. Sci. 16, 1349-1355 (1981)
(3) T. Yamamura, "Development of high tensile strength Si-Ti-C fiber using an organometallic polymer precursor", Am. Chem. Soc., Polymer Preprints, 25(1), 8-9 (1989).

(4) T. Yamamura, T. Ishikawa, M. Shibuya, T. Hisayuki and K. Okamura, "Development of a new continuous Si-Ti-C-0 fiber using an organometallic polymer precursor". J. Mater. Sci. 23, 2589-2594 (1988)
(5) Takeshi Isoda, "Silicon Nitride fibers", 1993, 161-167, Proceedings of Euro-Japanese Colloquium on Ceramic Fibers in The 6th European Conference on Composite Materials(ECCM-6)
(6) Pierre. J. Lamicq, Gerald. A. Bernhart, Martime. M. Dauchier and Jean. G. Mace, "SiC/SiC Composite Ceramics", Am. Ceram. Soc. Bull., 65, 336- 338 (1986)
(7) T. Ishikawa, T. Yamamura, T. Hirokawa, Y. Hayashi, Y. Noguchi and M. Matushima, " Strength and Fracture toughness properties of oxidation resistant high-temperature ceramic matrix composites", Proc ICCM 9, 2, 137-144 (1993)

Table and Figure

Table 1. Typical properties of methylhydrocooligosilazane (SNC)

Molecular Weight (Mn)	Density (g/cm^3)	Viscosity (mPa.s) at 30°C	Composition(%)		
			Si	N	C
800 - 1000	1.1 - 1.2	40 - 80	56	21.9	14.5

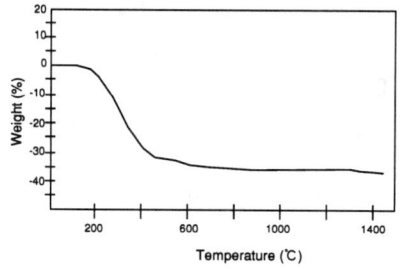

Fig.1 Thermogravimetric analysis of SNC

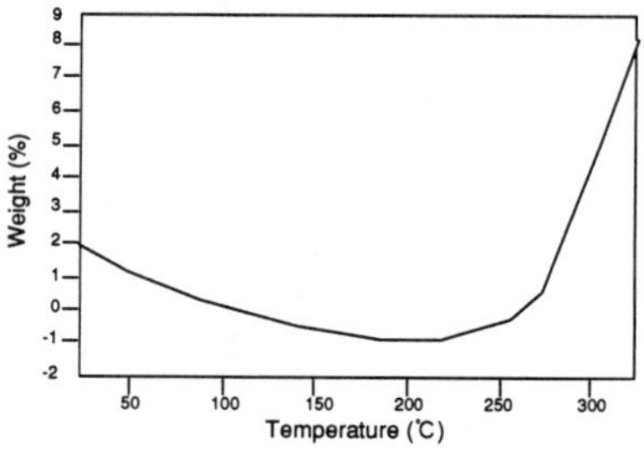

Fig.2. Viscosity vs temperature curve of SNC

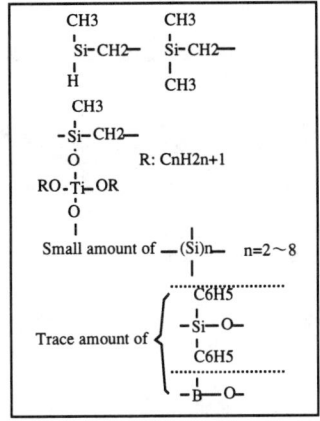
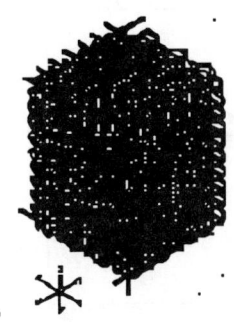

Fig.3 Unit Structures of PTC   Fig. 4 Structure of 3-D Woven Fabric

Table 2. Typical properties of ceramic fibers

Fiber		SNF	SNBF	Low-oxygen Si-C F	Low-oxygen Si-Ti-C-O F
Producer		TONEN	TONEN	Nippon Carbon	Ube Ind.
Tensile Strength (GPa)		2.2	2.2	2.8	3
Tensile Modulus(GPa)		200	200	270	196
Density(g/cm^3)		2.5	2.5	2.7	2.4
Composition	Si	59.6	48	62.4	55
(wt.%)	Ti	-	-	-	2
	B	-	5.5	-	-
	N	37.1	32.8	-	-
	C	0.4	5.4	37.1	38
	O	2.7	7.6	0.5	5

Table3. Physical properties of Si-Ti-C-o Fibers

Composition(wt%)	Si:48 - 57 C:30 - 32	Ti: 2.0,
	O:13 - 18	B:$\leq$0.1
Filament Diameter(μm)		8.5
Filaments/ Yarn		1600
Tex		200
Density at 25°C(g/cm^3)		2.3 - 2.4
Tensile Strength(GPa)		3.0 - 3.6
Tensile Modulus(GPa)		180 - 220
Tensile Strain-to-Failure(%)		1.5 - 2.0
Coefficient of Thermal Expansion(/°C)		3.1 - 3.6x10^{-6}
(along fiber axis, 0 - 500°C)		
Specific Heat(cal/deg/g)		~0.19(20°C)
		~0.28(400°C)

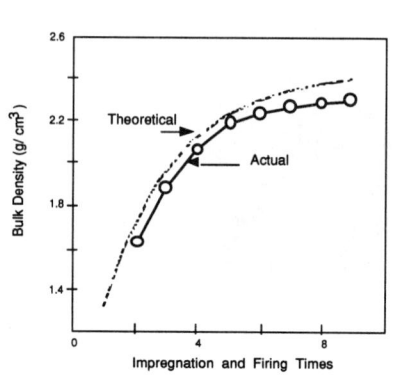

Fig. 5 Densification Curve of UD Composite

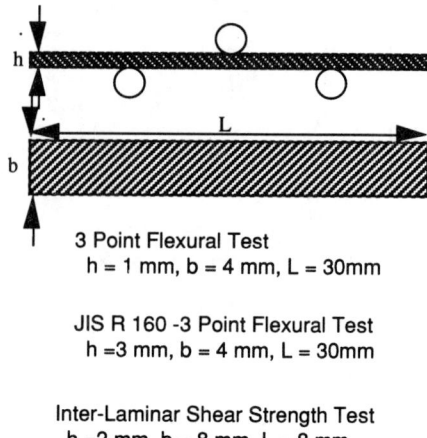

3 Point Flexural Test
h = 1 mm, b = 4 mm, L = 30mm

JIS R 160 -3 Point Flexural Test
h = 3 mm, b = 4 mm, L = 30mm

Inter-Laminar Shear Strength Test
h = 2 mm, b = 8 mm, L = 8 mm

Fig. 6. Specimens for Flexural Tests

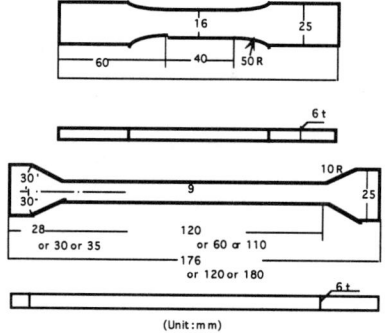

Fig. 7 Shape and Dimensions of Specimens at Ambient and High Temperature Test

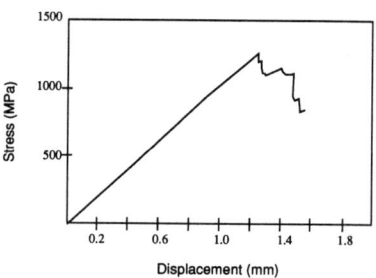

Fig. 8. Stress-Displacement Curve of UD SNF Composite

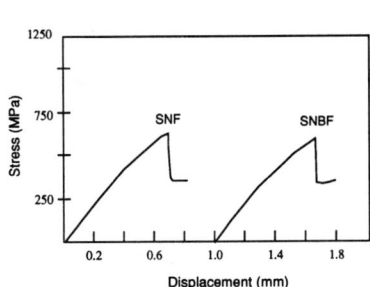

Fig. 9. Stress-Displacement Curves of 2D SNF Composite and 2D SNBF Composite

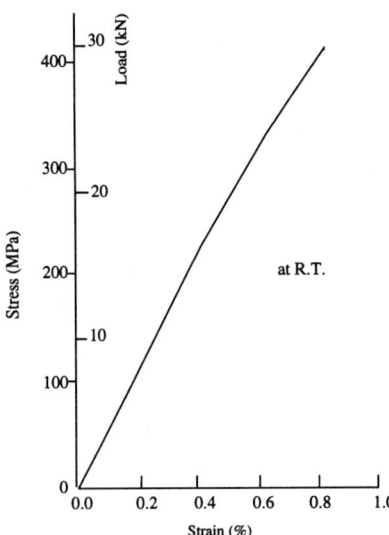

Fig. 10 Stress (Load) - Strain Relation for PCPI Material

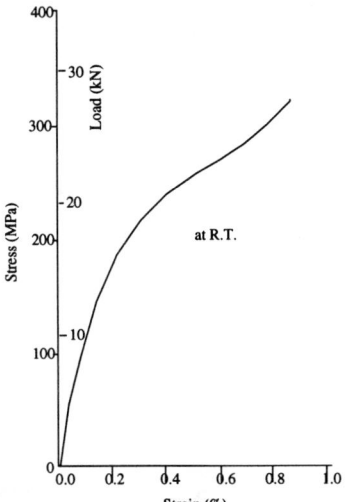

Fig. 11 Stress (Load) - Strain Relation for CVI Material

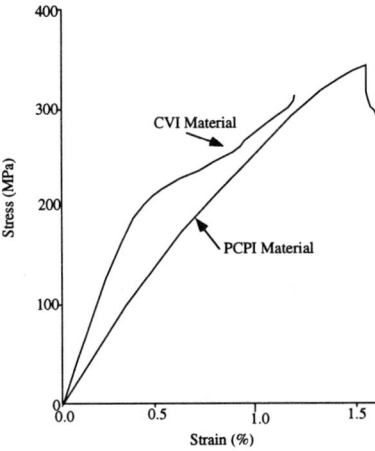

Fig. 12 Stress - Strain Relationships at 1200°C

OXIDATION KINETICS OF MgO-SiC COMPOSITES

M. E. F. Camey and D. W. Readey
Colorado Center for Advanced Ceramics
Colorado School of Mines
Golden, CO 804011

ABSTRACT

MgO-SiC composites were oxidized between 1200 °C and 1500 °C. The oxidation kinetics were followed by changes in thickness of the oxidized layer with time. The oxidized layer consisted of three distinct regions: a dark inner region, a porous intermediate region, and a dense, columnar grain surface layer. The oxidized layer morphology and the rate of oxidation suggest that magnesium diffusion in MgO is controlling the rate of oxidation. Sc^{+3}-doping increased the rate of oxidation as expected. Oxidation in ^{18}O-enriched atmospheres confirmed the mechanism of oxidation.

INTRODUCTION

Alumina,[1,2] mullite[2,3,4] and alumina-zirconia[5,6] have been used as matrices with both SiC particles or whiskers for composite oxidation studies. Most exhibit parabolic oxidation between 1200 and 1500 °C which suggests diffusion is rate controlling. Porosity is developed during oxidation and attributed to CO evolution. Hallum[7] studied the oxidation kinetics of MgO-SiC composites. In the temperature range from 1100 to 1500°C, parabolic oxidation was observed, strongly dependent on the amount of SiC present in the matrix. Comparisons between activation energies and diffusion coefficients calculated from oxidation data[7] and existing diffusion data[8,9,10] suggest magnesium diffusion controls the rate of oxidation. The most striking microstructural feature was the large columnar grain layer on the surface of the sample as shown in Figure 1. The presence of the columnar-grain surface layer reinforces the model of Mg^{2+} diffusion to the surface as the rate controlling step.[7]

The current research was to either confirm or reject the magnesium diffusion mechanism of oxidation. To this end, some experiments were carried out with MgO doped with Sc_2O_3. It is well established that Sc^{3+} additions generate cation vacancies[11] and increase the Mg^{2+} diffusion coefficient.[12] As a result, Sc_2O_3 additions to MgO would be expected to increase the rate of oxidation of MgO-SiC composites.

To the extent authorized under the laws of the United States of America, all copyright interests in this publication are the property of The American Ceramic Society. Any duplication, reproduction, or republication of this publication or any part thereof, without the express written consent of The American Ceramic Society or fee paid to the Copyright Clearance Center, is prohibited.

EXPERIMENTAL

Magnesium oxide-silicon carbide (MgO-SiC) composites with SiC as particles or whiskers dispersed in a MgO matrix were studied. Samples with SiC whiskers were prepared previously[7] and those with SiC powders were prepared in this study. SiC powder was mixed with pure MgO to give 5, 10, and 15 v/o SiC. Sc_2O_3-doped MgO powders previously prepared by calcining and milling were mixed with 10 v/o SiC powder. Sufficient Sc_2O_3 was added to ensure the presence of a second phase during oxidation. The MgO-SiC_p compacts were hot pressed to final density at 1635 °C for 90 minutes. All samples were greater than 98% percent of theoretical density and the grain size of the undoped samples was about 0.4 µm while those of doped sample was about 2 µm. Roughly 5 mm cube samples were cut and oxidized in air or ^{18}O-enriched oxygen.

RESULTS

Observation by optical microscopy shows two distinct layers at the sample surface after oxidation, Figure 2. At the surface there is a "white layer" and below that is a "dark layer." Higher magnification shows that the white layer consists of two regions, an outermost large-grain size layer and a porous region below this as shown in Figure 1. Neither optical or SEM microscopy could discern any significant difference between the white and dark layers. TEM found Mg_2SiO_4 in the porous part of the white region while in the dark region, particles of silicon, and not carbon, were found. Several differences were noted with doping. First, the thickness of the dark layer increased considerably compared to the thickness of the white layer. Second, the thickness of the large-grain layer increased considerably as shown in Figure 3. As this figure shows, the large-grain layer is clearly columnar and does not show any obvious continuous porosity

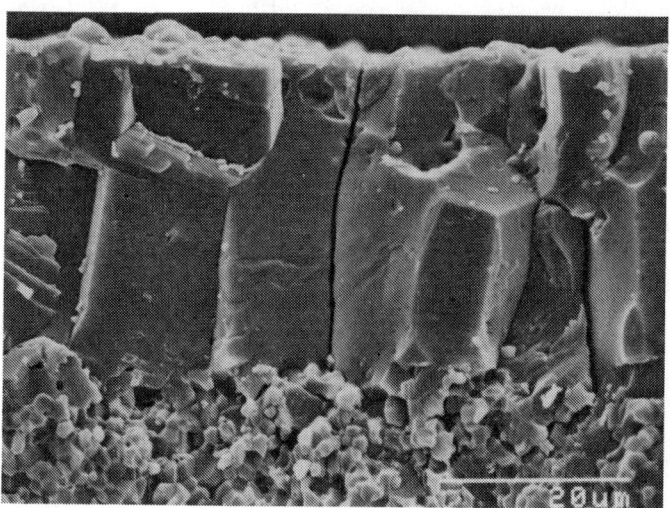

Figure 1. Columnar grains on the surface of an oxidized MgO-SiC composite.[7]

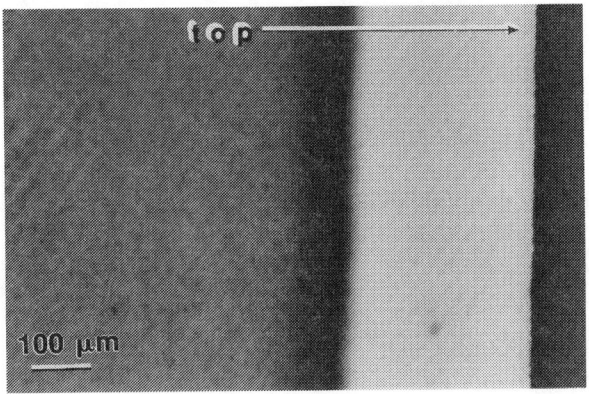

Figure 2. Optical micrograph of the oxidized surface of a composite.

through the layer. X-ray diffraction of this layer shows a strong (200) preferred orientation, Figure 4, and the presence of only MgO. The thickness of the oxidized layers increases with reaction time and temperature. Figure 5 shows typical behavior of both doped and undoped samples. The dark layer grows approximately parabolically while the white layer grows linearly. The columnar surface layer seems to associated with the growth of the dark layer and essentially stops growing as the white layer gets thicker, Figure 6. The ^{18}O signal obtained by imaging SIMS for a doped sample oxidized for 25 hours at 1400 °C in $^{18}O/^{16}O$ (50%/50%) is shown in Figure 7.

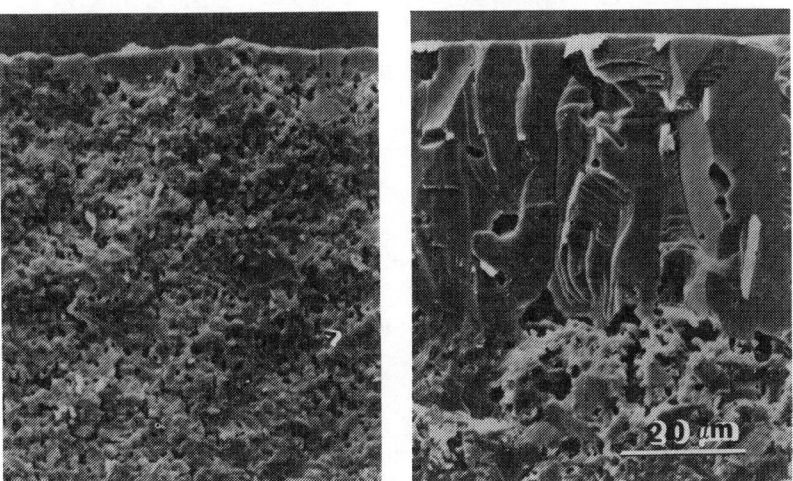

Figure 3. Columnar layer formed during oxidation at 1400 °C for 20 hrs. Left, undoped; right, Sc_2O_3 doped.

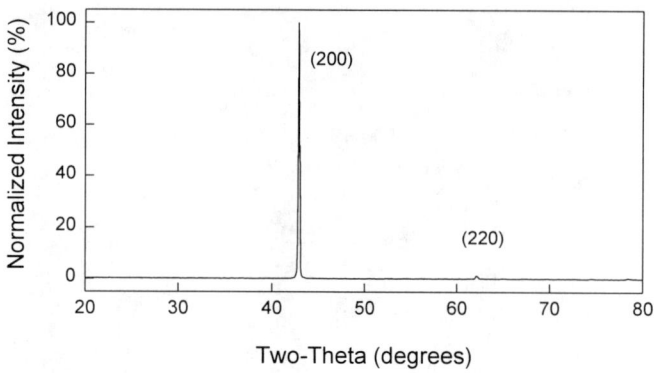

Figure 4.   X-ray pattern of Sc_2O_3-doped sample oxidized for 30 hrs. at 1400 °C.

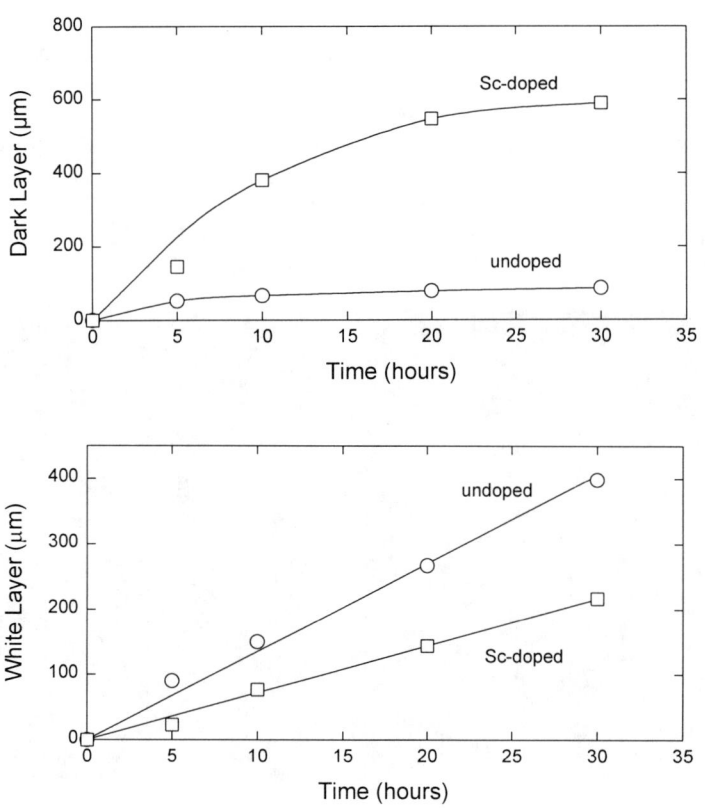

Figure 5.   Oxidation layer thicknesses versus time at 1400 °C.

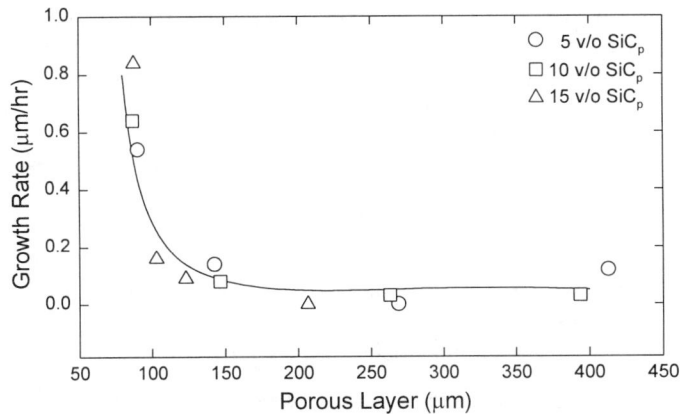

Figure 6. Columnar layer growth rate versus porous layer thickness; 1400 °C.

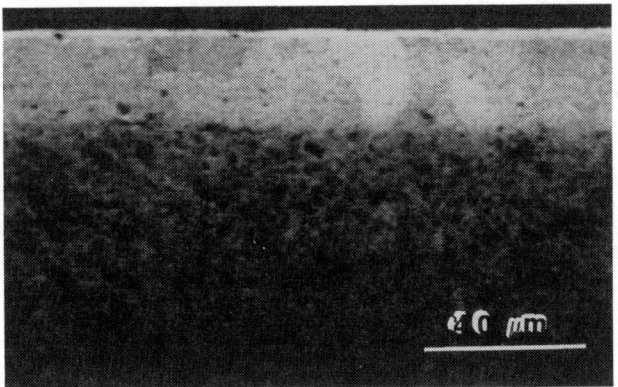

Figure 7. ^{18}O SIMS image of Sc_2O_3-doped columnar layer; 25 hrs. at 1400 °C.

DISCUSSION AND CONCLUSIONS

The microstructure, the preferred orientation, the absence of other phases, and the ^{18}O concentration suggest that the columnar grain surface layer grew over the original surface of the sample. These observations plus the more rapid and thicker growth of this layer with scandium doping is consistent with magnesium ion diffusion to the surface of the oxidizing sample. The presence of free silicon in the dark layer is a surprise and suggests that the dark layer is formed by a reaction such as:

$$SiC(s) + MgO(s) \rightleftharpoons Si(s) + CO^{-2} + Mg^{+2}.$$

The assignment of charges to the two species is purely arbitrary and only to balance the charge. Both the CO^{13} and the magnesium diffuse to the sample surface with the magnesium forming the columnar grain layer. In the white layer, silicon is oxidized to SiO_2 and forms forsterite while the CO pressure increases sufficiently to form porosity.

This research is supported by NSF grant number DMR-9110927.

REFERENCES

1. F. Lin, T. Marieb, A.A. Morrone, and S.R. Nutt, "Thermal Oxidation of Al_2O_3-SiC Whiskers Composites: Mechanisms and Kinetics," in Mater. Res. Soc. Symp. Proc., Vol. 120, (Materials Research Society, Pittsburgh) 323-332 (1982).

2. M.P. Borom, M.K. Brun, and L.E. Szala, " Kinetics of Oxidation of Carbide and Silicide Dispersed Phases in Oxide Matrices," Adv. Cer. Mat. 3[5] 491-497 (1988).

3. K.L. Luthra and H.D. Park, "Oxidation of Silicon Carbide-Reinforced Oxide-Matrix Composites at 1375° to 1575°C," J. Am. Ceram. Soc. 73[4] 1014-1023 (1990).

4. M.I. Osendi, "Oxidation Behavior of Mullite-SiC Composites," J. Mater. Sci. 25[8] 3561-3565 (1990).

5. P. Wang, G. Grathwohl, F. Porz, and F.Thummler, "Oxidation Behaviour of SiC Whisker-Reinforced Al_2O_3/ZrO_2 Composites," Powd. Met. Int. 23[6] 370-5 (1991).

6. M. Backhaus-Ricoult, "Oxidation Behavior of SiC-Whisker-Reinforced Alumina-Zirconia Composites," J. Am. Ceram. Soc. 74[8] 1793-1802 (1991).

7. G.W. Hallum, "High Temperature Effects of Oxidation on MgO-SiC Composites," Ph.D. Thesis, The Ohio State University (1990).

8. R. Lindner and G.D. Parafitt, "Diffusion of Radioactive Magnesium in Magnesium Oxide Crystals," J. Chem. Phys. 26[1] 182-185 (1957).

9. B.C. Harding and D.M. Price, "Cation Self-Diffusion in MgO up to 2350°C," Philos. Mag. 26[1] 253-260 (1972).

10. B.J. Wuensch, W.C. Steele, and T. Vasilos, "Cation Self-Diffusion in Single Crystal MgO," J. Chem. Phys. 58[12] 5258-5266 (1973).

11. W.H. Gourdin and W.D. Kingery, "The Defect Structure of MgO Containing Trivalent Cation Solutes: Shell Model Calculations," J. Mat. Sci. 14 2053-73 (1979).

12. D.R. Sempolinski and W.D. Kingery, "Ionic Conductivity and Magnesium Vacancy Mobility in Magnesium Oxide," J. Am. Ceram. Soc. 63[11-12] 664-669 (1980).

13. H. Kathrein, H. Gonska, and F. Freund, "Subsurface Segregation and Diffusion of Carbon in Magnesium Oxide," Appl. Phys. A30 33-41 (1983)

EFFECT OF INTERPHASE CARBON THICKNESS ON ENVIRONMENTAL RESISTANCE OF CONTINUOUS FIBER-REINFORCED CERAMIC MATRIX COMPOSITES

James D. Cawley
Department of Materials Science and Engineering
Case Western Reserve University
10900 Euclid Avenue, Cleveland OH 44106.

ABSTRACT

One variable in the design of ceramic-matrix composites is the thickness of the interphase material intentionally placed at the interface between fiber and matrix to induce debonding. For the case of interphase carbon in a SiC-SiC composite, it is shown that thickness does not significantly affect the rate of carbon oxidation, but it does strongly affect the susceptibility to fusion of the fibers to the matrix near matrix cracks.

INTRODUCTION

Environmental resistance continues to be a difficult characteristic to design into a ceramic matrix composite (CMC). A number of alternatives exist for high-toughness and high-strength ceramics, but essentially all of these suffer some type of degradation when the temperature is raised. This is particularly true for CMC's that employ interphase carbon to facilitate debonding [1-10].
There are three distinct degradation mechanisms that can occur in nonoxide continuous fiber-reinforced ceramic-matrix composites (CFCC's) when a matrix crack occurs exposing interphase carbon and fibers to the ambient: i) decoupling of the fiber from the matrix; ii) localized attack of the fiber; and iii) fusion of the fiber to the matrix. These are schematically depicted in Fig. 1.

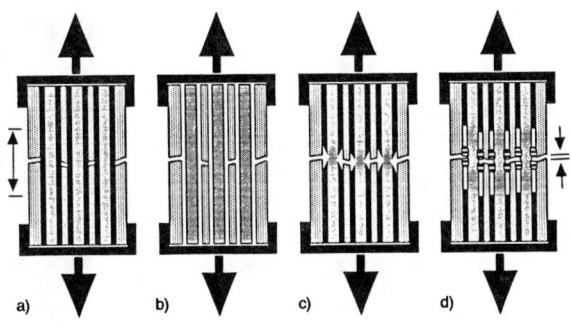

Figure 1. Schematic representations of the possible undesirable effects of oxidation in a SiC-SiC CMC with interphase carbon. a) A CMC loaded to the point of matrix cracking. b) Complete decoupling of the fibers from the matrix by carbon oxidation. c) Localized attack of the fiber by the atmosphere. d) Fusion of the fibers to the matrix by the formation of silica.

Decoupling of the fiber from the matrix removes the possibility of stress transfer between the matrix and the fiber. Since the external loads are nearly always applied through contact with the matrix, an uncoupled system behaves as if it was porous rather than reinforced. This compounds the fact that the matrix usually has properties less than those attainable when the same material is used in monolithic form.

Fiber degradation has obvious consequences. The entire load is carried by the fibers in the plane of a matrix crack. Thus, when corrosive effects of the atmosphere admitted by the matrix crack weaken individual fibers, the system will fail prematurely.

Fusion of the fibers can also lead to premature failure, but in this case by causing stress concentration rather than weakening the fiber. After the matrix has cracked an increase in load is accompanied by strain in the fiber. In a pristine composite, the section of fiber that elongates is on the order of twice the debond length (typically on the order of a millimeter). In contrast, after fusion, only the section of fiber between the faces of the open crack (on the order of a micron) can elongate (to first order). Therefore, for the same net displacement of the matrix, the local strain, and therefore stress, on the fibers may be a factor of 10^3 higher.

Design of an effective composite requires assessment of the balance between the propensity for each degradation mechanism, in the context of overall performance. In the following discussion, the role interlayer thickness will be discussed for SiC-SiC CFCC's that employ a thin carbon layer. In particular, it will be shown that reducing the thickness of a carbon layer does not appreciably reduce the rate of decoupling, but it does greatly reduce the time required for fusion. Degradation of the fiber will not be considered.

BACKGROUND

The problem to be considered has two components, as illustrated in Fig. 2. First, oxidation of the carbon involves recession of carbon away from the plane of the matrix crack parallel to the direction of the fiber reinforcement. Secondly, oxidation of the SiC-matrix and SiC-fibers forms SiO_2; the resultant volume expansion causes a radial outward migration of surface of the fiber and inward

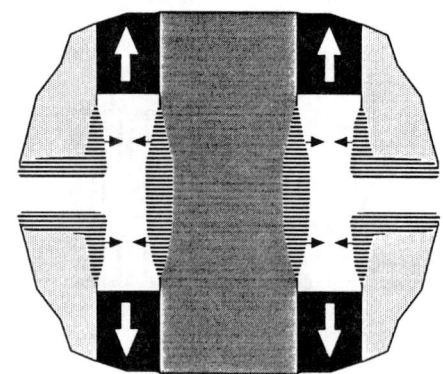

Figure 2. Schematic illustration of the geometry considered for oxidation reactions. The carbon recedes away from the matrix crack as CO is formed. The formation of silica by oxidation of the SiC-fiber and SiC-matrix more than compensates for the recession of silicon carbide and the free surfaces of the silica approach each other.

migration of the matrix surface. The rate of carbon oxidation determines the susceptibility to decoupling and the rate of the SiC oxidation determines susceptibility to fusion.

The temperature range to be considered is 700 to 1200°C. As will be discussed, it is not correct to consider materials like silicon carbide to be oxidation resistant, even at the low end of this temperature range.

ANALYSIS

In general, the rate of a chemical reaction in which one of the species is supplied by diffusion can be phenomenologically described using a linear-parabolic rate law, i.e.,

$$\frac{x^2}{k_p} + \frac{x}{k_l} = t \tag{1}$$

where t is time (s), k_p is a "parabolic" rate constant (m^2/s), and k_l is a "linear" rate constant (m/s). The same formalism can be applied to both carbon and SiC oxidation, but the meaning of the distance 'x' is different. For carbon oxidation, x refers to the recession distance, i.e., how far the carbon surface has moved via oxidation [2,5,11-13]. On the other hand, for SiC oxidation, x refers to the thickness of the SiO_2 scale after a given anneal [6,7,14]. The effect of fundamental variables such as temperature, pressure, and the relation to system parameters such as thickness of the interphase carbon are revealed through inspection of the functional form of k_p and k_l.

It is useful to recall that Eqn. 1 is derived assuming two processes take place is series. The slower process controls the overall rate.

EFFECT OF INTERLAYER THICKNESS ON RATE OF CARBON OXIDATION

As discussed in a recent analysis of the oxidation of interphase carbon [2], the rate constants can be expressed as

$$k_l = K_r \left(\frac{\chi c_T}{N} \right) \tag{2}$$

and

$$k_p = f(\chi) \frac{D c_T}{N} \tag{3}$$

where K_r is the reaction rate constant, conventionally written using an Arrhenius form, (m/s), D is the gas phase diffusion coefficient (m^2/s), c_T is total concentration of gas molecules in the atmosphere (mol/m^3), N is the molar density of carbon (mol/m^3), χ is the oxygen partial pressure, and $f(\chi)$ is a weakly varying function of value roughly 2.

None of the variables that affect k_l are dependent on the thickness of the interphase carbon; χ and c_T are external variables, and although both N and K_r depend on the density and crystallinity of the oxidizing carbon, neither depend on the thickness.

The parabolic constant can be affected by carbon thickness. The principal parameter that depends on carbon thickness is D. In addition, $f(\chi)$ may also depend on carbon thickness; the effect is always small compared to that on D, but, interestingly, in can be of opposite sign (D becomes smaller, but $f(\chi)$ becomes larger).

It is conventional to designate two regimes of gas phase diffusion, molecular and Knudsen. Molecular diffusion is obtained when the mean free path of the gas molecules is small compared characteristic dimension of the container, in this case the pore left behind as the carbon recedes. D is then calculated using an expression such as Chapman-Enskog [15,16]. Knudsen diffusion is the complement, i.e., it occurs when the container dimension is small compared to the molecular mean free path. In this regime D is calculated using a different functional form that includes a direct proportionality to the characteristic container dimension, d [15,16]. In reality, of course, these names are simply for convenience and the true behavior is a gradual transition from one limiting case to the other.

A range of calculated values for k_p are graphically depicted in Fig. 3. The three curves were calculated assuming the carbon reacts to form CO and $\chi = 1$. It is evident that in the molecular regime (large d), k_p is independent of pressure and only weakly dependent on temperature (i.e. less than a 10% change between 800 and 1000°C).

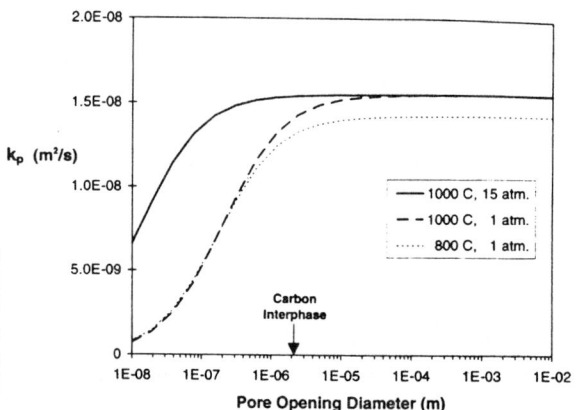

Figure 3. Calculated values for the parabolic rate constant, for several combinations of temperature and pressure, as a function of the characteristic dimension of the container (in this case, the width of the gap).

Most importantly, the reduction in k_p expected as carbon thickness is reduced is modest. Using k_p for carbon of 2 μm thickness as a baseline, the k_p expected for an interlayer of 0.1 μm is not even a factor of two smaller. Reducing the thickness of the carbon to 10Å (which is impractical) would only yield a k_p which is a factor of 7 smaller than the baseline.

The inability of effecting a practical reduction in oxidation rate made clear by considering the magnitude of oxidation rate predictions assuming k_p is small compared to k_l. (If k_p is not small compared to k_l changing the carbon thickness will have no effect.) For 1 atm. of O_2, and a temperature range of 800 - 1000°C, a recession of 1 cm will be expected in approximately 2 hours. Decreasing k_p by a factor of two (corresponding to a thickness of 0.1 μm) means that the same 1 cm of recession will occur in roughly 4 hours. Clearly, either recession rate would be unacceptably high for an engineering application and some other means of controlling oxidation must be employed.

(Incidentally, oxidation of pitch-based carbon fibers indicates that for the temperature range of interest the kinetics are essentially linear to large recession distances. This indicates that they are reaction controlled and, therefore, not sensitive to carbon thickness. It is presumed that dense and ordered interphase-carbon would also be reaction controlled for practical recession distances.)

FUSION OF FIBER TO MATRIX BY SILICA FORMATION

Under some circumstances, the rate at which a gap between the fiber and matrix is filled becomes a concern. For example, when a fugitive phase (*such as carbon*) was used and, after decoupling, sufficient mechanical interlocking is present for composite behavior to be observed [10]. With SiC-SiC systems filling such a gap is inevitable. If the only concern was to control the rate of decoupling by carbon oxidation, then such sealing with silica can prevent it [6,7]. However, experiments suggest that fusion of the fiber and matrix is undesirable [1].

The volume of vitreous silica (≈ 26 cm^3/mol) produced by oxidation is slightly more than twice the volume originally occupied by the SiC (≈ 12.5 cm^3/mol). Taking into account the recession of the silicon carbide and limiting consideration to gaps that are thin compared to the diameter of the fiber (so that parallel plate geometry applies) means that the free surfaces of the silica scales approach each other at a rate that is equal to the rate of scale growth. Thus, fusion of the fiber to the matrix will occur when the scale thickness grows to be equal to that of the original gap.

In most situations, the oxidation of silicon-based ceramics can be considered to be purely parabolic, although some exceptions are recognized (e.g., the short time linear regime during dry oxidation of silicon, slow- and fast-face oxidation of single crystal SiC, and the reduction in oxidation kinetics associated with devitrification to form cristobalite) they are not important for the material and temperature range being considered. In this case, k_p is given by

$$k_p = \frac{2D_{ox}\chi C_T}{N} \tag{4}$$

where D_{ox} is the effective diffusion coefficient for oxygen in vitreous silica [14,17].

As long as the scale formed is silica (i.e., it is not doped with impurities from the oxidizing material), k_p should be insensitive to the "type" of silicon carbide. This appears to be the case, for example, with Nicalon fibers and CVD-SiC. Figure 4 is an Arrhenius plot for the oxidation of various classes of SiC (slightly modified from the source [18]) and Fig. 5 is a plot of the thickness of the silica scale formed on Nicalon fibers after exposure to air at elevated temperature [19]. Using the 1200°C data in Fig. 4 yields an estimated k_p of 4×10^{-18} m^2/s. Correcting this by a factor of 5, to account for the difference in χ, yields 2×10^{-17} m^2/s at $10^4/T = 6.79$. This value is at the upper range of values summarized in Fig. 4, but is close to an extrapolation of the values for elemental Si.

The data presented in Fig. 5 thus allow a conservative estimate of the time necessary to fill a gap. For this case, 0.1 and 1 μm characteristic distances are compared.

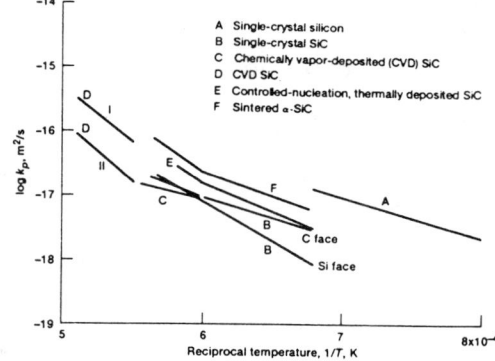

Figure 4. Parabolic rate constants for a variety of SiC materials referenced to single crystal silicon. (From ref. 18)

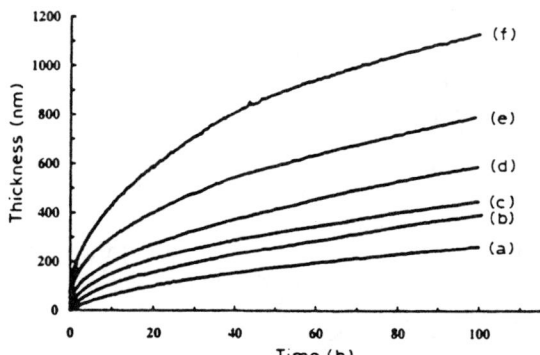

Figure 5. Thickness of silica scales grown on Nicalon NLM 202 fibers during aging in air at (a) 700°C, (b) 800°C, (c) 900°C, (d) 1000°C, (e) 1100°C, (f) 1200°C. (From Ref. 19)

For exposure in air, a 1 μm coating would be comfortably large compared to the thickness of the silica scale for 100 hour exposures up to 1100°C, but would only survive for ≈60 hours at 1200°C. A 2 μm coating would not be expected to fuse for times as long as 250 hours. Significant times would be predicted even with larger values of χ.

In contrast, very short time is required to fuse a 0.1 μm gap. Even at temperatures as low as 700°C, such a gap would fill within less than 20 hours. Referring to the curves for increasing temperature, the time to fill the gap becomes too small to resolve above 900°C. Thus at the higher end of the temperature range under consideration, a 0.1 μm gap cannot be expected to persist for any significant time, regardless of type of SiC.

CONCLUSIONS

The analysis presented indicates that using thinner carbon films does not reduce the rate at which a fiber becomes decoupled from a matrix as a result of carbon oxidation. However, the rate at which the fiber becomes fused to the matrix is very sensitive to the gap opening, and, thereby, to carbon thickness.

The use of multilayers, such as alternating layers of C and SiC, does not change the analysis.

REFERENCES

1. F. E. Heredia, J. C. McNulty, F. W. Zok, and A. G. Evans, "An Oxidation Embrittlement Probe for Ceramic Matrix Composites," J. Am. Ceram. Soc., in press.
2. A. J. Eckel, J. D. Cawley, and T. A. Parthasarathy, "Oxidation Kinetics of a Continuous Carbon Phase in a Nonreactive Matrix, J. Am. Ceram. Soc., 78 [4] 972-80 (1995).
3. J. D. Cawley, O. Unal, and A. J. Eckel, "Oxidation of Carbon in Continuous Fiber Reinforced Ceramic Matrix Composites" Ceramic Transactions, Advances in Ceramic Matrix Composites, Vol. 38, ed. by N.P. Bansal, p 541-552. (1994).
4. F. Lamouroux, G. Camus, and J. Thebault, "Kinetics and Mechanisms of Oxidation of 2D Woven C/SiC Composites: I, Experimental Approach," J. Am. Ceram. Soc., 77 [8] 2049-57 (1994).
5. F. Lamouroux, R. Naslain, and J.-M. Jouin, "Kinetics and Mechanisms of Oxidation of 2D Woven C/SiC Composites: II, Theoretical Approach," J. Am. Ceram. Soc., 77 [8] 2058-68 (1994).
6. L. Filipuzzi, G. Camus, R. Naslain, and J. Thebault, "Oxidation Mechanisms and Kinetics of 1D-SiC/C/SiC Composite Materials: I, An Experimental Approach," J. Am. Ceram. Soc., 77 [2] 459-66 (1994).
7. L. Filipuzzi and R. Naslain, "Oxidation Mechanisms and Kinetics of 1D-SiC/C/SiC Composite Materials: II, Modeling," J. Am. Ceram. Soc., 77 [2] 467 80 (1994).
8. P.F. Tortorelli, S. Nijhawan, L. Riester, and R.A. Lowden, Influence of Fiber Coatings on the Oxidation of Fiber-Reinforced SiC Composites", Ceram. Engr. and Sci. Proc. p. 358-66 (1993).
9. R. T. Bhatt, "Oxidation Effects on the Mechanical Properties of a SiC-Fiber-Reinforced Reaction-Bonded Si3N4 Matrix Composite", J. Am. Ceram. Soc., 75 [2] 406-12 (1992).
10. T. Mah, T.A. Parthasarathy, K. Keller and J. Guth, "Fugitive Interface Coating in Oxide-Oxide Composites: A Viability Study", Ceram. Engr. and Sci. Proc., 12 [9-10] 1802-18 (1991).
11. S. Drawin, M.P. Bacos, J.M. Dorvaux and O. Lavigne, "Oxidation Model for

Carbon-Carbon Composites", AIAA Paper 92-5016, Fourth International Aerospace Planes Conference, Orlando, FL, 7, 1112-1122 (1992).
12. K. L. Luthra, "Oxidation of Carbon/Carbon Composites -- A Theoretical Analysis", Carbon, 26, [2], 217-224 (1988).
13. J. Bernstein and T. B. Koger, "Carbon Film Oxidation-Undercut Kinetics", J. Electrochem. Soc., 135 [8] 2086-90 (1988).
14. B. E. Deal and A. S. Grove, "General Relationship for the Thermal Oxidation of Silicon", J. Appl. Phys., 36 [12] 3370-8 (1965).
15. W. Geankoplis, Mass Transport Phenomena, p. 27-28, 151-153, Ohio State University Bookstores, 1978.
16. R. B. Bird, W. E. Stewart, and E. N. Lightfoot, Transport Phenomena, p. 496-501, John Wiley and Sons, New York, NY, 1960.
17. J. A. Costello and R. E. Tressler, "Oxidation Kinetics of Silicon Carbide Crystals and Ceramics: I, In Dry Oxygen", J. Am. Ceram. Soc., 69 [9] 674-81(1986).
18. N. S. Jacobson, "Corrosion of Silicon-Based Ceramics in Combustion Environments," J. Am. Ceram. Soc., 76 [1] 3-28 (1993).
19. M. Huger, S. Souchard, and C. Gault, "Oxidation of Nicalon SiC Fibres," J. Mat. Sci. Lett. 12 414-420 (1993).

Keyword and Author Index

2,4-dichloro-2,4 disilapentane, 311

ABAQUS, 243
Al_2O_3-matrix CMCs, 41
Al_2O_3-SiO_2 fibers, 205
Alumina, 243
Alumina-based fibers, 85
Alumina fibers, 209
Ammonia, 281
Amorphism, 317
Analytical models, 343

Balagurova, N.M., 205., 209, 337
Baldus, H.-P., 75
Bashkirova, S.A., 261, 337
Behavior, 317
Berroth, Karl, 317
Besmann, Theodore M., 1, 119
Bezruchenko, V.I., 205
Binders, 205
Birot, M., 311
Biswas, Dhiman K., 105
Brennan, John J., 53
Bunsell, A.R., 85
Berger, M.H., 85
Bhatti, A.R., 181
BN-coated Nicalon fiber/glass-ceramic matrix composites, 53
Bodet, R., 305
Borosilazane, 75
Borosilicate glass, 349
Bourrat, X., 305
Braue, W., 275
Bullock, E., 125, 155

Cain, M.G., 41
Calcium fibers, 325
Camey, M.E.F., 371
Campaniello, J.J., 155
Capillary flow, 149
Carbon-carbon materials, 267
Carbon-carbon composites, 237
Carbon-ceramic materials, 267
Carbon coatings, 243, 249
Carbon-fiber/Si_3N_4, 125

Carbon fibers, 255
Carborundum, 331
C/C-SiC, 275
Carlsson, R., 143
Casadio, S., 193
Cawley, James D., 377
Ceramic fibers, 85, 325, 361
Ceramic processing, 161
Ceramic structure, 209
Characterization, 85
Chemical structure, 209
Chemical synthesis, 209
Chemical vapor deposition, 267, 349
Chemical vapor infiltration, 1, 187
Chermant, J.L., 355
Chernyshev, E.A., 205, 209, 261, 337
Chollon, G., 299, 305
Claussen, N., 167
Clements, N., 231
Coatings, 75, 267
Constrained sintering, 161
Continuous fiber ceramic composites, 1
Continuous fiber-reinforced CMCs, 377
Continuous fibers, 187
Crack deviation, 161
Creep, 299, 331
Cristallite size, 237
Crosslinking, 337
Crystallinity, 317
CVR, 267

David, P. 237
Debray, K., 161
Demin, A.V., 267
DiCarlo, J.A., 331, 343
Doleman, P., 41
Donato, A., 193
Doreau, F., 355
Dow Corning, 331
Dunoguès, J., 311

Eckerbom, Lena, 95
Electrical properties, 287
Electron beam curing, 65
Electron-probe microanalysis, 305

Electrophoretic infiltration, 155
Engeli, Georg, 317
Environmental resistance, 377

Fabrication, 105
Fiber/matrix interfacial microstructure stability, 53
Fibers, 75
    Distribution, 143
    -Matrix interface, 125, 131, 181, 275, 355
    -Reinforced ceramics, 175
    -Reinforced composites, 13, 155
    -Reinforced glass-ceramic matrix composites, 53
    Reinforcement, 167, 255
    Stability, 131
    Synthesis, 317
    Thermostructural properties, 343
Filament winding, 143
Finite element analysis, 243
Flexural testing, 361
Flow-through porous media, 105
Fracture numerical simulation, 161
Free radical, 293
Fugitive coating, 243
Fugitive interface, 243
Fujikura, M., 215
Funayama, O., 199
Functional composite materials, 325

Gas evolution, 293
Gas phase process, 317
Gatica, Jorge E., 105
Gent, J., 41
Gern, Frank H., 149
Gershkohen, S.L., 205, 209
Glass-ceramic composites, 137
Glass-ceramic matrix composites, 53
Goldsby, J.C., 331, 343
Goleki, I., 231
Grain sizes, 331
Grenet, C., 125
Guillet, F., 237

Hapke, J., 13
Heat treatment, 287, 299
Heintz, J.M., 161
Hémida, A. Tazi, 311

Hi-Nicalon, 65, 131, 305, 331
Hi-Nicalon S, 65

High-melting borides, 267
High strength, 361
Hochet, N., 85
Hoult, A.P., 349
Hot pressing, 181, 215
HREM, 275
Hybrid composites, 13
Hydridopolycarbosilane, 111

Ichikawa, Hiroshi, 65, 215
Image analysis, 143
Impregnated hydrated cellulose, 325
Impregnation, 325
Impurity segregation, 275
Infiltration, 75, 105
Influence of environment, 331
Interrante, L.V., 111
Interfaces, 41, 137, 187
    Properties, 199
Interphase carbon thickness, 377
Interphases, 355
Intrabundle microstructure, 275
Isoda, Takeshi, 199, 361
Ivanov, V.V., 261., 337

Janssen, R., 167

Kamimura, Seiji, 281
Kanka, B., 175
Kasai, Noboru, 281
Knowles, Kevin M., 137
Kooner, S., 155
Kostikov, V.I., 267
Kravetskii, G.A., 267
Kristoffersson, A., 143
Kroupa, J.L., 243
Kumar, Atul, 137

Laarz, E., 143
Laser processing, 349
Layered interphases, 23
Letullier, P., 161
Lewis, M.H., 41
Liquid infiltration, 193
Liquid silicon infiltration, 149
Low free carbon, 311
Lücke, J., 13
Lundberg, Robert, 95, 143

Manufacturing, 105
Martin, E., 161
Masaki, S., 187

Matlin, W.M., 119
Matrix cracking, 181
Matrix slurry, 243
Mechanical properties, 125, 199
Mechanisms, 281
Methylhydorocooligosilazane, 361
Methylhyrdrosilazane, 199
MgO-SiC composites, 371
Microstructure, 181, 299, 355
Modeling, 105
    Optimization, 243
Molybdenum disilicide, 181
Moore, E.H., 243
Moriya, K., 187
Morozumi, H., 199
Morris, R.C., 231
Morscher, G.N., 343
Mortimer, B., 181
Mouhamath, B., 249
Mullite, 243
Mullite/alumina, 243
Mullite/mullite-ceramic composites, 175
Multidirectional woven fabric, 361
Multilayered ceramic, 161

Nakano, K., 215
Nakashiba, Koji, 287
Nannetti, C.A., 193
Nano/microcomposites, 13
Nanosized mullite precursor, 175
Narasimhan, D., 231
Narcy, B., 237
Narisawa, Masaki, 287
Naslain, R., 23, 299, 311
Nd-YAG laser, 349
Near-stoichiometric ceramic fibers, 311, 331
Nicalon, 65, 119
Nitridation, 281
Numerical simulation, 149
Nutt, Steven R., 53

Ohnabe, H., 187
Okamura, Kiyohoto, 65, 281, 287, 293
Olry, P., 299
Organosilicon polymers, 261
Ortona, A., 193
Oxidation, 137
    Kinetics, 371
Oxide fibers, 85
Oxide interphase, 95

Oxide matrix CMCs, 41
Oxide-oxide, 243
Oxides, 95
Oxygen-free SiC fiber, 65

Pailler, R., 299, 305, 311
Pashby, I.R., 349
Passing, G., 75
Perhydrosilazane, 199
Phyllosiloxide, 23
Pickering, S., 155
Pillot, J.P., 311
Pleger, R., 275
Plunkett, L., 125
Polyak, L.G., 205, 337
Polyalumoxanes, 209
Polycarbosilane, 193, 281, 293
Polycarbosilane-derived ceramic-grade fibers, 305
Polymer impregnation, 187
Polymer precursors, 111, 205
Polymer pyrolysis, 243
Polymers, 255
Polyorganosilanes, 337
Polytitanocarbosilane, 361
Porous carbon-carbon preforms, 231
Powder processing, 131
Preannealing, 331
Preceramic polyalumoxane siloxane precursor, 205
Preceramic polymer, 199
    Impregnation, 199
        And pyrolysis method, 361
Precursors, 13, 209
Pressure infiltration, 125, 243
Pritchard, A.J., 181
Processing, 143
(PyC-SiC)n interphases, 23
Pyrocarbon, 23
Pyrolysis, 193

Quenisset, J.M., 161

Rabinovitch, R.A., 205, 209
Radiation curing, 281, 293
Rapid densification, 231
Ravel, F., 237
Razzell, A.G., 41
Reaction-bonded alumina, 95
Reaction-bonded oxide ceramics, 167
Reaction-bonded Si3N4, 131
Reaction bonding, 255

Reaction mechanism, 293
Readey, D.W., 371
Reese, O., 175
Reinforcement, 261
Rescio, M., 193
Residual stresses, 243
Rheology, 143
Rodionova, V.V., 267
Rupture behavior, 331

Saka, H., 215
Saphikon, 41
Sapphire, 95
Saruhan, B., 175
Sasaki, K., 215
Sato, K., 199
Scanning electron microscopy, 137, 305
Schneider, H., 175
Seguchi, Tadao, 65, 281, 293
Shamasundar, S., 243
Sherwood, W., 111
Shibuya, M., 187
Shobu, Kazuhisa, 255
Si-B-(N,C), 75
Si-B-N fiber, 361
Si-C-N (polycarbosilazanes), 13
Si-C-O fiber, 361
SiC, 255, 261
    -Based fibers, 305, 311
    Fibers, 65, 85, 131, 137, 287, 293, 299, 331, 337, 343, 349, 361
    Filler, 193
SiC/BN/SiC, 215
SiC/BN/SiC/Si3N4, 215
SiC/C/SiC, 215
SiC/C/Si3N4, 215
SiC(Nicalon)-Pyrex composites, 249
SiC/SiC, 23, 111, 187
SiCf-YMAS, 355
Silicate/nicalon CMCs, 41
Silicon capillary impregnation, 275
Silicon concentration, 149
Si3N4 fibers, 199, 281, 317
Si-N fiber, 361
Si-T-C-O fiber, 361
Slip casting, 267
Slurry, 181
    Infiltration, 125
Sporn, D., 75
Stinton, D.P., 119
Stoichiometric SiC fiber, 65
Sugimoto, Masaki, 293

Sun, Ellen Y., 53

Takeda, N., 249
Tani, Eiji, 255
Tensile strength, 317
Tensile testing, 361
Tewari, Surendra N., 105
Tezuka, A., 199
Thermal expansion mismatch, 181
Thermal gradient chemical vapor infiltration, 231
Thermally activated mechanisms, 343
Thierauf, A., 75
Tikchonovich, T.V., 261, 337
Titova, L.V., 325
Transmission electron microscopy, 137, 305
Tuersley, I.P., 349
Turbostratic BN, 23
Two-step forced chemical vapor infiltration process, 119

Ulyanova, T.M., 325
UV irradiation, 337

Vacuum impregnation, 243
Veyret, J.B., 125
Vicens, J., 355
Vogt, Ulrich, 317
Vygovskii, N.A., 205

Watanabe, Kiyoshi, 281
Weiss, R., 275
Wendorff, J., 167
Whitmarsh, C.W., 111
Wroblewska, G.H., 131

X-ray diffraction, 237

Yamamura, Takemi, 187, 361
Yun, H.M., 331, 343

Ziegler, G., 13, 131
Zirconium titanate fibers, 325